AF711739

Optimization Modelling

A Practical Approach

Optimization Modelling

A Practical Approach

Ruhul A. Sarker
Charles S. Newton

CRC Press
Taylor & Francis Group
Boca Raton London New York

CRC Press is an imprint of the
Taylor & Francis Group, an informa business

CRC Press
Taylor & Francis Group
6000 Broken Sound Parkway NW, Suite 300
Boca Raton, FL 33487-2742

© 2008 by Taylor & Francis Group, LLC
CRC Press is an imprint of Taylor & Francis Group, an Informa business

No claim to original U.S. Government works
Printed in the United States of America on acid-free paper
10 9 8 7 6 5 4 3 2 1

International Standard Book Number-13: 978-1-4200-4310-5 (Hardcover)

This book contains information obtained from authentic and highly regarded sources. Reprinted material is quoted with permission, and sources are indicated. A wide variety of references are listed. Reasonable efforts have been made to publish reliable data and information, but the author and the publisher cannot assume responsibility for the validity of all materials or for the consequences of their use.

No part of this book may be reprinted, reproduced, transmitted, or utilized in any form by any electronic, mechanical, or other means, now known or hereafter invented, including photocopying, microfilming, and recording, or in any information storage or retrieval system, without written permission from the publishers.

For permission to photocopy or use material electronically from this work, please access www.copyright.com (http://www.copyright.com/) or contact the Copyright Clearance Center, Inc. (CCC) 222 Rosewood Drive, Danvers, MA 01923, 978-750-8400. CCC is a not-for-profit organization that provides licenses and registration for a variety of users. For organizations that have been granted a photocopy license by the CCC, a separate system of payment has been arranged.

Trademark Notice: Product or corporate names may be trademarks or registered trademarks, and are used only for identification and explanation without intent to infringe.

Library of Congress Cataloging-in-Publication Data

Sarker, Ruhul A.
Optimization modelling : a practical introduction / Ruhul A. Sarker and Charles S. Newton.
p. cm.
Includes bibliographical references and index.
ISBN 978-1-4200-4310-5 (alk. paper)
1. Mathematical models. 2. Mathematical optimization. I. Newton, Charles S. (Charles Sinclair), 1942- II. Title.

QA401.S266 2007
658.4'0352--dc22 2007013747

Visit the Taylor & Francis Web site at
http://www.taylorandfrancis.com

and the CRC Press Web site at
http://www.crcpress.com

Table of Contents

Section II Modelling Techniques

Section III Model Solving

Section IV Practical Problems

List of Figures

List of Tables

List of Mathematical Notations

$\in$	Belongs to or subject to
$\cup$	Union
$\forall$	For all
$\exists$	There exists
$\sum$	Summation
$\leq$	Less than or equal to
$\geq$	Greater than or equal to
$\ni$	Such that
π	Pi (a constant)
Ω	Omega
θ	Theta

Preface

Because of the complexity of most real-world problems, it has been necessary for researchers and practitioners, when applying mathematical approaches, to reduce the complexity of the problem by either simplifying the problem or constraining it by making numerous assumptions. As a result, the solutions obtained from the modified model may differ significantly from an acceptable real practical solution to the original problem. To reduce the discrepancies between solutions obtained from a mathematical model approach and a realistic solution to the problem, one needs to apply appropriate modelling techniques and efficient solution approaches. As can be observed in most operations research, management science, and optimization books, journal articles, and conference proceedings papers, a tremendous amount of effort has been applied to the development of solution approaches over the past half a century. However, the appropriateness of particular modelling approaches to certain categories of problems and the modelling techniques used have received very little attention.

Mathematical modelling is an art. It is a discipline in its own right, but it is not as widely appreciated by problem-solving and decision-making practitioners as it should be. Although some modelling techniques are introduced in many operations research, management science, and optimization books, they have not been systematically covered in these texts nor applied in detail to real-world problem situations. This book provides an opportunity to discern the importance of modelling, come to grips with a wide range of modelling techniques, and illustrate the important influence of modelling on the decision-making process. This book also demonstrates the use of available software packages in solving optimization models without going into difficult mathematical details and complex solution methodologies. In addition, the book discusses the practical issues of modelling and problem solving.

This book emphasizes the modelling aspects of optimization problems. Different modelling techniques are presented in a very simple way illustrated by various examples. The formulation and modelling of a number of well-known theoretical and practical problems are provided and analyzed. Solution approaches are briefly discussed. The use of optimization packages is demonstrated in the solution of various mathematical models and an interpretation of some of these solutions is provided. The practical aspects and difficulties of problem solving and solution implementation are presented. In addition, a number of practical problems are studied.

The book is based on the authors' extensive teaching and consulting experiences in decision making and problem solving. Some of the material presented in the book has been compiled from teaching notes prepared in the 1980s and 1990s, and every effort has been made to identify the sources of this material. Any unintentional omission will be rectified in possible future editions of the book, if brought to the attention of the authors.

What Is Different in This Book?

In general, the emphasis of the current book is on modelling techniques rather than solution algorithms. Most books in the field address the solution aspects of mathematical models with very little coverage of the modelling approaches. The specific features of this book include the following:

- Describes the importance of modelling and demonstrates the appropriateness of mathematical modelling to the decision-making process.
- Deals with a wide range of model-building techniques that can be applied to problems ranging from simple and small to complex and large. The alternative modelling approaches for certain problem areas are also introduced.
- Discusses briefly the existing solution approaches and the appropriate use of software packages in solving optimization models without going into difficult mathematical details and complex solution methodologies.
- Presents different data-collection and data-preparation methods. The influence of data availability on mathematical modelling and problem solution is also discussed.
- Provides the modelling of a number of well-known theoretical problems and several interesting real-world problems. A brief review of some practical problems, with their modelling and solution approaches, is presented.
- Discusses the difficulties and practical issues of modelling, problem solving, and implementation of solutions.

The book would work as a single source for a variety of modelling techniques, classic theoretical and practical problems, and data collection and input-preparation methods, use of different optimization softwares, and practical issues for modelling, model solving, and implementation.

Benefits for the Potential Reader

Readers may benefit from the wide range of modelling techniques presented in the book, from the illustrations of the usage of various computer packages in solving developed models without going through complex solution methodologies, and from the lessons learnt by the authors through their own experiences relating to practical problem-solving and implementation difficulties. The specific benefits of the book are as follows:

- Provides a useful source for a wide range of modelling techniques. To the best of our knowledge, no other book covers modelling in as systematic a way and with similar detail.
- Presents different modelling techniques in a comprehensive way illustrated by various examples.
- Provides the formulation and modelling approaches of a number of well-known theoretical and practical problems frequently mentioned in the literature.
- Discusses existing solution approaches briefly. The application of optimization packages to solving mathematical models is demonstrated and suggested interpretations of solutions are provided.
- Presents the data-collection and data-preparation methods for model solving and discusses their relevant issues.
- Presents practical aspects and difficulties of problem solving. In addition, a number of case problems are provided.

Organization

The table of contents for *Optimization Modelling: A Practical Introduction* is laid out in a fairly traditional format; however, topics may be covered in a variety of ways. The book is divided into four sections as follows:

Section I: Introduction to Optimization and Modelling
Section II: Modelling Techniques
Section III: Model Solving
Section IV: Practical Problems

Section I contains three chapters (Chapters 1 through 3). Chapter 1 provides a general introduction to modelling and optimization. Chapter 2 describes the process of optimization and discusses its components. In Chapter 3, an introduction to mathematical modelling of optimization problems is given.

Section II contains five chapters (Chapters 4 through 8). Chapters 4 and 5 cover various modelling techniques frequently used in practice. Chapters 6 and 7 present a number of well-known problems frequently mentioned in the literature and that have arisen in practice. Chapter 8 discusses the alternative formulations of real-world problems.

Section III contains four chapters (Chapters 9 through 12). Chapter 9 provides an overview of existing optimization techniques and of using optimization software. Chapter 10 discusses the data-collection and data-preparation methods. Chapter 11 presents the problem solutions and discusses practical issues in problem solving. Few basic optimization algorithms are demonstrated in Chapter 12.

Section IV contains three chapters (Chapters 13 through 15). Chapters 13 and 14 provide full-scale mathematical models for a number of real-world problems. Chapter 15 provides the solutions of some of the models presented in earlier chapters.

For the Instructor

To use this book as a text, instructors should cover the material in Chapters 1 through 5 and Chapters 9 through 11 first. Then they can choose material from the remainder of the book based on their personal preferences.

An instructor's manual, prepared by the authors, containing PowerPoint slides and solutions to all the text problems, will be available from the publisher.

Acknowledgments

We are indebted to many of our colleagues and friends for their helpful comments and useful suggestions during the development of this book. Among these are

Hussein Abbass, UNSW, at ADFA, Australia
Rezaul Begg, Victoria University, Australia
Gopinath Chattopadhyay, CQU, Australia
Frantz Clermont, J.P. French Associates & University of York, United Kingdom
Graham Freeman, UNSW, at ADFA, Australia
Eldon Gunn, Dalhousie University, Canada
Aman Haque, Pennsylvania State University, United States of America
Anwarul Haque, VC, RUET, Bangladesh
Zohrul Kabir, IIU, Dhaka, Bangladesh
Bob McKay, Seoul National University, Korea
Abu Mamun, Qantas Airlines, Australia
M. Quaddus, CUT, Australia
Tapabrata Ray, UNSW, at ADFA, Australia
K.C. Tan, National University of Singapore, Singapore
Xin Yao, Birmingham University, United Kingdom
Sajjad Zahir, University of Lethbridge, Canada

We would like to express our gratitude to the unknown reviewers who reviewed the book proposal and provided constructive comments. A special note of thanks must go to all the staff at CRC Press–Taylor & Francis Group, whose contributions throughout the whole process from the proposal submission to the final publication have been invaluable. In fact, this book would not have been possible without the ongoing professional support from Senior Acquisitions Editor Ms. Cindy Renee Carelli and Project Coordinator Mr. David Fausel at CRC Press–Taylor & Francis Group.

Finally, we like to thank our families for their love, support, and patience throughout the entire book project. This book is dedicated to Ruhul Sarker's family (his wife Smriti, son Rubai, and daughter Shajoti) and to Maria Newton.

Ruhul A. Sarker
School of ITEE, UNSW, at ADFA
Canberra, Australia

Charles Newton
DPU International College
Dhurakij Pundit University
Bangkok, Thailand

and

School of ITEE, UNSW, at ADFA
Canberra, Australia

Authors

Ruhul A. Sarker obtained his PhD in operations research (1991) from DalTech (formerly Technical University of Nova Scotia, TUNS), Dalhousie University, Halifax, Canada.

He is currently a senior lecturer in operations research at the School of Information Technology and Electrical Engineering, University of New South Wales (UNSW), Australian Defence Force Academy (ADFA) campus, Canberra, Australia. Before joining UNSW, at ADFA in 1998, he worked with Monash University and Bangladesh University of Engineering and Technology. He has published more than 125 refereed technical papers in international journals, edited reference books, and conference proceedings. He has edited six reference books and several proceedings, and served as guest editor and technical reviewer for a number of international journals. One of his edited books *Evolutionary Optimization* was published by Kluwer (now Springer) in 2001. His research interests include applied mathematical modelling, optimization, and evolutionary computation.

He was a technical cochair of IEEE-CEC2003 and served many international conferences in the capacity of chair, cochair, or as a PC (program committee) member. He is a member of INFORMS, IEEE, and ASOR. He is the editor of *ASOR Bulletin*, the national publication of the Australian Society for Operations Research.

Charles S. Newton is an emeritus professor in the School of Information Technology and Electrical Engineering, University of New South Wales (UNSW), ADFA Campus, Canberra, Australia. Currently, he also holds the position of Dean at DPU International College, Dhurakij Pundit University, Bangkok, Thailand.

He obtained his PhD in nuclear physics from the Australian National University, Canberra, in 1975. He joined the School of Computer Science in 1987 as a senior lecturer in operations research. In May 1993, he was appointed head of the school and became professor of computer science in November 1993. Professor Newton was also the Deputy Rector (Education) at UNSW, at ADFA. Before joining ADFA, he spent 9 years in the analytical studies branch of the department of defense. During 1989–1991, Professor Newton was the National President of the Australian Society for Operations Research. His research interests encompass group decision support systems, simulation, war-gaming, evolutionary computation, data mining, and operations research applications. He has published extensively in national and international journals, books, and conference proceedings.

Section I

Introduction to Optimization and Modelling

1

Introduction

1.1 General Introduction

A model is an abstraction or mathematical representation of a problem of interest and is an essential part of the process of solving that problem optimally. However, it is difficult, and sometimes impossible, to develop a mathematical model that addresses all aspects of the problem and its planning environment, since most real-world problems are too complex and involved. As a result, researchers and practitioners attempt to formulate either a simplified version of the problem or make numerous assumptions and approximations. As the modelling approach provides solutions to the simplified or approximated problem, there may exist a significant discrepancy between those solutions and the subjectively expected realistic solution to the original problem. This discrepancy may lead to an inappropriate decision being made if the decision is made based solely on the solutions of the simplified model. This may happen in many practical decision-making or design processes. As can be seen in most books, journal articles, and conference proceedings on optimization, a tremendous effort has been put into the development of solution approaches over the past half-a-century. However, the appropriateness of modelling and appropriate techniques have received little attention. In fact, mathematical modelling may be considered an art that has its own domain and has not been generally explored by problem-solving practitioners. So, instead of solution techniques, the emphasis of this book is on the modelling aspects such as

- Importance of modelling in the decision-making process
- Modelling techniques
- Influence of modelling in decision making
- Linking of the mathematical model to the other components of a decision-making process

In this introductory chapter, we present a brief history of optimization, the nature of the optimization problem, the nature of the mathematical model used in optimization problem-solving approaches, the basic concept of optimization, and the classification of optimization problems. The general structure of the book is also discussed.

1.2 History of Optimization

Optimization techniques have been available for more than a century. In the beginning, differential calculus was the basic tool applied for finding maxima or minima of functions, which arose in many practical situations and theoretical problems. There is a clear evidence of the use of mathematical models and optimization techniques at the turn of the twentieth century such as (1) in 1900, when H.L. Gantt used charts to efficiently schedule jobs on machines, which are known as Gantt charts today; (2) in 1915, when F.W. Harris derived the mathematical formulation for the most economic quantity of an item to order from a vendor, which is the well-known economic order quantity in inventory management today; and (3) in 1917, when A.K. Erlang derived the mathematical formula for analyzing problems encountered by callers to an automated telephone switchboard, which has led to the present queuing/waiting line analysis.

During World War II, the British government organized civilian scientific groups to assist field commanders in solving complex, strategic, and tactical problems. The purpose was to maximize their war effort with the limited resources they had. The success of the British groups leads the United States to institute similar efforts in 1942, although a small-scale project dated back to 1937 has been reported. The British scientific community described the activities the groups conducted as "operational research" whereas in the United States it was termed "operations research." Following the successes of those activities, operations research has been recognized and established as a separate discipline within the academic arena. It must be mentioned here that optimization is considered a subset of operations research discipline.

After World War II, a dramatic development and refinement of operations research techniques occurred with a corresponding expansion from singularly addressing military problems to problems encountered in almost all areas of public and private industry as well as in government services. The managers and decision makers realized that the savings incurred from applying operations research approaches to solving problems were very significant because even a cent saved per unit on a large production run could total up to millions of dollars.

In 1947, George B. Dantzig developed the simplex algorithm for solving linear programming problems, which established him as one of the forefathers of the discipline. Linear programming is one of the basic techniques used in optimization. Dantzig has since stated that "The tremendous power of the simplex method is a constant surprise to me." The systematic development of practical computing algorithms for addressing linear programming problems began in 1952 at the Rand Corporation in Santa Monica, United States, under the direction of Dantzig. He worked intensively on this project until late 1956, by which time great progress had been made on first-generation computers. However, the importance of linear programming methods was described, in 1980, by computer scientist Laszlo Lovasz* who wrote "If one collected statistics about which mathematical problem is using up most of the computer time in the world, then... the answer would probably be linear programming." Also that year, Eugene Lawler* of Berkeley wrote "Linear programming is used to allocate resources, plan production, schedule workers, plan investment portfolios and formulate marketing (and military) strategies. The versatility and economic impact of linear programming in today's industrial world is truly awesome."

In addition to many other conventional optimization techniques developed over the past half-a-century (as will be discussed later), the recent development of modern heuristic techniques such as simulated annealing, tabu search, genetic algorithms, neural computing, fuzzy logic, and ant colony optimization are providing practitioners with some sophisticated tools to address more complex situations.

1.3 Optimization Problems

Problems that seek to maximize or minimize a mathematical function of a number of variables, subject to certain constraints, form a unique class of problems, which may be called optimization problems. Many real-world and theoretical problems can be modelled in this general framework.

A common term optimize is usually used to replace the terms maximize or minimize. The mathematical function that is to be optimized is known as the objective function, containing usually several variables. An objective function can be a function of a single variable for some practical problems; however, a single variable function may not challenge from an optimization point of view. Optimization problems may involve more than one objective function and are known as multi-objective optimization problems.

Depending on the nature of the problem, the variables in the model may be real or integer (pure integer or binary integer) or a mix of both. The optimization problem could be either constrained or unconstrained. In the constraint part of a mathematical model, the left-hand side of the constraint function (or a single variable) is separated from the right-hand-side value by one of the three signs: (1) equal to ($=$), (2) less than or equal to ($\leq$), or (3) greater than or equal to ($\geq$).

In this book, we mainly discuss deterministic modelling. The functions, either objective or constraints, may be from either the linear or nonlinear domain. As per the function properties, they could follow any pattern such

*Quotations taken from Freund (1994) SIAM News as referenced later.

as continuous or noncontinuous, differentiable or nondifferentiable, convex or nonconvex, or unimodal or multimodal. These properties are discussed later under the problem classification.

1.4 Mathematical Model

A general structure of a mathematical model (also known as a mathematical programming model) can be represented as follows:

Find x to

$$\begin{aligned} &\text{Maximize } f(x) \\ &\text{Subject to } g_i(x) \le gb_i, \quad i = 1, \ldots, m \\ &\qquad h_j(x) = hb_j, \quad j = 1, \ldots, p \\ &\qquad x \ge 0 \end{aligned} \qquad \text{Model (1.1)}$$

where the objective function f is a function of a single variable x, and the constraint functions g_i and h_i are general functions of the variable (otherwise expressed as an unknown, decision variable or sometimes as a parameter) $x \varepsilon R^n$. The right-hand sides, gb_i and hb_j, are usually the known constants for deterministic problems. The non-negativity constraint, $x \ge 0$, is necessary for many practical problems (since many variables cannot be negative) and for many solution approaches (assumption by default). The above standard model may vary as follows: (1) contains upper and lower bounds of x instead of a non-negativity constraint, (2) contains upper and lower bounds of x instead of any other constraint, and (3) the above standard model, with or without (1) and (2), with multiple variables.

Let us assume $\overline{x}$ represents a set of variables, where $\overline{x} = (x_1, x_2, \ldots, x_n)$, then the above model can be rewritten for multiple variables as follows:

$$\begin{aligned} &\text{Maximize } f(\overline{x}) \\ &\text{Subject to } g_i(\overline{x}) \le gb_i, \quad i = 1, \ldots, m \\ &\qquad h_j(\overline{x}) = hb_j, \quad j = 1, \ldots, p \\ &\qquad \overline{x} \ge 0 \end{aligned} \qquad \text{Model (1.2)}$$

1.4.1 Characteristics and Assumptions

The general characteristics of a mathematical model can be described as follows:

- A limited quantity of resources (usually represented by the right-hand side of a constraint equation) is described by a parameter.
- The resources are used for some activity (usually represented by a decision variable) such as to produce something or to provide some service.

- There are a number of alternative ways in which the resources can be used.
- Each activity in which the resources are used yields a return in terms of the stated goal (contribution to the objective function).
- The allocation of resources is usually restricted by several limitations (known as constraints).

Suppose $g_i(\overline{x})$ and $f(\overline{x})$ in Model 1.2 are linear functions and they can be represented as follows:

$$
\begin{aligned}
&f(\overline{x}) = c_1x_1 + c_2x_2 + \cdots + c_nx_n \\
&\text{and} \\
&g_1(\overline{x}) = a_{11}x_1 + a_{12}x_2 + \cdots + a_{1n}x_n \leq gb_1 \\
&g_2(\overline{x}) = a_{21}x_1 + a_{22}x_2 + \cdots + a_{2n}x_n \leq gb_2 \\
&\vdots \\
&\vdots
\end{aligned}
\qquad \text{Model (1.3)}
$$

In the constraint, $g_1(\overline{x})$, a_{11} is the resource required from gb_1 for each unit of activity x_1, a_{12} is the resource required from gb_1 for each unit of activity x_2, and so on. In the objective function, $f(\overline{x})$, c_1 is the return per unit of activity x_1, c_2 for activity x_2, and so on. Here, c_i and a_{in} are known as the coefficients of the objective function and the constraint functions, respectively.

The general assumptions for formulating a mathematical model can be outlined as follows:

- Returns from different allocations of resources can be measured by a common unit (such as dollars, kilograms, or utility) and can be compared.
- Resources are to be used in the most economical manner.
- All data are known with certainty for deterministic problems (note that this book mainly considers deterministic cases).
- Decision variables are either real or integer or a mix of both.
- Function type is general (that means not restricted to any particular type).

Example 1.1: A simple example

A small retail shop receives the supply of units of one particular product, directly from either a manufacturer or a supplier, and sells them to individual customers in an open market. The demand for the items is approximately constant over time. It is convenient for the retailer to order the items in a batch at regular time intervals and store them in the shop/warehouse, until they are sold. The retailer faces the basic questions of how many items should be ordered in each order and how often?

Suppose x is the quantity (amount) of the items (or batch size) to be ordered each time. There are some costs involved with the ordering, storing, and administrative activities. To keep the example simple, it will be discussed very briefly. The annual demand, D, can be calculated as the demand rate is known and constant. The number of orders (n) required per year is therefore equal to D/x. There is a fixed cost associated with each order, F \$/order, which is independent of the quantity to be ordered. As x decreases, n as well as the total ordering cost (TOC) per year increases. With the increase of x, the cost of the total inventory holding (TIC) increases almost linearly. Suppose h is the inventory holding cost (\$) per unit per year. Let us now write down the cost equation for the problem:

$$\text{TOC per year} = \frac{D}{x}F \tag{1.1}$$

$$\text{TIC per year} = \frac{x}{2}h \tag{1.2}$$

where $x/2$ represents the average inventory per year or per any cycle of the ordering process. If we assume p as the unit item price, the total purchase cost (TPC) $= Dp$.

So, the total cost of the system (TCS) is equal to

$$\begin{aligned} \text{TCS} &= \text{TPC} + \text{TOC} + \text{TIC} \\ &= Dp + \frac{D}{x}F + \frac{x}{2}h \end{aligned} \tag{1.3}$$

Generally speaking, the order size (x) should be the one for which TCS is minimum. The mathematical model of the problem can then be written as

$$\begin{aligned} &\text{Minimize TCS} = Dp + \frac{D}{x}F + \frac{x}{2}h \\ &\text{Subject to} \\ &\qquad \text{lb} \le x \le \text{ub} \end{aligned} \qquad \text{Model (1.4)}$$

The constraint indicates that x should be within the lower bound (lb) and the upper bound (ub) set by the retailer. Depending on the type of problem, x can be either real or integer.

1.5 Concept of Optimization

The basic concept of optimization is to find the best possible solution (point or alternative) to a given model/problem. To find it, one should examine all alternatives and prove that the one selected is indeed the best. In the above example, it is to find the value of x that minimizes TCS while satisfying the upper and lower bounds/constraints. For a better understanding of the optimization task, let us analyze TOC, TIC, and TCS in regard to x. We can exclude TPC as it is independent of x and it would not affect the optimization process. So the revised optimization model is the following:

$$\text{Minimize TCS} = \frac{D}{x}F + \frac{x}{2}h$$

$$\text{Subject to} \qquad\qquad \text{Model (1.5)}$$

$$\text{lb} \le x \le \text{ub}$$

For ease of analysis, we need an instance of the problem. Informally, in an instance we are given the "input data" and should have enough information to obtain a solution; a problem is a collection of instances, usually all generated in a similar way.

Suppose for a given instance of the problem, $D = 100$ units, $F = 4$ \$/order, $h = 8$ \$/unit/year, lb = 1, and ub = 15, then the TCS for the problem instance can be calculated as

$$\begin{aligned} \text{TCS} &= \frac{D}{x}F + \frac{x}{2}h \\ &= \frac{100 \times 4}{x} + \frac{8}{2}x \\ &= \frac{400}{x} + 4x \end{aligned} \qquad (1.4)$$

The range of x is 1–15, which contains an infinite number of alternative values (for real x). If x is an integer, there exist only 15 alternative numbers. For the integer case, it is not difficult to find all 15 TCS function values by simple calculation as shown in Table 1.1.

TABLE 1.1

TOC, TIC, TCS versus x for a Given Problem Instance

x (Amounts)	TOC $= \frac{400}{x}$	TIC $= 4x$	TCS (\$) = TOC + TIC
1	400.00	4.00	404.00
2	200.00	8.00	208.00
3	133.33	12.00	145.33
4	100.00	16.00	116.00
5	80.00	20.00	100.00
6	66.67	24.00	90.67
7	57.14	28.00	85.14
8	50.00	32.00	82.00
9	44.44	36.00	80.44
10	**40.00**	**40.00**	**80.00**
11	36.36	44.00	80.36
12	33.33	48.00	81.33
13	30.77	52.00	82.77
14	28.57	56.00	84.57
15	26.67	60.00	86.67

From Table 1.1, it is clear that TCS = \$80.00 is the optimal system cost with a corresponding value of $x = 10$ units. In fact, this is the optimal solution for the given problem instance. That means, the retailers should order 10 items each time they require a resupply. This problem is easy and simple in regard to finding the optimal solution. Interestingly, it is not possible to find the optimal solutions for all models/problems by using an enumeration method as above because of a possible large number of alternative solutions involved.

To apply the optimization techniques/algorithms, which find the optimal solutions by analyzing only a subset of all possible alternative solutions, we need some knowledge regarding the function properties. The plot of TOC, TIC, and TCS against the continuous variable x (from 6 to 15) is shown in Figure 1.1 to get a feeling of the function behavior. As we can see, TIC is linearly increasing, TOC is decreasing, and TCS (which is a sum of TIC and TOC) first decreases and then increases as x increases.

We are more interested in TCS and the point of the function where TCS is a minimum. A more detailed plot of TCS against x is shown in Figure 1.2.

From the plot, it is easy to see that the function TCS is smooth, continuous, and convex with one clear minimum at $x = 10$, which represents the optimal solution of the problem instance. It is impossible to visualize a function with more than two variables. In that case, we are fully dependent on the mathematical properties of the function and its analysis. The function TCS is differentiable for real x. After differentiating the function TCS with respect to x, we get

$$\frac{\mathrm{d(TCS)}}{\mathrm{d}x} = -\frac{D}{x^2}F + \frac{h}{2} = -\frac{400}{x^2} + 4 \tag{1.5}$$

Mathematically, $\frac{\mathrm{d(TCS)}}{\mathrm{d}x}$ is known as the gradient of the function TCS. If we calculate the gradients of TCS at points $x = 8$, 10, and 12, we get

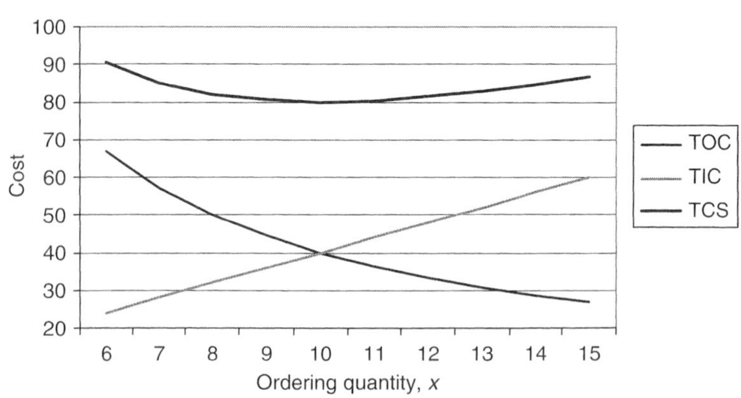

FIGURE 1.1
Plot of TOC, TIC, TCS versus x for the given problem instance.

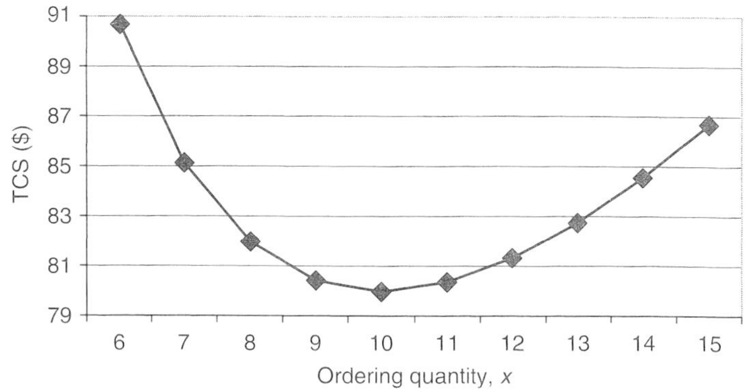

FIGURE 1.2
Plot of TCS versus x for the given problem instance.

$$\text{for } x = 8, \quad \frac{d(\text{TCS})}{dx} = -\frac{400}{8^2} + 4 = -2.25$$
$$\text{for } x = 10, \quad \frac{d(\text{TCS})}{dx} = -\frac{400}{10^2} + 4 = 0$$
$$\text{for } x = 12, \quad \frac{d(\text{TCS})}{dx} = -\frac{400}{12^2} + 4 = +1.22 \tag{1.6}$$

It can be seen here that the gradient is zero at the optimum (minimum) point, negative when x is lower than the optimum value, and positive when x is greater than the optimum value. That means if we like to find an optimum point we must search for a point on the function with zero gradient.

1.6 Classification of Optimization Problems

The general optimization problems can be classified as shown in Figure 1.3.

We have discussed the objective classification (single or multiple) and the objective type (maximization or minimization) in an earlier section. In case of multiple objectives, the objectives usually contradict each other. If they do not, the multiple objectives can be converted into a single-objective problem. The problem classification (in the next page) indicates whether the problem contains constraints or not. Some people believe that there are no unconstrained optimization problems in the real world, as these all will have either constraint functions or variable bounds (upper or lower) or both. The study of unconstrained problems is very important since many optimization algorithms solve constrained problems by converting them into an unconstrained or a sequence of unconstrained problems. In addition, several unconstrained optimization techniques can be extended in a natural way to provide and motivate solution procedures for constrained problems.

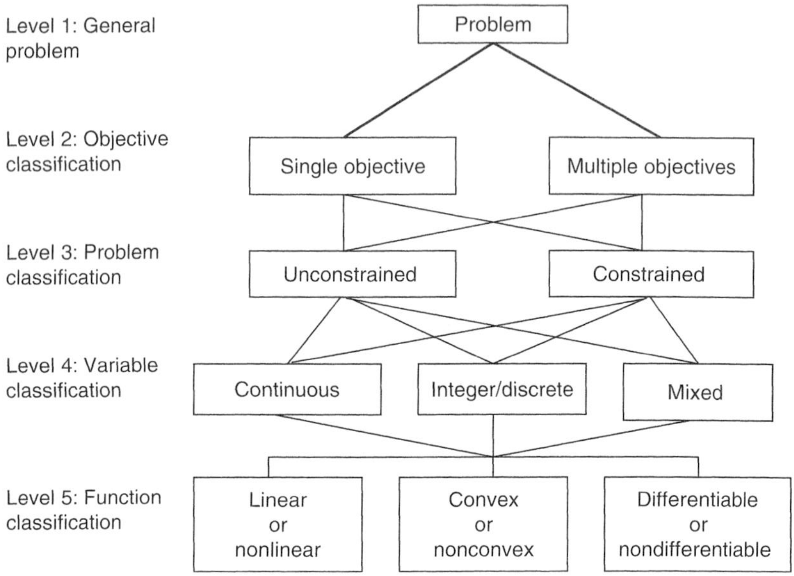

FIGURE 1.3
Classification of optimization problems.

We have already discussed the variable classification as real, integer, or mixed integer. However, many practitioners recognize them as continuous, integer, discrete, or mixed. In problems with continuous (real) variables, we are generally looking for a set of real numbers. The optimization problem with integer or discrete variables is termed a combinatorial problem. In combinatorial problems, we are looking for an object from a finite or infinite set—typically an integer, set, permutation, or graph. These two kinds of problems generally have quite different flavors, the methods for solving them have become quite divergent.

The function classification mainly deals with functions' mathematical properties, which are very important from the solution approach point of view. The objective or constraint functions may be either linear, nonlinear, or both. If all the functions are linear in a given model, we call it a linear programming model or linear model. If one or more of the functions of a model involve nonlinearity, we call it a nonlinear model. The solution approaches of nonlinear models are quite different and more complex than those of linear models. An unconstrained problem with a single linear objective function does not raise any interest from an optimization point of view. However, unconstrained nonlinear optimization problems attract many interesting research studies and applications.

Convexity is considered as an important property in classical optimization as many optimization techniques/algorithms are developed based on the assumption that the function is convex. In optimization, the solution approaches can be divided into two major groups: (i) those with derivatives and (ii) those without derivatives (/derivative free). Differentiability of the function is necessary when using derivative-based techniques. Differentiability is closely related to the continuity of functions. For continuity, differentiability and convexity properties of functions, consult any first year mathematics book. A brief description is also available in works by Hillier and Lieberman (2001) and Bazaraa et al. (1990).

As an example, the function properties of a single-objective constrained problem with continuous variables could be nonlinear, convex, and differential (see Figure 1.3). In addition to the above general classification, the optimization problem domain also considers function properties such as unimodal versus multimodal, static versus dynamic, and constraint properties such as soft versus hard constraints. A function with only one peak (/optimum solution) is known as a unimodal function whereas a function with more than one peak (either local or global optima) is recognized as a multimodal function. If a function changes over time, it is known as a dynamic function. In this book, we restrict ourselves to static functions with a minor variation. The constraints that must be satisfied, in the final solution, are known as hard constraints. Soft constraints are those constraints that can be violated with a certain penalty or under certain conditions. These constraints are discussed in detail in a later chapter.

1.7 Organization of the Book

The book is divided into four sections as follows:

Section I: Introduction to Optimization and Modelling
Section II: Modelling Techniques
Section III: Model Solving
Section IV: Practical Problems

Section I contains three chapters (Chapters 1 through 3). Chapter 1 provides a general introduction to modelling and optimization, a brief history of optimization, concept of optimization, the basic structure of mathematical model, and a general classification of optimization problems. Chapter 2 describes the process of optimization and discusses its components and the role of optimization within that process. In Chapter 3, we provide an introduction to mathematical modelling for optimization problems. This chapter discusses the basic components of a mathematical model and provides simple modelling examples. It also introduces various types of models such as linear programming, integer programming, multi-objective optimization, goal programming, and nonlinear programming.

Section II contains five chapters (Chapters 4 through 8). Chapter 4 covers a number of simple modelling techniques frequently used in practice. In this chapter, we also introduce the use of subscripts in variables and provide several examples to demonstrate the simple modelling techniques as well as the use of subscripts. In Chapter 5, we discuss simple integer and nonlinear modelling techniques frequently used by modellers. In addition, a few other modelling techniques are presented. A number of examples are provided to demonstrate the use of integer and nonlinear variables in modelling practical problems. In Chapter 6, we introduce the use of the summation sign when expressing a number of variables together and the use of the subset sign to separate certain variables—which should make the model even more compact specifically when we are dealing with large-scale modelling. We also present a number of well-known problems frequently mentioned in the literature and that have arisen in practice. In Chapter 7, we present a few more well-known problems frequently mentioned in the literature. The mathematical models of these problems are analyzed and their practical applications are briefly discussed with reference to the literature. Chapter 8 discusses the alternative formulations of real-world problems based on the assumptions and decision process. Two different alternative models are presented with examples of practical problems.

Section III contains four chapters (Chapters 9 through 12). Chapter 9 provides an overview of existing conventional optimization techniques, heuristic techniques, and optimization software. A number of appendices instructing the use of well-known optimization software are also provided. Chapter 10 discusses the data requirement for the models, data collection methods, and data preparation as required by the software packages and solution approaches. Chapter 11 presents the solutions of models and the outputs provided by different software packages, and discusses practical issues faced by users when dealing with optimization models and software packages. A number of examples to demonstrate these issues are provided. As indicated earlier, the purpose of this book is to cover the basic modelling techniques and to demonstrate the use of software packages when solving optimization models. However, to give a feeling about how the optimization packages solve the complex models, we present and analyze a few basic optimization algorithms in Chapter 12.

Section IV contains three chapters (Chapters 13 through 15). Chapter 13 provides full-scale mathematical models for a number of linear programming-based real-world problems. Chapter 14 presents few more complex practical optimization models. Chapter 15 provides the solutions of some of the models presented in Chapters 13 and 14, and discusses a number of practical issues in solving those models.

Exercises

1. Jenny will run an ice cream stand in the coming week-long multicultural event. She believes the fixed cost per day of running the stand is $60. Her best guess is that she can sell up to 250 ice creams per day at $1.50 per ice

cream. The cost of each ice cream is \$0.85. Find an expression for the daily profit, and hence find the break-even point (no profit–no loss point)?

2. The total cost of producing x items per day is $(45x + 27)$ dollars, and the price per item at which each may be sold is $(60 - 0.5x)$ dollars. Find an expression for the daily profit, and hence find the maximum possible profit?
3. A stone is thrown upward so that at any time x (in second) after throwing, the height of the stone is $y = 100 + 10x - 5x^2$ (in meter). Find the maximum height reached?
4. A manufacturer finds that the cost $C(x) = 2x^2 - 8x + 15$, where x is the number of machines operating. Find how many machines should one operate to minimize the total cost of production. What is the optimal cost of production?
5. A string 72 cm long is to be cut into two pieces. One piece is used to form a circle and the other a square. What should be the perimeter of the square in order to minimize the sum of two areas?
6. Find the maximum or minimum values of the following quadratic functions, and the values of x for which they occur:

   (i) $f(x) = x^2 - 4x + 7$

   (ii) $f(x) = 3 - 8x - x^2$

References

Bazaraa, M., Jarvis, J., and Sherali, H., *Linear Programming and Network Flows*, John Wiley & Sons, New York, 1990.

Freund, R. (contributed extensively), Professor George Dantzig: Linear programming founder turns 80, SIAM News, November 1994 (also available in http://www.stanford.edu/group/SOL/dantzig.html).

Hillier, F. and Lieberman, G., *Introduction to Operations Research*, 8th ed., McGraw-Hill, Boston, 2005.

2

The Process of Optimization

2.1 Introduction

A key step in many decision-making and design processes is the optimization phase, which itself contains several stages. The purpose of the optimization process is to help determine realistic and practical outcomes of management decision-making and design processes. So, it is appropriate at this point to provide an overview of the optimization process.

In this chapter, we discuss the general structure of the decision-making process, giving a brief description of its components and the role of optimization within that process.

2.2 Decision Process

The decision-making process may be initiated whenever an individual or group becomes concerned about some issues or matters they have encountered. In addressing this concern, the individual may be able to transpose that concern into a problem that demands a solution. The problem solvers then have basically two avenues open to them, either the qualitative approach or the quantitative approach. Using only a qualitative approach, the problem solver, when making a decision, relies on personal judgment or past experience acquired in dealings with similar problems in the past. In some cases this approach may be adequate; however, there are many situations where a quantitative approach to the problem provides a better structured and logical path through the decision-making process.

Decision makers apply the quantitative approach when they feel that the problem facing them can be solved by applying techniques that have already been developed and applied to similar problems, when the problem is recurrent, or when the problem is very complex involving many variables.

Quantitative analysis skills are usually acquired through the study of mathematical tools, both pure and applied, and practical on-the-job experience. Optimization techniques are applied to real-life systems that have perceived complex management problems. The premise of decision making is that there are several different ways to address a problem and that the best way may not be obvious or necessarily unique. The process of finding the best (optimal) way is called optimization. Most analysts break a decision-making process down into six major steps or phases:

1. Identifying and clarifying the problem
2. Defining the problem
3. Formulating and constructing a mathematical model
4. Obtaining a solution to the model
5. Testing the model, evaluating the solution, and carrying out sensitivity analysis
6. Implementing and maintaining the solution

Some analysts, however, combine some of the six stages above to fewer stages, whereas others expand the six phases into seven or more phases, in order to facilitate the process. In the above decision-making process, steps 2–5 represent the optimization process. No one phase or step can be said to be the most critical element in the process, although if the problem has not been correctly identified, the analysis will produce erroneous results.

The steps of the decision-making process are shown in Figure 2.1.

A study may start by proceeding through the phases in a sequential order; however, it is rare for an analyst to get the first two stages absolutely correct, and therefore they need to revisit the first two stages during the model-development stage and later stages for further clarification and definition. In most situations, it may be necessary to also apply this recursive approach to other stages by revisiting previous phases to modify the outcome or solution of the present phase. Because of this iterative nature of most decision-making processes, they do not necessarily have to finish in the same order. In fact, several phases may be ongoing simultaneously until the project reaches completion.

To be able to identify a successful formulation of the problem, the analyst and sponsor need to consider all phases of the project. This is because the interrelationships between various aspects of the problem may suggest a form of model, which in turn may dictate what data is needed for problem formulation, testing, and implementation. After the initial consideration of the problem and the development of a model, it may be necessary to modify the model so that the results can easily be implemented. Following the implementation and testing of the solution, it may be necessary to go right back and reformulate the problem to achieve a better solution.

Sometimes, the analysis of the process is accompanied by a feasibility study that addresses issues such as technical, economical, and operational factors. The feasibility study should inform the sponsors of the analysis what techniques are available for the solution of the problem, whether the

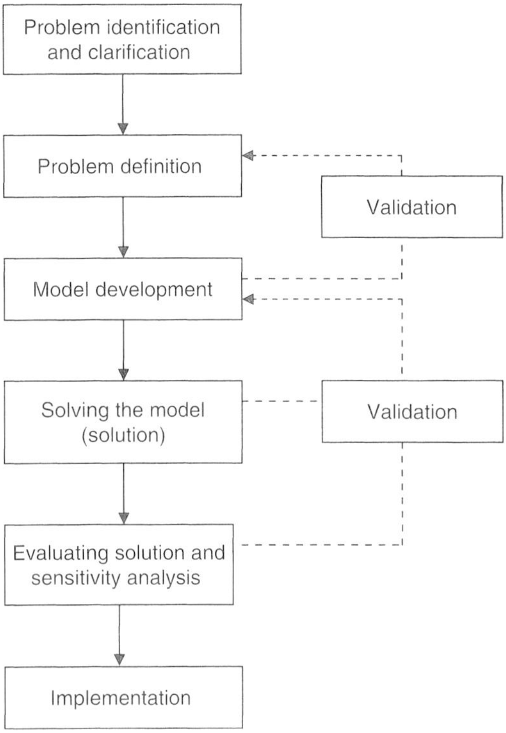

FIGURE 2.1
The decision-making process.

cost of the analysis is realistic, and whether the probability of a successful implementation of the results is high or low. We briefly describe the phases of the decision-making process below.

2.3 Problem Identification and Clarification

The first step in the process is to develop a clear understanding of the problem, usually by carrying out a detailed observation of the real-world system. Through the observation of various symptoms, a problem becomes apparent, although it is not always clear what the problem is. In many cases, managers are those who become aware of the existence of the problem through having misgivings or concerns. They may develop concerns over the activities and operations of their work unit or project. The work unit may not be achieving all the goals and objectives set for it, staff may not be performing to their full capacity, staff may be disillusioned and unhappy, or the project may not be meeting its deadlines or milestones. It is this dissatisfaction or concern that could be a symptom of a serious underlying problem.

In order to establish the existence of a problem and to bring a structured approach to addressing it

- There must be an individual or a group of individuals who through their concerns perceive that they have a problem that merits a solution that achieves some objectives.
- There must be a number of alternative ways of achieving the objectives and there must be some doubt in the decision maker's mind as to which alternative is best in terms of achieving those objectives.
- There is a relevant environment within which the concerns have arisen and have led to the perception that there exists a problem to be solved.

Consider the following example. During the last 5 years, a major wholesale-distributing company (ADC), distributing certain products in a state capital city in Australia, has captured about 35% of the market share. Senior management of the company did not make any major changes to their operational policies during that 5 year period. They have been maintaining a constant number of employees and were using similar resources each year. The company has been experiencing incrementally increasing profits each year without major management influences. As there were no new competitors in the market and the other competitors had not challenged them, the management had automatically assumed that they were doing pretty well.

However, one of their competitors has recently announced price discounts and another is preparing to do so. The management of ADC has become concerned about the news and has anticipated that they may lose some of their customers. It seems that ADC management is facing a problem, which may show the symptoms of customer loss. So a customer loss problem has been identified by management.

2.4 Problem Definition

It is sometimes difficult to distinguish between symptoms and problems as real-world problems are complicated by many interrelated factors. By defining the problem, we mean recognizing that a problem (or an opportunity) exists, determining its magnitude, defining it precisely, and noting what the symptoms are. Sometimes, what is described as a problem may only be a symptom. The problem definition phase is the phase where a clear and concise statement of the problem has to be derived from the observations carried out in the problem identification and clarification stage. This clear problem definition must define the objective, the measures of effectiveness (MoE) to be used to compare alternatives, stakeholder interests, assumptions, restrictions and limitations, and possible available information on resources and parameters involved with the problem. Experience has shown that bad problem definition has been a major cause of failure of analysts arriving at realistic solutions to their organizations' problems, thereby failing to serve their organizations well.

The analysts and manager may find that a large proportion of the time spent in a study is directed toward identifying the real problem. It is this phase of the approach that is most important and usually the most difficult. It is necessary to identify the underlying causes of the problem, so that an accurate definition of the problem can be established.

The formulation of a problem may often require several attempts or refinements. The process of formulation starts with a tentative formulation followed by a procession of progressive reformulations and refinements as the project proceeds and the team members gain deeper insights into the problem.

Consider the situation of ADC discussed in the previous section. Apparently, ADC identified that they either had to match the discount offer or do nothing. For matching the discount offer, ADC would need to collect data and carry out further analysis of the impact of offering discounts. A quick analysis shows that ADC would not survive offering an equivalent level of discount as their competitors, due to their operating costs being higher than the expected revenue for the same level of market share. However, if they do not offer similar discounts, they will be out of business within a short period. What should ADC do now? Suppose ADC wants to stay in the business. Can you think of any other option that may help ADC to survive?

Under the current situation, consider the concern (based on new information derived from the analysis) that operating costs are higher than the expected revenue. If we assume that the revenue will remain constant, then the only option available is to reduce the operating costs. So ADC's objective could be the minimization of operating costs. ADC performs two different operations: (1) receives the supply of goods and stores them in the warehouse and (2) distributes/delivers the supplies to customers via a fleet of vehicles. Now it seems that the possible problems are (1) the efficient use of the vehicles and (2) the efficiency of the storage/inventory system. In other words, minimization of operating costs through efficient utilization of storage space and vehicles. Now, one can probe further into the concerns to determine the assumptions, unknown parameters and constraints, and collect the necessary information.

2.5 Development of a Mathematical Model

Once we have defined the problem to be analyzed, the next step is to develop a mathematical model. The mathematical model is a mathematical representation of the real system or object and is able to present the important features of the system in a form that is easy to interpret. The mathematical model is the experimental test bed on which one can test new decision alternatives' impact on the existing system without disturbing the real system. Initially, model building may be qualitative in nature, involving a rather informal descriptive approach. From this informal qualitative model, a formal quantitative model can be developed. Mathematical models are developed to represent specific attributes of interest of a real-life system and they usually do not include features that are not relevant to the problem at hand.

There exist three major types of models: *iconic*, *analog*, and *symbolic* or *mathematical* models. An iconic model is basically a realistic scale model of the system, such as a scale model of an aircraft for use in a wind tunnel. An analog model utilizes one physical property to represent another physical property, such as using a damped electronic circuit to investigate a mechanical hydraulic suspension system. A symbolic or mathematical model uses a set of mathematical symbols and functional relationships to represent some physical situation. The major kinds of models of interest to optimization are the symbolic or mathematical models that allow manipulation of the entity modelled. The purpose of a model is to explain, predict, or control the behavior of the entity modelled.

There are four broad categories of mathematical models available to the analyst. The first category of model is the *descriptive* model, which allows the analyst to represent some physical situation in a visual mode. Simulation models and queuing models belong to this category. A simple type of descriptive model is the influence diagram. The second category of model is the normative or optimization model, also called the prescriptive model. Most of the techniques, such as linear programming, transportation algorithms, and project management techniques fall into this category. These types of models endeavor to optimize some objective function subject to certain limitations. The third type of model used in operations research is the heuristic model. Heuristic models that basically employ rules of thumb or intuitive rules, typically guided by common sense. Heuristic models provide good solutions, but not necessarily optimal. The fourth type of model is the predictive model, which has been developed so that estimates of future trends can be made, for example, in areas such as manpower planning and financial planning. This category includes models of regression and time series analysis.

In the case where all the functional relationships and parameters are known with certainty, one is able to develop a deterministic model, whereas stochastic models address those situations where there is uncertainty involved; data is usually described in terms of probability distributions.

Models, when developed, should follow certain guidelines such as those described below:

Simple: Models should be kept as simple as possible so that they can be described in terms that can be understood by nonspecialists, who often do not have a scientific background.

Robust: Models should be developed so that they are general enough to give valid answers over a wide range of input values.

Adaptive: If a model is developed to represent the real world, which is a very dynamic system, then the model must be sufficiently adaptive to be able to incorporate those rapid changes, which may be experienced in the present or future worlds.

Complete: As models are to represent the total problem and its environment, the model should address all aspects of the problem and not just some components of the problem.

User friendly: The user must be able to change inputs easily and to obtain answers quickly without having to rely on expert support.

In a mathematical model, the properties of the objects under consideration usually take the form of decision variables whose optimal values will be determined by the model. Starting values for the model are usually chosen by the decision maker from a specified set of values known to provide a feasible solution to the problem. Those variables that are beyond the control of the decision maker are represented as random variables, or by parameters and constants. The objective function or measure of effectiveness expresses the contribution each decision variable makes toward the decision maker's objectives. Constraints are represented by relationships between the variables, which may be in the form of functions, equations, or inequalities.

A general purpose model is a mathematical model that can be used to solve a whole category of problems that satisfies certain assumptions as to the form of the decision variables, and includes the nature of the functional relationships between them. It is a special purpose model if its structure is specific to a unique problem.

With most general purpose models, one may associate one or more specialized techniques for finding the optimal solution, such as linear programming, goal programming, or critical path method. For this reason, general purpose models are often referred to as techniques. Many problems can be solved by several techniques, each offering certain advantages. The expert analyst should be aware of these advantages and can thus determine the technique that is more appropriate for the problem at hand.

In order for a model to identify the preferred solution, it is necessary for the model to differentiate between the different alternatives in their ability to meet the objective of the decision.

2.5.1 Measure of Effectiveness

A quantitative measure, used to compare the effectiveness of alternatives in achieving the objective, is called a measure of effectiveness. A measure of effectiveness is some scale that can be used to compare the performance of alternative solutions to the problem being studied. The measure of effectiveness, in order to be a suitable scale, has to relate in a direct way to the objectives of the study. An initial choice of a measure of effectiveness should have occurred during the problem-formulation stage, when the objectives of the study were being defined. Investigating measures of effectiveness used previously in similarly related studies provides a good starting point in determining appropriate measures of effectiveness for the current study. In some problems, the measure of effectiveness may be expressed as an objective function, which can be in the form of a mathematical relationship between decision variables and cost coefficients.

The Military Operations Research Society has developed a four-level hierarchy of measures from high-level force effectiveness to low-level rudimentary measures of physical entities to address a wide variety of complexity in objectives of studies. The four levels proposed are the following:

- Measures of policy effectiveness (MoPE), which focus on policy and societal outcomes
- Measures of force effectiveness (MoFE), which focus on how a force performs its mission or the degree to which it meets its objectives
- Measures of C2 (command and control) effectiveness (MoCE), which focus on the impact of C2 systems within the operational context
- Measures of performance (MoP), which focus on internal system structures, characteristics, and behavior

The analyst is often interested, when carrying out quantitative analysis, in measures that can be used to express the response or performance of the system in terms of the decision maker's objectives, or those measures that are needed to verify if constraints imposed on the decision choices are satisfied. The objectives and constraints thus determine measures of performance or low-level measures of effectiveness that are relevant for evaluating alternative courses of action. There is often a one-to-one correspondence between objectives and measures of performance. There can also be instances where no suitable set of performance measures can be defined to evaluate the performance.

Measures of performance can be aggregated to form measures of effectiveness that evaluate the response of the system, in terms of each of the decision maker's objectives, to different courses of action. For a single objective, the measure of effectiveness to be optimized is usually referred to as the objective function. The objective function may be expressed in numerical form as a function of decision variables and parameters. However, it is not essential that the function be of a numerical form. A function may be designed to use any one of the four scales: nominal, rank or ordinal, relative or interval value, and absolute value or ratio. A better–worse, success–failure, or nonnumerical ranking of each alternative action may often suffice to identify the best decision. If only a single objective is to be met then the measures of effectiveness may be expressed as a single value, whereas in the case of multiple objectives a complex array or matrix of values may be required. The measure of effectiveness may be more conveniently expressed in analytical terms as a functional relationship of the decision variables, if the number of alternative courses of action is very large or even infinite.

In cases where decision makers wish to achieve multiple objectives, they may encounter conflict between the objectives. One possible way of overcoming conflicts is to prioritize the objectives and solve the problem with respect to the most important objective while setting minimal performance targets (goals) on all other objectives. Indeed, one could reduce all the objectives into goals (goal programming approach). In the case where objectives can be measured in comparable units, then one may be able to aggregate all the objectives into some common measure, such as costs and returns, or into units of the most important objective. An aggregate of objectives may show how much the variation in one objective's values is worth in terms of other objectives' values. When analyzing the optimal solution, the analyst has to examine carefully how changes in these objectives, which reflect alternative policy decisions, affect the solution. Sometimes, the various objectives can be expressed in the form of goals to be achieved as near as possible. If the objectives are highly conflicting, it may not be possible to reach each goal exactly. A mathematical technique derived to cater for such problems is the goal programming technique.

2.6 Deriving a Solution

Once the problem has been formulated and a mathematical model developed to investigate the problem then a solution has to be found using the model. Usually, the optimal solution to a model is found by following a sequence of mathematical evaluations. A starting solution, which is usually a feasible solution to the problem, starts the sequence of mathematical operations to derive a new and ideally better-suited solution. This sequence of operations starting from an initial solution as input to the model and the generation of an improved solution as an output is called an iteration. The improved output solution is then resubmitted to the model as a new input, and the process is repeated until certain conditions—referred to as stopping rules—are satisfied, indicating that an optimal solution has been reached with some predetermined desired degree of accuracy. For some problems, there may not exist feasible or bounded solutions.

Initially, the analyst will use the model to carry out a *dry run* using data that has not necessarily been collected for the specific study, but is representative enough to validate the model's or algorithm's behavior. These dry runs will show whether the assumptions made were adequate, whether the model/algorithm provides the necessary in-depth detail required for the study, and whether the model provides the answers required.

For an algorithm to be a practical solution method, each iteration should lead to an improved solution, converging to the optimal solution in a finite number of steps.

After being satisfied that the model is performing correctly, one then applies data, which has been collected specifically for the problem at hand, to the model to obtain primary (baseline) results. This stage is sometimes called a *wet run*. These results should demonstrate to the sponsor that the model developed is an adequate description of the problem and will be able to provide useful and usable results.

2.7 Sensitivity Analysis

One important phase of a study is the sensitivity analysis phase. It is important to understand how the optimal solution depends on the input data. Carrying out a sensitivity analysis allows one to determine the accuracy required of the input data and also to identify those decision variables that have the most significant impact on the solution. Sensitivity analysis also allows the analyst to see how sensitive the choice of the preferred option is to changes in the assumptions and changes in the data. By carrying out sensitivity analysis, one is able to determine how robust the preferred option is, and how much the input data needs to change for an alternate option to become the preferred option.

In carrying out sensitivity analysis, the analyst modifies the assumptions or data to promote the discarded options at the expense of the previously determined optimal option. The amount of modification of the assumptions or of the data is a measure of robustness of the model. If the original solution still remains optimal under these conditions, then the solution is very robust.

2.8 Testing the Solution

The reason for undertaking a quantitative investigation of a system is to improve the performance of the system. The decision maker will need to be convinced that the proposed solution performs better than the present alternative before deciding on whether to implement the derived solution or keep the status quo. By giving presentations and making demonstrations of the possible improvements that can be achieved by implementing the derived solution may help the analyst substantially in convincing the decision maker to implement the proposed changes. When the analysts test the solution, they normally want to be certain that the decision rules derived from the optimal solution will perform as anticipated and also to be able to stipulate what the expected net benefits of implementing the solution will be. A solution should be tested against previous behavior and also against future behavior. In a military environment, the solution in many cases may be tested in field exercises, for either a short period or over extended periods. When testing the proposed solution, it must be validated against observations of actual performance, it must be independent of how the optimal solution was derived, the data used should be representative of future behavior, covering the entire range of behaviors likely to be observed in the future, and the tests should be carried out over a sufficiently long period to ensure that the model is robust over time.

2.9 Implementation

The final phase of an operations research project is the implementation of the solution. In order to help the decision makers introduce and use the optimal outputs from the study, it may be necessary for the analyst to convert the mathematical findings into a set of easily understood operating procedures or decision rules. It may also be necessary to educate the decision makers in the proper application of the findings and help them introduce the changes required to take them from the present situation to the new desired mode of operations, and support them in establishing control mechanisms to maintain and update the solution. The analysts jointly with the decision makers must continually monitor the performance until the new mode of operation becomes routine. In many cases, it is often not possible to implement all the recommendations resulting from a study. In those cases, it is more appropriate to talk about the degree of implementation achieved. The analyst must be fully involved in the implementation to ensure that most of the potential benefits that have been identified in the study's outcomes are implemented. In most cases, the analyst will have to fine-tune the model and solution to totally satisfy the decision maker. The analysts' participation in the implementation phase is important if the analyst is to be able to identify and incorporate future refinements into the model.

There are three major factors that can cause problems in the implementation of the findings. First, the degree of change required to move from the present situation to the proposed new situation can have an impact on the acceptance of the proposed solution. If this change is major, then that may deter the decision makers from introducing it. Second, the ability of the users using the solution, such as the personalities of the users, their motivation and pride in the job, their ages and backgrounds, level of education, and the impact of activities on their jobs, can influence the degree of acceptance of the proposed changes. Third, the changes may disrupt the present allocation of support or power of stakeholders, such as the support given to the solution by higher echelons in the hierarchy and organizational implications of the solution, and those changes that may be perceived as threatening to employees or customers may create resistance to the proposed changes.

Generally, analysts focus their attention to the first factor, as analysts often believe that problems only involve technical issues. There has been a tendency in the past to neglect the human factor aspects of the situation, which are qualitative in nature, and therefore are not conducive to quantitative treatment. As these factors are very difficult to incorporate into the analysis, they are often ignored. However, ignoring these factors will only lead to a solution that is unusable, because not all factors of the problem have been considered.

To enhance the probability of implementing the solution, it is necessary for the sponsor to have ownership of the solution. If the communication between the analysts and sponsors has been open and contact kept to a maximum, the sponsors will feel that they have contributed significantly to the project and will feel a strong feeling of ownership, guaranteeing the implementation and continued use of the solution. The analyst must establish contact with the sponsor at the beginning of the project and maintain it throughout the duration of the project. It is no use trying to convince the sponsor of the merits of the solution at the implementation stage, if the sponsor has not been heavily involved in the whole process.

Given that the present-day environment in which most organizations operate is very dynamic and is constantly undergoing change, it is necessary to monitor whether the solution remains optimal. Management needs to put in place procedures that will monitor quantitative and qualitative changes in the environment, and processes that will enable corrective action to be taken when such changes become significant. A change is considered significant if the cost of making the adjustment is less than the benefits achieved from making the adjustment.

Even after the solution has been implemented, the analysts need to remain involved. The solution needs to be continually checked to ensure that the actual performance of the solution does not deteriorate after a while. Also, the actual benefits gained need to be compared to those that were identified in the first two stages of the process. If major differences are observed then reasons for those differences need to be examined and adjustments made. Also, those people involved in the implementation must have the appropriate training and if they do not, then further training may have to be provided to give them the required skills. Only then can the success of the project be judged.

2.10 Summary

Resolving a decision maker's concerns through the identification and formulation of the underlying problem, developing a modelling test bed, selecting the optimal solution, analyzing the solution, and implementing the preferred alternative has been shown in this chapter as a logical and structured approach. Although it has been suggested that the stages of identification and formulation of the problem are vital for the successful analysis and implementation of the problem, the choice of modelling technique is also a very important stage. Analysts must ensure that they choose the right mathematical approach for the problem at hand. Modelling approaches are addressed in the next few chapters.

Exercises

1. The elevator users of a multistoried building have complained about the delay in getting an elevator. Being the property manager, how do you define the problem in order to solve it? In other words, what is your problem precisely which you intend to solve?

   The problem definition may vary from person to person for such a situation. If you cannot define the problem appropriately it is unlikely that it will be solved. For example, the problem may be thought as

   a. Minimizing the waiting time by using better and efficient elevator, which would require an expensive reengineering of the elevator system.
   b. Minimizing the people movement by studying the reasons for frequent elevator usage and reduce them by taking appropriate action. For example, having laundry in each floor instead of a common laundry at the basement.
   c. Minimizing or eliminating the complaints using simple but innovative means such as putting mirrors on the walls around the lobby of the building. This would not change the waiting time of the elevators and people movement, but will change the perception, because people become occupied with another activity. So the complaints will disappear.
   d. You may add another option.

   Most people would choose the first one as the problem definition (minimizing waiting time) and suggest an expensive reengineering as the solution. Which one you would choose and why?

2. A manufacturing company produces several products in its shopfloor and sells them directly to their customers, through its retail section. Although the production capacity is fixed and known, the demand of each product varies from period to period. As a result, few products are experiencing shortages during some periods whereas some other products are having excess stocks. The retail manager knows that the overall performance of the company can be improved by applying optimization techniques. The company is currently performing very well financially. The top management is neither familiar with optimization techniques

nor intended to make any changes in its current production schedule. As the retail manager, how would you convenience the top management to study the current system using optimization techniques?

3. Consider Problem 2 in Exercises. As you are a key personnel in the retail team, after your repeated requests, suppose the top management has agreed to do the study. Although the study shows significant improvement in company's performance, the top management has no idea how this result was obtained. As a consequence, the top management is hesitant to implement the resulting production schedule. What would you do now?
4. In most major airports, it is always a complaint that it takes too long to get the arriving baggage. Being a key member of the airport baggage-handling team, how would you define the problem in order to solve it? In other words, what is your problem precisely which you intend to solve?

3

Introduction to Modelling

3.1 Introduction

Mathematical model is the name given to a family of tools designed to help solve managerial, planning, and design problems in which the decision maker must allocate scarce (or limited) resources among various activities to optimize a measurable goal. The basic characteristics of a problem that can be solved through mathematical modelling and the assumptions required to develop a mathematical model have already been discussed in Chapter 1. A number of different categories of mathematical models have been established in the literature based on the classification tree presented in Chapter 1, for example, linear programming, integer programming, goal programming, and nonlinear programming. The details of these techniques will be discussed later.

The use of mathematical models, especially linear programming, is so common today that optimization software can be found in just about any organization equipped with a computer. However, this has led in some cases to the use of inappropriate modelling techniques, because of the availability of specific software to uninformed users, resulting in an inappropriate problem solution, thereby presenting the decision maker with an unrealistic decision alternative. In some cases, such a bad or wrong decision alternative could be disastrous.

In this chapter, we discuss the different components of a mathematical model, introduce various types of mathematical models, and formulate mathematical models for several simple problems.

3.2 Components of a Mathematical Model

Mathematical models consist of three major components: *decision variables* (unknowns of the model), an *objective function* (which needs to be optimized), and *constraints* (restrictions or limitations of the model). These components are briefly discussed below.

3.2.1 Decision Variables

The decision variables depend upon the type of problem considered. For example, the decision variables can be the quantities of the resources to be allocated, the number of units to be produced, or both. The decision maker searches for the value set of these unknown variables that will provide an optimal solution to the problem. The decision variables are usually denoted by $x_1, x_2, \ldots$ or x, y, and z. However, model developers are free to define the names of variables. Although some software products restrict the length of variable names, others allow any length of alphabetic or alphanumeric characters. Sometimes, it is useful to define meaningful names for variables. Shorter names are usually preferred as (1) using shorter names reduces the probability of making mistakes in writing and typing and (2) the model looks more compact.

3.2.2 Objective Function

The objective function represents the goal/objective of the problem in terms of decision variables. The decision maker endeavors to either maximize or minimize this function, such as to either maximize the total profit or minimize the total cost of production when producing or selling certain products. Data such as profit (for maximization) or cost (for minimization) per unit product are parameters required in association with the decision variables to form the objective function. These parameters are known as coefficients (profit or cost) of the objective function.

3.2.3 Constraints

The constraints are known as restrictions or limitations of the problem. A constraint has two components, usually a function and a constant, related by either an equality or inequality sign. For a resource constraint, the function represents the total resource required in terms of the decision variables and the constant specifies the total resource availability. Data such as the resources required per unit product are required to form the constraint functions. These data are known as coefficients associated with the constraints or technological coefficients. Note that most optimization software products adhere to the convention of having the variable expression on the left-hand side (LHS) of the constraint equation and a constant on the equation's right-hand side (RHS).

3.3 Simple Examples

Let us consider five small but simplified examples which will be analyzed and formulated as mathematical models later.

Example 3.1: A product-mix problem

A furniture manufacturer produces tables and chairs. The process involves machining, sanding, and assembling the pieces to make the tables and chairs. It takes 5 h to machine the pieces for a table, 4 h to sand the pieces, and 3 h to assemble a table. A chair requires 2 h to machine the pieces, 3 h to sand the pieces, and 4 h to assemble a chair. There are 270 h available for machining the pieces, 250 h for sanding the pieces, and 200 h for assembling. If the profit for a table is \$100 and for a chair \$60, how many tables and chairs should the manufacturer produce in order to maximize the overall profit? What if there is an additional requirement to produce four chairs for each table?

Example 3.2: A diet problem

A sports dietician is planning a food menu consisting of three major food stuffs A, B, and C. Each gram of A contains 3 units of protein, 2 units of carbohydrates, and 4 units of fat. Each gram of B contains 1, 3, and 2 units of protein, carbohydrates, and fat, respectively; and each gram of C contains 3, 1, and 4 units of protein, carbohydrates, and fat, respectively. The dietician wants the meal to provide at least 440 units of fat, at least 150 units of carbohydrates, and at least 320 units of protein. If 1 kg of A costs \$15.60, 1 kg of B costs \$18.90, and 1 kg of C costs \$12.70, how many grams of each foodstuff should be served to minimize the cost of the meal and yet satisfy the dietician's requirements?

Example 3.3: A vehicle mix problem

A commanding officer (CO) has been tasked to move the battalion out into the field for a major exercise. The CO has been informed that only a maximum of 100 vehicles is allowed, each to be used for a single round trip. There are two types of vehicles available to the CO, a Mack vehicle that can carry 24 m^3 and a Mercedes vehicle that can carry 16 m^3. The Mack vehicle is estimated to use 50 L of fuel on a round trip, whereas the Mercedes vehicle will only use 25 L of fuel per round trip. The battalion has a total of 4000 L of fuel available. The vehicles will require maintenance after each round trip; each Mack vehicle requires 3 h and each Mercedes 9 h. The maintenance squadron has 720 h available. To move a maximum amount of cubic meters, what mix of Mack and Mercedes vehicles should the CO choose?

Example 3.4: A curtain material trim loss problem

A curtain manufacturer receives three orders for curtain material with widths and lengths as follows:

Order Number	Width (m)	Length (Number of Rolls)
1	2.5	30
2	3.8	50
3	4.9	10

Rolls of curtain material are produced in two standard widths, 5 and 10 m. These can be cut to the sizes specified by the order. There is no practical length limitation as rolls can be joined together. Determine the production plan that minimizes the curtain material trim loss.

Example 3.5: A usage/supply problem

A conference organizer has to cater for lunches for the 5 days of a conference. The organizer is faced with the problem of deciding on the daily supply of clean serviettes. The organizer knows that the number of conference attendees requiring lunch for the 5 days are 130, 220, 180, 120, and 100. The alternatives are

- to buy new serviettes at 25¢ a piece, or
- to send used serviettes to the laundry where they can receive either a 48 h service at 10¢ a serviette, or a 24 h service at 15¢ a serviette.

If this conference is a once-off activity with respect to the provision of serviettes, how should the organiser provide serviettes to minimize the total cost?

3.4 Analyzing a Problem

Not all problems are amenable to a mathematical modelling approach, in that a mathematical model may not be able to represent the problem realistically or the solution approach would be computationally expensive. Consider Example 3.1. Is this problem suitable to be solved using mathematical modelling? To answer this question, we need to examine whether the problem meets the basic characteristics of mathematical modelling as discussed in Chapter 1.

The problem requires three different activities: machining, sanding, and assembly. These three activities are required to produce two products: tables and chairs. The resources to be shared for these activities are limited as indicated previously to 270 h of machining, 250 h of sanding the pieces, and 200 h in the assembly department. The amount of a given resource required by one unit of a product could be different from that of another product. In Example 3.1, it takes 5 h to machine the pieces for a table, 4 h to sand the pieces, and 3 h to assemble a table. A chair requires 2 h to machine the pieces, 3 h to sand the pieces, and 4 h to assemble a chair. So, there are many alternative ways of sharing these three resources to produce the proposed two products, for example, producing 5 chairs and 2 tables, 10 chairs and 5 tables, 10 chairs and 10 tables, and so on. Each product produced would earn a profit—the profit for a table is $100 and for a chair $60.

TABLE 3.1

Alternative Resource Usage

Alternative Number	Number of Tables (x_T)	Number of Chairs (x_C)	Machining (Limit: 270)	Sanding (Limit: 250)	Assembling (Limit: 200)
1	1	1	7	7	7
2	1	2	9	10	11
3	2	2	14	14	14
:	:	:	:	:	:
:	10	10	70	70	70
:	11	10	75	74	93
:	:	:	:	:	:
:	50	10	270	230	190
k	50	11	272	233	194
:	:	:	:	:	:

Finally, no violation of the above resource limitations is allowed. That means the problem meets all the general characteristics of a mathematical modelling approach. So, we will be able to develop all three components required for a formal mathematical model of the above problem.

Although we have no intention of solving the problem in this chapter, we would like to elaborate on the alternate options of resource use.

Suppose,

x_T = the number of tables to be produced

x_C = the number of chairs to be produced

The total resources required were calculated using the following equations. The details of the equations can be found in the next section.

$$\text{Total machining hours required} = 5x_T + 2x_C \tag{3.1}$$

$$\text{Total sanding hours required} = 4x_T + 3x_C \tag{3.2}$$

$$\text{Total assembly hours required} = 3x_T + 4x_C \tag{3.3}$$

As one can see in Table 3.1, it is possible to share the resources in many different ways. However, the alternative k violates the machining hour limitation. That means this alternative does not represent a feasible solution.

3.4.1 A Nonmathematical Programming Problem

Let us consider a different example. In a drive-through McDonalds restaurant, the cars arrive at an average rate of λ per hour and the staff can serve at an average rate of μ customers per hour. The arrival and service rates are usually stochastic (probabilistic) and the customers are served on the basis of a first-in first-out (FIFO) rule. The service times required by different customers may be different but are fixed for the individual customer. If $\lambda \gg \mu$, the customers will wait for long periods in the queue, which is a negative indicator for customer satisfaction. However, for $\mu \gg \lambda$, the staff utilization rate will be low leading to higher service costs. What is the exact issue facing McDonalds? It could be that they need (1) to minimize the customer waiting time (for better customer satisfaction), (2) to minimize the service cost (saving costs), or (3) to minimize both waiting time and service cost (a compromise solution).

Let us examine whether the problem meets the general characteristics of a mathematical modelling approach. The staff can be treated as resources that provide (produce) services. The number of staff is usually limited; however, more staff could be hired if necessary. Once the number of staff is fixed, there is no flexibility in the usage of those resources, thereby removing possible decision alternatives. That means, under the current situation, the problem cannot be solved appropriately using mathematical modelling. However, if we decide to determine the optimum number of staff required, then we may have to analyze a limited number of decision alternatives by varying the resource usages, which can be analyzed easily (even by hand calculation) without modelling. In addition, we need to rely on nonmathematical programming approaches such as simulation or queuing analysis to determine the average waiting time for a given number of staff. So, this is not an interesting problem from the mathematical modelling point of view.

3.5 Modelling a Simple Problem

Consider the product-mix problem as stated in Example 3.1. Sometimes, it is convenient in modelling if the problem is summarized in tabular form. The tabular summary of Example 3.1 is as follows:

Resource/Item	Per Unit Product		Resource Availability (h)
	Table	Chair	
Machining	5	2	270
Sanding	4	3	250
Assembly	3	4	200
Profit per unit ($)	100	60	

In modelling the problem, we need to define the variables and develop the mathematical functions associated with the objective and the constraints.

3.5.1 Defining the Variables

The decision variables in this case are

$T =$ the number of tables to be produced
$C =$ the number of chairs to be produced

T and C are the unknown (decision) variables of the problem, which we would like to determine. Alternatively, one can define the variables as x_1 and x_2, or x and y, instead of T and C.

3.5.2 Objective Function

As indicated in the problem, the objective is to maximize the profit from the selling of produced tables and chairs. The objective function can be developed as follows:

Profit from any product = (the profit per unit) × (the number of units).
Total profit from tables = $100T$
Total profit from chairs = $60C$
Total profit from tables and chairs = $100T + 60C$
So, the objective function is

$$\text{Maximize } Z = 100T + 60C \tag{3.4}$$

The letter Z represents the objective function and the word Maximize or Max indicates that the objective function is to be maximized.

3.5.3 Constraints

There are three constraints in this problem: the hours available for the machining, sanding, and assembly facilities.

Limitation for the machining stage: There are 270 machining hours available in the machining facility. Each unit of table requires 5 machining hours, whereas each unit of chair requires 2 machining hours. So the total amount of machining hours required can be calculated as follows:

Total machining hours required by any product = (the machining hour per unit) × (the number of units).
Total machining hours required by tables = $5T$
Total machining hours required by chairs = $2C$
Total machining hours required by tables and chairs = $5T + 2C$
So, the machining constraint is

$$5T + 2C \leq 270 \tag{3.5}$$

The LHS of the constraint indicates the total machining hour requirement, while the RHS represents the availability of machining hours. The less than or equal to sign (between the LHS and RHS) indicates that the total requirement must not exceed the total availability.

Limitation for the sanding stage: There are 250 sanding hours available in the sanding facility. Each unit of table requires 4 sanding hours, whereas each unit of chair requires only 3 sanding hours. So the sanding constraint can be written as

$$4T + 3C \leq 250 \tag{3.6}$$

Limitation for the assembly: There are 200 h available for assembly in the assembly shop. Each unit of table requires 3 h to assemble it, whereas each unit of chair requires 4 h to be assembled. So the assembling constraint can be written as

$$3T + 4C \leq 200 \tag{3.7}$$

Nonnegativity conditions: Finally, both T and C must be nonnegative (zero or positive) as we cannot produce any negative quantity of products in reality. As indicated earlier, many software packages require this condition to solve the model. So the nonnegativity constraints of the problem are

$$T \geq 0 \quad \text{and} \quad C \geq 0 \tag{3.8}$$

The final form of the mathematical model is therefore

$$\begin{aligned}
&\text{Maximize } Z = 100T + 60C \\
&\text{Subject to} \\
&\qquad 5T + 2C \leq 270 \quad \text{Machining} \\
&\qquad 4T + 3C \leq 250 \quad \text{Sanding} \\
&\qquad 3T + 4C \leq 200 \quad \text{Assembly} \\
&\qquad T \geq 0 \text{ and } C \geq 0 \quad \text{Nonnegativity}
\end{aligned} \qquad \text{Model (3.1)}$$

The model indicates that Z is to be maximized subject to three constraints such as machining, sanding, and assembly. The sign $\in$ is sometimes used instead of "subject to." In solving the above model, we mean to determine the numerical values of T and C, which maximizes the objective function while not violating the constraints.

Suppose the definitions of T and C were given. We are sure that you would come up with the same mathematical model for the above problem given the same information. However, can you make up a story of the underlying problem from the above mathematical model? You know the model is a mathematical representation of the problem. One should note here that the mathematical model of a problem is not unique. It is fully dependent on the decision variables considered and the assumptions made. Sometimes you may feel that you need additional variables, for example, a variable for determining the total machining hours required or the unused machining hours. You may agree that these additional variables can be calculated easily from the machining constraint if T and C are known. That means you can avoid such variables in order to keep the model compact. It is expected that a mathematical model will be developed with a minimum number of variables and constraints since the computational complexity is a function of the number of variables and constraints used in the model.

3.6 Linear Programming Model

In the above model, the variables were assumed real (any number to be greater than or equal to zero), and the objective function and all three constraint functions were developed as linear functions. This type of model is known as a linear programming model or linear programming or linear program (LP) in the optimization literature. The general linear programming model can be described as follows:

Given a set of m linear inequalities or equations in n variables, we wish to find the nonnegative values of these variables, which will satisfy the constraints and maximize or minimize some linear function of variables.

The basic assumptions of linear programming are

- Certainty: The values of the parameters (data) are known and are constant.
- Proportionality: Any function (objective or constraint) is proportional to the level of the activity (with consistent unit of measure).
- Additivity: The total activity is the sum of all individual activities.
- Divisibility: The decision variables could be either real or integer.
- Nonnegativity: Only positive values of variables are allowed.

3.7 More Mathematical Models

In this section, we will revisit Example 3.1 and develop mathematical models for Examples 3.2 and 3.3.

In the last sentence of Example 3.1, it was indicated that there was a requirement to produce four chairs for each table.

This is an additional constraint, which will ensure that for every table produced there will be four chairs available. This condition can be expressed mathematically as follows:

$$T = C/4 \tag{3.9}$$

To check the condition, one can assume that $T = 5$ tables produced, then the above expression gives us that $C = 20$ chairs. Note that this is an equality type constraint. Since most optimization software packages require the variables to be in the LHS and constants in the RHS of the constraint, we rearrange the above condition as follows:

$$4T - C = 0 \tag{3.10}$$

So the revised mathematical model is

$$\begin{aligned}
&\text{Maximize } Z = 100T + 60C \\
&\text{Subject to} \\
&\qquad 5T + 2C \leq 270 \quad \text{Machining} \\
&\qquad 4T + 3C \leq 250 \quad \text{Sanding} \\
&\qquad 3T + 4C \leq 200 \quad \text{Assembly} \\
&\qquad 4T - C = 0 \quad \text{Additional} \\
&\qquad T \geq 0 \text{ and } C \geq 0 \quad \text{Nonnegativity}
\end{aligned} \qquad \text{Model (3.2)}$$

We can summarize the problem in Example 3.2 as follows:

	Foodstuff			
Item	**A**	**B**	**C**	**Requirements**
Fat	4	2	4	440+
Carbohydrates	2	3	1	150+
Protein	3	1	3	320+
Cost ($/kg)	15.60	18.90	12.70	

We will formulate this problem without showing the detailed calculations as we did in Example 3.1.

Defining variables:

x = quantity of foodstuff A in kg

y = quantity of foodstuff B in kg

z = quantity of foodstuff C in kg

Objective function:
This is to minimize (as opposed to maximize in Example 3.1) the total cost, which is $15.60x + 18.90y + 12.70z$. So the objective function is

$$\text{Minimize } Z = 15.60x + 18.90y + 12.70z \tag{3.11}$$

Constraints:
There are three constraints in this problem: limitations (minimum requirements) on fat, carbohydrates, and protein. All the constraints would be of a greater than equal to type as they must be greater than the given lower limits. So the constraints can be written as

$$\text{Fat limit: } 4x + 2y + 4z \geq 440 \tag{3.12}$$

$$\text{Carbohydrate limit: } 2x + 3y + 1z \geq 150 \tag{3.13}$$

$$\text{Protein limit: } 3x + 1y + 3z \geq 320 \tag{3.14}$$

So the complete linear programming formulation is

$$\begin{aligned}
&\text{Minimize } Z = 15.60x + 18.90y + 12.70z \\
&\text{Subject to} \\
&\qquad 4x + 2y + 4z \geq 440 \quad \text{Fat limit} \\
&\qquad 2x + 3y + 1z \geq 150 \quad \text{Carbohydrate limit} \\
&\qquad 3x + 1y + 3z \geq 320 \quad \text{Protein limit} \\
&\qquad x, y, z \geq 0
\end{aligned} \qquad \text{Model (3.3)}$$

The problem in Example 3.3 is straightforward and similar to the above two examples. The summary of the problem is shown below:

	Number of Macks	Number of Mercedes	Availability/Limitation
Fuel (L)	50	25	4000
Maintenance (h)	3	9	720
Load carried (m^3)	24	16	—
Maximum number of vehicles allowed	—	—	100

Defining variables:

x_1 = number of Mack vehicles to be used

x_2 = number of Mercedes vehicles to be used

Objective function:
Maximize the volume transported.

$$\text{Maximize } Z = 24x_1 + 16x_2 \tag{3.15}$$

Constraints:

$$\text{Fuel limit: } 50x_1 + 25x_2 \leq 4000 \tag{3.16}$$

$$\text{Maintenance hour limit: } 3x_1 + 9x_2 \leq 720 \tag{3.17}$$

$$\text{Number limit on vehicles: } x_1 + x_2 \leq 100 \tag{3.18}$$

So the overall linear programming model is

$$\begin{aligned}
&\text{Maximize } Z = 24x_1 + 16x_2 \\
&\text{Subject to} \\
&\qquad 50x_1 + 25x_2 \leq 4000 \quad \text{Fuel limit} \\
&\qquad 3x_1 + 9x_2 \leq 720 \quad \text{Maintenance hour limit} \\
&\qquad x_1 + x_2 \leq 100 \quad \text{Number limit on vehicles} \\
&\qquad x_1, x_2 \geq 0
\end{aligned} \qquad \text{Model (3.4)}$$

3.8 Integer Programming

Integer programming (IP) is an extension of the general linear programming problem. The decision variables of an optimal solution to a general linear programming problem may take on either nonnegative fractional values or integer values. Unfortunately, fractional values are neither practical nor very meaningful in certain types of business, manufacturing, or defense problems. For example, two-thirds of a tank or half an aircraft cannot be bought or used in reality. One can assume integer programming as being a linear program with indivisibility requirements. There are three types of integer programming models:

- Integer: Where all the decision variables are integers.
- Binary integer: Where all the decision variable values are binary (either zero or one) only. There are some situations, where variables can assume only one of two values (e.g., yes or no) that can be designated as zero and one.
- Mixed integer or mixed integer linear: Linear programs with some integer and some real decision variables.

Recall Examples 3.1 and 3.3 and their linear programming models. The solutions to these models may provide fractional values for the decision variables such as 3.6 tables, 14.2 chairs, 51.3 Mack vehicles, and 49.7 Mercedes vehicles. The fractional values are not meaningful for such items in practice. Note that fractional values are not an issue for Example 3.2. To avoid fractional values, we need to define the variables as integer. We will rewrite the model of Example 3.1 here with an integer declaration.

$$
\begin{aligned}
&\text{Maximize } Z = 100T + 60C \\
&\text{Subject to} \\
&\qquad 5T + 2C \leq 270 \quad \text{Machining} \\
&\qquad 4T + 3C \leq 250 \quad \text{Sanding} \\
&\qquad 3T + 4C \leq 200 \quad \text{Assembly} \\
&\qquad T \text{ and } C \geq 0 \text{ and integer}
\end{aligned}
\qquad \text{Model (3.5)}
$$

The above model can be called an integer program. If you remove the integrality condition, this reverts to a linear program.

Example 3.6: A capital budgeting problem

Local councils and organizations frequently face situations where they have to select one or more projects (investment opportunities) from a number of competing projects. Consider the following list of projects. If \$30 million is available, which projects should be selected?

Project Number	Project	Cost (\$ million)	Expected Utility
1	After-school program	6	18
2	Road security	18	16
3	Crime reduction	10	12
4	Road extension	9	25
5	Child care facility	4	14

The problem is to maximize the total expected utility not exceeding the budget limitation. Here, the decision needs to determine whether a project should be chosen or not. So using binary variables would be a suitable approach to make the yes/no decision.

Defining variables:

$$x_1 = \begin{cases} 1 & \text{if project 1 is selected} \\ 0 & \text{otherwise} \end{cases}$$

$$x_2 = \begin{cases} 1 & \text{if project 2 is selected} \\ 0 & \text{otherwise} \end{cases}$$

$$x_3 = \begin{cases} 1 & \text{if project 3 is selected} \\ 0 & \text{otherwise} \end{cases}$$

$$x_4 = \begin{cases} 1 & \text{if project 4 is selected} \\ 0 & \text{otherwise} \end{cases}$$

$$x_5 = \begin{cases} 1 & \text{if project 5 is selected} \\ 0 & \text{otherwise} \end{cases}$$

The integer programming model is

$$\begin{aligned} &\text{Maximize } Z = 18x_1 + 16x_2 + 12x_3 + 25x_4 + 14x_5 \\ &\text{Subject to} \\ &\qquad 6x_1 + 18x_2 + 10x_3 + 9x_4 + 4x_5 \leq 30 \\ &\qquad x_1, x_2, x_3, x_4, x_5 \text{ are either 1 or 0} \end{aligned}$$

Model (3.6)

Example 3.7: A knapsack problem
Four items are considered for loading onto a small military aircraft, of carrying capacity 12 tonnes, involved in relief operations. The weights and volumes of the items are given below:

Item	Weight (tonnes)	Volume (m^3)
I-A	3	20
I-B	5	30
I-C	4	25
I-D	2	18

Which items should be loaded onto the aircraft to maximize the volume of the items transported?

Variables:

$$x_A = \begin{cases} 1 & \text{if item I-A is selected} \\ 0 & \text{otherwise} \end{cases}$$

$$x_B = \begin{cases} 1 & \text{if item I-B is selected} \\ 0 & \text{otherwise} \end{cases}$$

$$x_C = \begin{cases} 1 & \text{if item I-C is selected} \\ 0 & \text{otherwise} \end{cases}$$

$$x_D = \begin{cases} 1 & \text{if item I-D is selected} \\ 0 & \text{otherwise} \end{cases}$$

The integer programming model is

$$\text{Maximize } Z = 20x_A + 30x_B + 25x_C + 18x_D$$

Subject to

$$3x_A + 5x_B + 4x_C + 2x_D \leq 12$$

$$x_A, x_B, x_C, x_D \text{ are either 1 or 0}$$

Model (3.7)

3.9 Multi-Objective Problem

So far, we have discussed optimization problems with only one objective or goal in mind. Nowadays, having only one objective or goal to achieve is not enough to survive in a dynamic (continuously changing) business environment, there are usually many competing goals. In addition, it is usually difficult to obtain an explicit statement of what the organization's goals are. As a result, managers usually want to attain several simultaneous objectives or goals, some of which may conflict with each other. Consider a profit-making firm. In addition to making money, the company wants to grow, to develop its products and its employees, to provide job security for its workers, and to serve the community. Here, the goal of serving the community means spending money on different community programs, which contradicts with the goal of making money.

Example 3.8: A production planning problem

The manufacturing unit of a large company has received a production order of 300 units that must be supplied within a week. Two production lines are available, each for 25 h during the week. Production line A can produce 5 units per hour. Using production line B, it takes 15 min to produce each unit. Line A costs $8 per hour to operate, and line B costs $11 per hour. Overtime is available; up to 30 h for line A at $10 per hour and up to 30 h for line B at $12 per hour. The company wants to develop a production plan (1) by minimizing the overall production cost and (2) at the same time, by maximizing the utilization of the regular working hours.

The tabular summary of the problem is as follows:

	Units per Hour		Total Hours		$/Unit	
Time	Line A	Line B	Line A	Line B	Line A	Line B
Regular	5	4	25	30	8	11
Overtime	5	4	25	30	10	12

Defining variables:

X_{AR} = units produced in line A during regular hours

X_{BR} = units produced in line B using regular hours

X_{AO} = units produced in line A during overtime
X_{BO} = units produced in line B using overtime

Objective function 1: minimizing the total cost

$$\text{Minimize } Z_1 = 8X_{AR} + 11X_{BR} + 10X_{AO} + 12X_{BO} \tag{3.19}$$

Objective function 2: maximizing the utilization of the regular working hours

$$\text{Maximize } Z_2 = X_{AR}/5 + X_{BR}/4 \tag{3.20}$$

The following are the constraints:

1. Production requirements

$$X_{AR} + X_{BR} + X_{AO} + X_{BO} = 300 \tag{3.21}$$

2. Regular hour limitation

$$X_{AR}/5 \leq 25 \tag{3.22}$$

$$X_{BR}/4 \leq 25 \tag{3.23}$$

3. Overtime hour limitation

$$X_{AO}/5 \leq 30 \tag{3.24}$$

$$X_{BO}/4 \leq 30 \tag{3.25}$$

4. Nonnegativity constraint

$$X_{AR}, X_{BR}, X_{AO}, X_{BO} \geq 0 \tag{3.26}$$

So a mathematical model with two objectives is

$$\begin{aligned}
&\text{Minimize } Z_1 = 8X_{AR} + 11X_{BR} + 10X_{AO} + 12X_{BO}\\
&\text{Maximize } Z_2 = X_{AR}/5 + X_{BR}/4\\
&\text{Subject to}\\
&\qquad X_{AR} + X_{BR} + X_{AO} + X_{BO} = 300\\
&\qquad X_{AR}/5 \leq 25 \qquad\qquad \text{Model (3.8)}\\
&\qquad X_{BR}/4 \leq 25\\
&\qquad X_{AO}/5 \leq 30\\
&\qquad X_{BO}/4 \leq 30\\
&\qquad X_{AR}, X_{BR}, X_{AO}, X_{BO} \geq 0
\end{aligned}$$

3.9.1 Objective versus Goal

In mathematical modelling, an objective is treated differently from a goal. An objective is represented by a function, which is to be either maximized or minimized. As an example, consider the objective function of Example 3.1 (Equation 3.4):

$$\text{Maximize } Z = 100T + 60C$$

A goal indicates a target value for a function. If we define a target for profit, say \$2500, then we mean the function $100T + 60C$ should be as close to \$2500, while still meeting all the constraints. However, in goal programming, the goal constraints are used to represent each of the goals to be met. That means multiple goals can be handled by one model. As opposed to the constraints in linear programming and integer programming, the goal constraints can be violated with certain penalties set by the decision makers. That is why, goal constraints are also known as soft constraints.

Example 3.9: A goal programming problem
Consider Example 3.8. Suppose the company has set a target of \$TPC for the overall production cost as part of its annual budget and a target for the utilization of the regular working hours (equal to 75% of the sum of available regular working hours). Here, although the company expects to achieve the target values some variations may be allowed under certain conditions.

3.10 Goal Programming

The above problem can be formulated as a goal programming model. Goal programming can be either a linear, integer, or nonlinear model. For ease of understanding, we will refer to it as linear goal programming in this section. The components of any goal programming are

1. Decision variables: same as linear programming
2. Deviational/goal variables: new but important for goal programming (discussed below)
3. System constraints: identical to linear programming constraints with no deviations allowed
4. Goal constraints: target values to be achieved, but deviations are allowed (also known as soft constraints)
5. Objective function: minimize the weighted sum of the undesirable deviations

In goal programming, a specific numeric goal is established for each goal function (constraint), and then a solution is derived that minimizes the (weighted) sum of deviations of these goal functions from their respective goals. There are three types of goals in goal programming as described below:

- A lower, one-sided goal sets a lower limit that we do not want to fall under (but exceeding the limit is fine).
- An upper, one-sided goal sets an upper limit that we do not want to exceed (but falling under the limit is fine).
- A two-sided goal sets a specific target range that we do not want to fall outside.

For Example 3.9, the decision variables (component 1) and the system constraints (component 3) are the same as for Example 3.8. The deviational variables (component 2) can be defined as follows:

For the overall production cost goal

d_1^+ = Overachievement deviational variable (if required more than the production cost target value)

d_1^- = Underachievement deviational variable (if achieved under the production cost target value)

For the regular working hour goal

d_2^+ = Overachievement deviational variable (if exceeded the target working hours)

d_2^- = Underachievement deviational variable (if achieved under the target working hours)

Now the goal constraints (component 4) can be derived. The production cost goal is an upper, one-sided goal that the company does not want to exceed. Also, the total number of regular working hours is a lower, one-sided goal that the company does not want to fall under.

The overall production cost function

$$Z_1 = 8X_{AR} + 11X_{BR} + 10X_{AO} + 12X_{BO} \tag{3.27}$$

can be rewritten as goal constraint

$$8X_{AR} + 11X_{BR} + 10X_{AO} + 12X_{BO} + d_1^- - d_1^+ = \text{TPC} \tag{3.28}$$

As a requirement of the formulation, both d_1^- and d_1^+ are greater than or equal to zero and at least one of them must be equal to zero. Note that it is necessary to write d_1^- with positive sign and d_1^+ with negative sign in the goal constraint.

Similarly, the regular hour function

$$Z_2 = X_{AR}/5 + X_{BR}/4 \tag{3.29}$$

can be rewritten as goal constraint

$$X_{AR}/5 + X_{BR}/4 + d_2^- - d_2^+ = \text{SFPTRH} \tag{3.30}$$

where SFPTRH = 75% of total regular hours.

Both d_2^- and d_2^+ are greater than or equal to zero and at least one of them must be equal to zero.

The objective function (component 5) of the goal programming is always to minimize the weighted sum of the undesirable deviations. In the production cost goal constraint, the overachievement factor (i.e., d_1^+) is undesirable. The underachievement factor (i.e., d_2^-) is undesirable for the available working hour goal constraint. We want to minimize the sum of these two variables. However, one can set the weight to reflect the importance of one variable as compared to the other. So the goal programming objective function can be written as follows:

$$\text{Minimize } Z = w_1 d_1^+ + w_2 d_2^- \tag{3.31}$$

So the overall goal programming model becomes

$$
\begin{aligned}
&\text{Minimize } Z = w_1 d_1^+ + w_2 d_2^- \\
&\text{Subject to} \\
&\qquad X_{AR} + X_{BR} + X_{AO} + X_{BO} = 300 \\
&\qquad 8X_{AR} + 11X_{BR} + 10X_{AO} + 12X_{BO} + d_1^- - d_1^+ = \text{TPC} \\
&\qquad X_{AR}/5 + X_{BR}/4 + d_2^- - d_2^+ = \text{SFPTRH} \\
&\qquad X_{AR}/5 \le 25 \\
&\qquad X_{BR}/4 \le 25 \\
&\qquad X_{AO}/5 \le 30 \\
&\qquad X_{BO}/4 \le 30 \\
&\qquad X_{AR}, X_{BR}, X_{AO}, X_{BO}, d_1^-, d_1^+, d_2^-, d_2^+ \ge 0
\end{aligned}
\qquad \text{Model (3.9)}
$$

Note that the goal programming model may not produce optimal solutions; however, it produces workable but satisfactory solutions.

3.11 Nonlinear Programming

Consider the inventory batch sizing example provided in Chapter 1. From Figures 1.1 and 1.2, it is clear that the function is nonlinear as the cost function value decreases until the order quantity reaches the optimum ordering point and then increases with the increase in ordering quantity. Such a problem is known as a nonlinear problem. A nonlinear programming model is similar to a linear programming model with one exception; it contains nonlinear terms either in the objective function or in the constraint equations or both. If a model contains all linear terms except one, it is still recognized as nonlinear model.

Example 3.10: A location problem

A New South Wales based radio channel provides radio services in most southern Australian cities and towns. The channel is planning to expand its service to four northern and western cities. To provide a good quality service, the channel needs to establish a new transmission tower, which will transmit radio frequency to the preexisting smaller towers in those cities. The new tower to be constructed can cover areas within a radius of K km. Thus, the new tower must be located within K km of each of those existing towers. The problem is to determine the tower location that minimizes the total distance from the new tower to each of the existing towers.

The location of each city can be calculated using two dimensional coordinates (x, y) from a given reference point as follows:

City	x	y
1	10	45
2	15	25
3	20	10
4	55	20

Defining the variables:
We need to determine the location of the new tower. So, the coordinates of the new tower would represent the decision variables as follows:

X = distance of the new tower with respect to the X-axis
Y = distance of the new tower with respect to the Y-axis

Objective function:
The objective is to minimize the total distance from the new tower to each of the existing towers.

$$\text{Minimize } Z = \sqrt{(10 - X)^2 + (45 - Y)^2} + \sqrt{(15 - X)^2 + (25 - Y)^2} + \sqrt{(20 - X)^2 + (10 - Y)^2} + \sqrt{(55 - X)^2 + (20 - Y)^2} \quad (3.32)$$

The first term in the objective function calculates the distance from the tower in city 1 to the location of the new tower. The remaining three terms perform similar calculations for the towers in cities 2, 3, and 4.

Constraints:
The new tower must be built within a distance of K km from the existing towers. That means, the distance from the new tower to any of the existing towers cannot exceed K km.

$$\sqrt{(10 - X)^2 + (45 - Y)^2} \leq K \tag{3.33}$$

$$\sqrt{(15 - X)^2 + (25 - Y)^2} \leq K \tag{3.34}$$

$$\sqrt{(20 - X)^2 + (10 - Y)^2} \leq K \tag{3.35}$$

$$\sqrt{(55 - X)^2 + (20 - Y)^2} \leq K \tag{3.36}$$

So the complete mathematical model is as follows:

$$\text{Minimize } Z = \sqrt{(10 - X)^2 + (45 - Y)^2} + \sqrt{(15 - X)^2 + (25 - Y)^2} + \sqrt{(20 - X)^2 + (10 - Y)^2} + \sqrt{(55 - X)^2 + (20 - Y)^2}$$

Subject to

$$\begin{aligned} \sqrt{(10 - X)^2 + (45 - Y)^2} &\leq K \\ \sqrt{(15 - X)^2 + (25 - Y)^2} &\leq K \\ \sqrt{(20 - X)^2 + (10 - Y)^2} &\leq K \\ \sqrt{(55 - X)^2 + (20 - Y)^2} &\leq K \\ X, Y &\geq 0 \end{aligned} \qquad \text{Model (3.10)}$$

Note that the objective function and all the constraints are nonlinear, and K must be known.

3.12 Summary

In this chapter, we have covered the basic concepts of mathematical modelling and the different types of mathematical models with a number of simple examples. Developing the mathematical formulation of these simple models was straightforward. However, in many cases the process is not that straightforward, one needs some mathematical agility to derive the mathematical formulation. The following chapter will discuss some of the not so straightforward approaches required when deriving the mathematical formulation.

Exercises

1. A furniture manufacturer employs 6 skilled and 11 semiskilled workers and produces two products: study table and computer table. A study table requires 2 h of a skilled worker and 2 h of an unskilled worker. A computer table requires 2 h of a skilled worker and 5 h of an unskilled worker. As per the industrial laws, no one is allowed to work for more than 38 h a week. The manufacturer can sell as many tables as he can produce. If the profit for a study table is $100 and for a computer table $160, how many study and computer tables should the manufacturer produce in a week in order to maximize the overall profit? Formulate a linear programming model.
2. Consider Problem 1 in Exercises. Suppose that the demands of the study and computer tables are at least 40 and 45, respectively, and the manufacturer pays $900 and $600 per week for each skilled and unskilled worker, respectively. If the manufacturer intends to fulfill the demand in full, what objective function would you suggest to the manufacturer's production planning problem? Justify your suggestion and formulate the problem as a linear programming model.
3. Consider Problems 1 and 2 in Exercises. Suppose the manufacturer is interested in maximizing his overall profit rather than fulfilling the demand. What objective function would you suggest to the manufacturer's production planning problem? Justify your suggestion and formulate the problem as a linear programming model.
4. A marketing manager wishes to allocate his annual advertising budget of $1.5 million in three media: television (TV), radio, and daily newspaper. The unit cost of an advertisement in TV is $10,000, in radio $5,000, and in newspaper $3,000. The company advertises in one TV channel, one radio station, and one newspaper only. The number of advertisements in each media must be at least 20. The expected effective

audience for each advertisement for TV is 30,000, for radio 18,000, and for newspaper 10,000. Develop a mathematical model.

5. A farmer in Queensland owns 100 acres of land and is planning to grow wheat and potato. Each acre cultivated with wheat yields a $400 profit, requires four workers, and 3 tonnes of fertilizer; each acre with potato yields $550 profit, requires three workers, and 5 tonnes of fertilizer. The farmer can hire up to 300 workers and manage 400 tonnes of fertilizer. Formulate a mathematical programming model that would maximize the total profit.
6. A food production and retail chain is considering several projects that have varying capital requirements over the next 3 years. The projects are (1) possible plant expansion, (2) possible warehouse expansion, (3) possible addition of a transport unit, and (4) possible purchase of new machinery. The estimated net present value for each project, the investment requirements, and the available capital over the next 3 years are shown below. All figures are in million dollars.

	Project				
	Plant	Warehouse	Transport	Machinery	Capital Available
Year 1 (IR)	0.30	0.15	0.10	0.15	0.60
Year 2 (IR)	0.25	0.20	0.06	0.12	0.65
Year 3 (IR)	0.20	0.15	0.08	0.10	0.40
Present value	1.0	0.60	0.30	0.50	

Which projects the company should choose in order to maximize the total net present value?

7. The central intelligence branch is considering the relocation of several intelligence units in Canberra to obtain better information from several new high-crime areas. The locations under consideration together with the areas that can be covered from these locations are given below:

Potential Locations for Units	Areas Covered
L1	A, C, F
L2	B, D, G
L3	D, E, G
L4	A, C, E, F
L5	C, E, G
L6	B, D, F

Formulate an integer programming model that could be used to find the minimum number of locations necessary to cover all the specified areas.

8. Eastern Aviation Company organizes summer charter flights from Sydney to Auckland for students. The company uses three types of aircraft whose operating cost and capacities are given below:

Aircraft Type	Capacity (Passenger)	Cost per Flight ($)	Maximum Number of Flights	Air Crew Required
A1	120	25,000	20	6
A2	180	35,000	12	14
A3	280	45,000	5	20

The company can spare only 200 crews for the entire mission. A total of 4,000 students signed up for the summer trip, each paying $450. Considering all students must be flown, find how many flights of each type should be used in order to maximize profit. Formulate a mathematical programming model.

9. CPS Investment Services must develop an investment portfolio for a new customer. Initially, the new customer would like to restrict portfolio to three stocks as shown below:

Stock	Price per Share ($)	Estimated Annual Return (%)
ACT cable	50	6
CTR retail	35	10
PCL petroleum	120	8

The customer wants to invest $80,000 and established two investment goals.

Goal 1: Obtain an annual return of at least 9%.

Goal 2: Limit the investment in CTR retail, the riskier investment, to no more than 50% of the total investment.

Formulate a goal programming model for the CPS investment problem.

10. Consider Problem 4 in Exercises. Now suppose the marketing manager established three goals for the advertising campaign as follows:

Goal 1: Reach at least 3 million audiences.

Goal 2: The number of TV advertisements should be at least 30% of the total number of advertisements.

Goal 3: The cost of advertisements should be within the budget.

Formulate a goal programming model for this problem.

11. After a major bushfire in 2003, the local government is planning to establish a new fire station to cover four major town centers, under

frequent fire threat, located in Town-1 (10, 20), Town-5 (50, 20), Town-8 (40, 40), and Town-14 (70, 60). Town-1 averages 20 fires per year; Town-5, 30 fires; Town-8, 40 fires; and Town-14, 25 fires. The local government wants to build the fire station in a location that minimizes the average distance that a fire engine must travel to respond to a fire. Develop a mathematical model to determine the location of the fire station.

12. An entertaining company is planning to spend $20,000 on advertising. It costs $5,000 per minute to advertise on TV and $3,000 per minute to advertise on radio. If the company buys x min of TV advertising and y min of radio advertising, its revenue in thousands of dollars can be expressed as $-2x^2 - y^2 + xy + 9x + 4y$. Formulate a mathematical model to maximize the company's revenue.

Section II

Modelling Techniques

4

Simple Modelling Techniques I

4.1 Introduction

After introducing the concept of a mathematical model in simple terms in Chapter 3, we would like to discuss some basic modelling techniques frequently used in practice. In this chapter, we also introduce how to use subscripts in variables to make the mathematical model formulation compact. A number of examples are provided to demonstrate the simple modelling techniques as well as the use of subscripts.

This chapter is organized as follows. First, we introduce the use of subscripts in variables. In Section 4.3, we discuss a number of simple modelling techniques. The techniques include additional work requirement in formulation, variables as fraction of other variables, maintaining certain ratios among different variables, one constraint is a fraction of another constraint, maxi–min or mini–max objective function, multi-period problem, infeasible to satisfactory solutions, transportation problem, assignment problem, network flow problem, and project management problem.

4.2 Use of Subscripts in Variables

To demonstrate the use of subscripts in variables, let us consider Example 3.6. The problem formulation of Example 3.6 is reproduced here for convenience.

Variables:

$$x_1 = \begin{cases} 1 & \text{if project 1 is selected} \\ 0 & \text{otherwise} \end{cases}$$

$$x_2 = \begin{cases} 1 & \text{if project 2 is selected} \\ 0 & \text{otherwise} \end{cases}$$

$$x_3 = \begin{cases} 1 & \text{if project 3 is selected} \\ 0 & \text{otherwise} \end{cases}$$

$$x_4 = \begin{cases} 1 & \text{if project 4 is selected} \\ 0 & \text{otherwise} \end{cases}$$

$$x_5 = \begin{cases} 1 & \text{if project 5 is selected} \\ 0 & \text{otherwise} \end{cases}$$

The integer programming model is

$$\begin{aligned} &\text{Maximize } Z = 18x_1 + 16x_2 + 12x_3 + 25x_4 + 14x_5 \\ &\text{Subject to} \\ &\qquad 6x_1 + 18x_2 + 10x_3 + 9x_4 + 4x_5 \leq 30 \\ &\qquad x_1, x_2, x_3, x_4, x_5 \text{ are either 1 or 0} \end{aligned} \qquad \text{Model (4.1)}$$

Instead of defining each variable at a time, we can define them together using subscripts as follows:

$$x_i = \begin{cases} 1 & \text{if project } i \text{ is selected} \\ 0 & \text{otherwise} \end{cases} \quad \text{for all } i$$

By "for all i," here we mean $i = 1, 2, 3, 4$, and 5. Usually we use the notation $\forall$ instead of writing "for all." So one can rewrite the variable definition to

$$x_i = \begin{cases} 1 & \text{if project } i \text{ is selected} \\ 0 & \text{otherwise} \end{cases} \quad \forall i$$

So the integer programming model becomes

$$\begin{aligned} &\text{Maximize } Z = 18x_1 + 16x_2 + 12x_3 + 25x_4 + 14x_5 \\ &\text{Subject to} \\ &\qquad 6x_1 + 18x_2 + 10x_3 + 9x_4 + 4x_5 \leq 30 \\ &\qquad x_i \text{ either 1 or 0} \quad \forall i \end{aligned} \qquad \text{Model (4.2)}$$

This model can be compacted further by using a summation sign, which will be discussed in the next chapter.

4.3 Simple Modelling Techniques

As we have seen in Chapter 3, it is not difficult to define the variables and develop the objective function and constraints once the problem has been defined properly and adequate descriptions are provided. However, this is usually not the case for real-world problems. In some problems, additional work may be required to define the variables and develop the functions. In many cases, some simple but interesting tricks need to be applied to formulate the constraints. The nature of the additional work and the tricks usually used in practice are discussed with illuminating examples in this section.

4.3.1 Additional Work Requirement in the Formulation

Let us now formulate the problems described as Examples 3.4 and 3.5 using the concept of defining the variables with subscripts. These examples require some additional work to define the variables and set up the constraint functions.

Example 4.1: A curtain material trim loss problem
Consider Example 3.4 here. The problem summary is given below:

Order Number	Width (m)	Length (in Terms of Number of Rolls)
1	2.5	30
2	3.8	50
3	4.9	10

Two standard widths are available: 5 and 10 m.
Objective is to minimize curtain material trim loss.

This problem looks simple but it is a bit harder than the linear programming (LP) problems we have formulated so far. This requires more thought and some additional work to define the variables and to set up the objective function and constraints.

Defining the variables:
The variables can be defined using subscripts of notation such as i, where i has a range of values ($i = 1, 2, \ldots, n$).

X_{1i} = number of rolls produced from a 5 m width roll of orders combination i

X_{2i} = number of rolls produced from a 10 m width roll of orders combination i

The 5 m width roll can be cut in three different ways and the 10 m width roll can be cut in six different ways to produce 2.5, 3.8, and 4.9 m width rolls. The variables are explained below to provide a better understanding.

Using a 5 m width roll, you can produce

- two 2.5 m width rolls with no trim loss (X_{11}),
- one 3.8 m width roll with 1.2 m trim loss (X_{12}), or
- one 4.9 m width roll with 0.1 m trim loss (X_{13}).

Using a 10 m width roll, you can produce

- four 2.5 m width rolls with no trim loss (X_{21}),
- two 2.5 m and one 3.8 m width rolls with 1.2 m trim loss (X_{22}),
- two 2.5 m and one 4.9 m width rolls with 0.1 m trim loss (X_{23}),

- two 3.8 m width rolls with 2.4 m trim loss (X_{24}),
- one 3.8 m and one 4.9 m width rolls with 1.3 m trim loss (X_{25}), or
- two 4.9 m width rolls with 0.2 m trim loss (X_{26}).

One, therefore, needs nine variables to model this problem. The different cut combinations with their trim losses are also shown in the following table. This table would make the modelling task easier.

Width Required ↓	Decision Variable									Required Rolls
	X_{11}	X_{12}	X_{13}	X_{21}	X_{22}	X_{23}	X_{24}	X_{25}	X_{26}	
2.5	2	0	0	4	2	2	0	0	0	30
3.8	0	1	0	0	1	0	2	1	0	50
4.9	0	0	1	0	0	1	0	1	2	10
Trim → Loss	0	1.2	0.1	0	1.2	0.1	2.4	1.3	0.2	

Objective function:
The objective is to minimize the total trim loss. The trim loss involved with each decision variable is shown in the bottom row. So the objective function can be expressed as

$$\text{Minimize } Z = 1.2X_{12} + 0.1X_{13} + 1.2X_{22} + 0.1X_{23} + 2.4X_{24} + 1.3X_{25} + 0.2X_{26} \tag{4.1}$$

Constraints:
There are three constraints: requirements of curtains with (1) 2.5 m width, (2) 3.8 m width, and (3) 4.9 m width. The number of rolls with 2.5 m width, which can be produced from one unit of each decision variable, is shown in the first row in the table. The figures for 3.8 and 4.9 m are given in the second and third rows, respectively. From these figures, we can calculate the possible number of rolls produced in each type.

2.5 m rolls: $2X_{11} + 4X_{21} + 2X_{22} + 2X_{23}$

3.8 m rolls: $X_{12} + X_{22} + 2X_{24} + X_{25}$

4.9 m rolls: $X_{13} + X_{23} + X_{25} + 2X_{26}$

Then the LP formulation is

$$\text{Minimize } Z = 1.2X_{12} + 0.1X_{13} + 1.2X_{22} + 0.1X_{23} + 2.4X_{24} + 1.3X_{25} + 0.2X_{26}$$

$\in$

$$\begin{aligned} &2X_{11} + 4X_{21} + 2X_{22} + 2X_{23} = 30 \\ &X_{12} + X_{22} + 2X_{24} + X_{25} = 50 \\ &X_{13} + X_{23} + X_{25} + 2X_{26} = 10 \\ &X_{11}, X_{12}, X_{13}, X_{21}, X_{22}, X_{23}, X_{24}, X_{25}, 2X_{26} \geq 0 \end{aligned} \qquad \text{Model (4.3)}$$

After solving the model, we can easily find the number of 5 and 10 m rolls required to fulfill the customer's demand with a minimum possible trim loss.

$$\text{Number of 5 m rolls required} = X_{11} + X_{12} + X_{13} \quad (4.2)$$

$$\text{Number of 10 m rolls required} = X_{21} + X_{22} + X_{23} + X_{24} + X_{25} + X_{26} \quad (4.3)$$

Example 4.2: The conference organizer problem

Consider Example 3.5. The problem is to manage serviettes for a 5 day conference's lunches at minimum possible cost. This problem requires some additional work in defining the variables and setting up the objective and constraint functions. In this problem, the options of using serviettes in any 1 day are

- Buy new serviettes
- Receive used serviettes from cleaning
  - 24 h cleaning and/or
  - 48 h cleaning
- Send used serviettes for cleaning
  - 24 h cleaning and/or
  - 48 h cleaning
- Store used serviettes—do not send for cleaning now

Assume that there is no serviette in stock at the beginning of day 1.

Decision variables:

x_i = number of serviettes bought on ith day

y_i = number of serviettes at end of ith day sent for 24 h cleaning

z_i = number of serviettes at end of ith day sent for 48 h cleaning

v_i = number of used serviettes not sent for cleaning at end of ith day

At the beginning of the day, the fresh serviettes available for use are

Day	Buy	From 24 h Cleaning	From 48 h Cleaning	Required for the Day
1	x_1	—	—	130
2	x_2	—	—	220
3	x_3	y_1	—	180
4	x_4	y_2	z_1	120
5	x_5	y_3	z_2	100
Cost	25	15	10	

The used serviettes sent for 24 h cleaning at the end of day 1 will be available for use on day 3 as they would return at the end of day 2. Similarly, the used serviettes sent for 48 h cleaning at the end of day 1 will be available for use on day 4 as they would return at the end of day 3.

At the end of the day, the used serviettes available for cleaning are

Day	For 24 h Cleaning	For 48 h Cleaning	Used—Not Sent for Cleaning	Available at the End of the Day
1	y_1	z_1	v_1	130
2	y_2	z_2	v_2	$220 + v_1$
3	y_3	z_3	v_3	$180 + v_2$
4	y_4	z_4	v_4	$120 + v_3$
5	y_5	z_5	v_5	$100 + v_4$

From the above two tables, we can easily formulate an LP model

Objective function:

$$\text{Minimize cost} = 25(x_1 + x_2 + x_3 + x_4 + x_5) + 15(y_1 + y_2 + y_3 + y_4 + y_5) + 10(z_1 + z_2 + z_3 + z_4 + z_5) \tag{4.4}$$

Constraints:
Beginning of the day constraints:

$$\text{Day 1:} \quad x_1 = 130 \tag{4.5}$$

$$\text{Day 2:} \quad x_2 = 220 \tag{4.6}$$

$$\text{Day 3:} \quad x_3 + y_1 = 180 \tag{4.7}$$

$$\text{Day 4:} \quad x_4 + y_2 + z_1 = 120 \tag{4.8}$$

$$\text{Day 5:} \quad x_5 + y_3 + z_2 = 100 \tag{4.9}$$

End of the day constraints:

$$\text{Day 1:} \quad y_1 + z_1 + v_1 = 130 \tag{4.10}$$

$$\text{Day 2:} \quad y_2 + z_2 + v_2 = 220 + v_1 \tag{4.11}$$

$$\text{Day 3:} \quad y_3 + z_3 + v_3 = 180 + v_2 \tag{4.12}$$

$$\text{Day 4:} \quad y_4 + z_4 + v_4 = 120 + v_3 \tag{4.13}$$

$$\text{Day 5:} \quad y_5 + z_5 + v_5 = 100 + v_4 \tag{4.14}$$

Nonnegativity constraint:

$$x_i, y_i, z_i, v_i \geq 0 \quad \text{for all } i$$

4.3.2 Variables as Fractions of Other Variables

Sometimes in practice, a certain condition may imply that the value of one or more variables must be within a certain percentage (fraction) range of the sum of the values of some or all other variables. Further in the chapter, examples are provided with such a condition and their mathematical model formulations are discussed.

Example 4.3: A financial management problem

ABC Credit Union offers five types of loans. The titles of these loans, with their respective yearly interest rates charged to customers, are presented in the table below:

Loan ID	Type of Loan	Interest Charged (%)
1	Industrial/commercial loans	9.00
2	Home extensions	8.00
3	First time home loans	6.50
4	Home loans–supplemented	7.50
5	Personal loans	10.00

The credit union has $50 million available for these loans. ABC's objective is to maximize the yield on investment in loans. The credit union maintains the following policies for their loan investments:

1. Home extension loan investments cannot be greater than 25% of first time home loan investments.
2. Industrial loan investments must be less than or equal to home loan–supplemented investments.
3. The credit union invests at least 70% of the funds in home loans (first time and supplemented).
4. For technical reasons, there must be at least $3 invested in first time home loans for every dollar invested in home loans–supplemented.

Decision variables:
X_i = Dollars invested in loan ID i ($i = 1, \ldots, 5$)

Objective function:

$$\text{Maximize } Z = 0.09X_1 + 0.08X_2 + 0.065X_3 + 0.075X_4 + 0.10X_5 \qquad (4.15)$$

Constraints:

1. Total available funds for lending is $50 millions:

$$X_1 + X_2 + X_3 + X_4 + X_5 \leq 50 \qquad (4.16)$$

2. Home extensions loans cannot be higher than 25% of first time home loans:

$$X_2 \leq 0.25X_3 \qquad (4.17)$$

3. Industrial loans must be smaller than or equal to home loans–supplemented:

$$X_1 \leq X_4 \tag{4.18}$$

4. The credit union invests at least 70% of the funds in home loans (first time and supplemented):

$$X_3 + X_4 \geq 0.7(X_1 + X_2 + X_3 + X_4 + X_5) \tag{4.19}$$

5. For technical reasons, there must be at least \$3 invested in first time home loans for every dollar invested in home loans–supplemented:

$$X_3 \geq 3X_4 \tag{4.20}$$

6. Nonnegativity constraint: $X_i \geq 0 \quad \forall i$

   After rearranging the variables, one obtains the following formulation:

$$\text{Maximize } Z = 0.09X_1 + 0.08X_2 + 0.065X_3 + 0.075X_4 + 0.10X_5$$

$\in$

$$\begin{aligned}
X_1 + X_2 + X_3 + X_4 + X_5 &\leq 50 && \text{Availability of funds} \\
X_2 - 0.25X_3 &\leq 0 && \text{Policy 1} \\
X_1 - X_4 &\leq 0 && \text{Policy 2} \\
-0.7X_1 - 0.7X_2 + 0.3X_3 + 0.3X_4 - 0.7X_5 &\geq 0 && \text{Policy 3} \quad \text{Model (4.4)} \\
X_3 - 3X_4 &\geq 0 && \text{Policy 4} \\
X_i &\geq 0 && \forall i
\end{aligned}$$

Example 4.4: An oil blending problem

A multinational oil supplier produces two grades of gasoline, U (unleaded) and L (leaded), which it sells for \$1.10 and \$1.00 per liter, respectively. The refinery can buy three different types of refined oil, from three different sources, with the following constituents and prices:

Refined Oil	Constituents			Price/L
	A	B	C	
1	0.70	0.20	0.10	\$0.60
2	0.25	0.25	0.50	\$0.50
3	0.60	0.10	0.30	\$0.45

The U grade gasoline must have at least 50% of constituent A and not more than 35% of constituent C. The L grade gasoline must not have more than

30% of constituent C. Determine how the refined oils should be mixed so as to maximize the profit.

Decision variables:
X_{ij} = amount used of refined oil i ($i = 1, 2, 3$) in producing the gasoline grade j (j = U, L)

Objective function:

The total amount of grade U gasoline produced = $X_{1U} + X_{2U} + X_{3U}$
The total amount of grade L gasoline produced = $X_{1L} + X_{2L} + X_{3L}$
The total amount of refined oil 1 used = $X_{1U} + X_{1L}$
The total amount of refined oil 2 used = $X_{2U} + X_{2L}$
The total amount of refined oil 3 used = $X_{3U} + X_{3L}$
Total profit = (total revenue from the gasoline grades U and L) − (the total cost of refined oils 1, 2, and 3 used in production of the gasoline grades U and L)

The objective is to maximize the profit:

$$\text{Maximize } Z = 1.10(X_{1U} + X_{2U} + X_{3U}) + 1.00(X_{1L} + X_{2L} + X_{3L}) - 0.60(X_{1U} + X_{1L}) - 0.50(X_{2U} + X_{2L}) - 0.45(X_{3U} + X_{3L}) \quad (4.21)$$

After simplification, one obtains

$$\text{Maximize } Z = 0.50X_{1U} + 0.60X_{2U} + 0.65X_{3U} + 0.40X_{1L} + 0.50X_{2L} + 0.55X_{3L} \quad (4.22)$$

Constraints:

1. Grade U gasoline must contain at least 50% of constituent A.
   The total amount of grade U gasoline produced = $X_{1U} + X_{2U} + X_{3U}$
   The total amount of constituent A in the refined oils 1, 2, and 3 used in the production of grade U gasoline = $0.70X_{1U} + 0.25X_{2U} + 0.60X_{3U}$
   So the constraint is

$$0.70X_{1U} + 0.25X_{2U} + 0.60X_{3U} \geq 0.5(X_{1U} + X_{2U} + X_{3U}) \quad (4.23)$$

   After simplification, the constraint becomes

$$0.20X_{1U} - 0.25X_{2U} + 0.10X_{3U} \geq 0 \quad (4.24)$$

2. Grade U gasoline must not contain more than 35% of constituent C.

The total amount of grade U gasoline produced $= X_{1U} + X_{2U} + X_{3U}$

The total amount of constituent C in the refined oils 1, 2, and 3 used in the production of grade U gasoline $= 0.10X_{1U} + 0.50X_{2U} + 0.30X_{3U}$

So the constraint is

$$0.10X_{1U} + 0.50X_{2U} + 0.30X_{3U} \leq 0.35(X_{1U} + X_{2U} + X_{3U}) \tag{4.25}$$

After simplification, the constraint becomes

$$-0.25X_{1U} + 0.15X_{2U} - 0.05X_{3U} \leq 0 \tag{4.26}$$

3. Grade L gasoline must not contain more than 30% of constituent C.

   The total amount of grade L gasoline produced $= X_{1L} + X_{2L} + X_{3L}$

   The total amount of constituent C in the refined oils 1, 2, and 3 used in the production of grade U gasoline $= 0.10X_{1L} + 0.50X_{2L} + 0.30X_{3L}$

   So the constraint is

$$0.10X_{1L} + 0.50X_{2L} + 0.30X_{3L} \leq 0.30(X_{1L} + X_{2L} + X_{3L}) \tag{4.27}$$

After simplification, the constraint becomes

$$-0.20X_{1L} + 0.20X_{2L} \leq 0 \tag{4.28}$$

The final overall LP model is

$$\begin{aligned} \text{Maximize } Z = {} & 0.50X_{1U} + 0.60X_{2U} + 0.65X_{3U} + 0.40X_{1L} \\ & + 0.50X_{2L} + 0.55X_{3L} \end{aligned}$$

$\in$

$$\begin{aligned} -0.20X_{1U} - 0.25X_{2U} + 0.10X_{3U} &\geq 0 \quad \text{At least 50\% of A in U} \\ -0.25X_{1U} + 0.15X_{2U} - 0.05X_{3U} &\leq 0 \quad \text{Not more than 35\% of C in U} \\ -0.20X_{1L} + 0.20X_{2L} &\leq 0 \quad \text{Not more than 30\% of C in L} \\ X_{ij} &\geq 0 \quad \forall i, j \end{aligned}$$

Model (4.5)

4.3.3 Maintaining Certain Ratios among Different Variables

In Example 4.3, the constraints given by Equations 4.17, 4.18, and 4.20 indicate that one variable must be less than or equal to (or greater than or equal to) n times that of another variable, where $n > 0$. There are also many practical situations where the quantities of different products (or items) must be of specific ratios to one another. For example, in a sales department, products A, B, and C could be required in a ratio of 2:3:4. An example of modelling such a situation will be provided here.

Example 4.5: A product mix problem

ABC International Manufacturing Company produces and markets three products P1, P2, and P3. Time to manufacture product P1 is twice that for P2 and thrice that for P3 and the products are to be produced in the ratio of 3:4:5. The raw material requirements for each product and the availability of the raw materials are given in the following table. If the whole labor force is engaged in only producing product P1, time will permit 1600 units of P1 to be produced. There is a demand for at least 185, 250, and 200 units of products P1, P2, and P3 to be produced and the profit earned per unit is \$50, \$40, and \$70, respectively. Find the quantities of P1, P2, and P3 to be produced?

Raw Materials	Requirements per Unit of Product (kg) P1	P2	P3	Total Availability (kg)
R1	6	4	9	5000
R2	3	7	6	6000

Decision variables:
X_1, X_2, X_3 denote the number of units of products P1, P2, and P3 to be manufactured.

Objective function:
The objective is to maximize the profit.

$$\text{Maximize } Z = 50X_1 + 40X_2 + 70X_3 \tag{4.29}$$

Constraints:

1. Raw material limitations:

$$\text{For raw material 1:} \quad 6X_1 + 4X_2 + 9X_3 \leq 5000 \tag{4.30}$$

$$\text{For raw material 2:} \quad 3X_1 + 7X_2 + 6X_3 \leq 6000 \tag{4.31}$$

2. Capacity constraint: As product P2 requires one-half and product P3 requires one-third of the time required by product P1, the constraint on the number of units manufactured can be expressed as below:

   Assume each product P1 takes t amount of time, so

$$tX_1 + (t/2)X_2 + (t/3)X_3 \leq 1600t,$$

   which can be rewritten as

$$X_1 + (1/2)X_2 + (1/3)X_3 \leq 1600 \tag{4.32}$$

3. Market demand constraints:

$$\text{For product P1:} \quad X_1 \geq 185 \tag{4.33}$$

$$\text{For product P2:} \quad X_2 \geq 250 \tag{4.34}$$

$$\text{For product P3:} \quad X_3 \geq 200 \tag{4.35}$$

4. Product ratio constraints: Since the products P1, P2, and P3 are to be produced in the ratio 3:4:5, $X_1{:}X_2{:}X_3 = 3{:}4{:}5$ or

$$\frac{X_1}{3} = \frac{X_2}{4} = \frac{X_3}{5}$$

$$\text{or} \quad \frac{X_1}{3} = \frac{X_2}{4} \quad \text{and} \quad \frac{X_2}{4} = \frac{X_3}{5}$$

$$\text{or} \quad 4X_1 - 3X_2 = 0 \quad \text{and} \quad 5X_2 - 4X_3 = 0$$

The corresponding constraints are

$$4X_1 - 3X_2 = 0 \tag{4.36}$$

$$5X_2 - 4X_3 = 0 \tag{4.37}$$

So the final LP model becomes

Maximize $Z = 50X_1 + 40X_2 + 70X_3$

$\in$

$6X_1 + 4X_2 + 9X_3 \leq 5000$	Raw material 1 availability
$3X_1 + 7X_2 + 6X_3 \leq 6000$	Raw material 2 availability
$X_1 + (1/2)X_2 + (1/3)X_3 \leq 1600$	Production capacity
$X_1 \geq 185$	Demand constraint for product P1
$X_2 \geq 250$	Demand constraint for product P2
$X_3 \geq 200$	Demand constraint for product P3
$4X_1 - 3X_2 = 0$	Ratio constraint 1
$X_2 - 4X_3 = 0$	Ratio constraint 2
$X_1, X_2, X_3 \geq 0$	Nonnegativity

Model (4.6)

4.3.4 One Constraint Is a Fraction of Another Constraint

In Examples 4.3 through 4.5, we discussed some relationships that may be found between different variables or groups of variables. In some cases, similar relationships may be necessary for the constraint functions. For example, a balanced (or uniform) work load among different sections of an organization may be required. See the following examples.

Example 4.6: A crop mix problem

Northern Queensland Agriculture Incorporated (NQAI) operates three agricultural farms with similar soil and weather conditions. Each farm has a limited amount of usable land and limited amount of water for irrigation. The data for usable land and available water are given below:

Farm	Land Available (Acres)	Water Available (Acre Feet)
1. Lakefield	500	1600
2. Blackbull	600	2000
3. Buchanan	400	1000

NQAI is considering three crops for cultivation, which differ greatly in their expected profit per unit area and in their usage of water. Furthermore, the total area that can be used for each of the crops is limited by the amount of appropriate harvesting equipments and personnel available.

Crop	Maximum Acreage	Water Consumption in Acre Feet per Acre	Expected Profit per Acre ($)
Wheat	600	5	450
Potato	700	4	350
Canola	300	3	150

In order to maintain a balance work load among the farms, it is the policy of NQAI that the percentage of usable area cultivated must be equal at each farm. However, the individual farm is free to choose any combination of crops to be cultivated in its land. NQAI wishes to find how much of each crop should be produced at each farm to maximize the expected profit. Formulate the problem as an LP model.

Decision variables:
Let X_{ij} ($i = 1, 2, 3$ for farm in Lakefield, Blackbull, and Buchanan; $j = \text{A, B, C}$ for crop wheat, potato, and canola) represent the number of acres of ith farm to be allotted to the jth crop.

Objective function:

$$\text{Total area used for wheat by all three farms} = X_{1A} + X_{2A} + X_{3A}$$
$$\text{Total area used for potato by all three farms} = X_{1B} + X_{2B} + X_{3B}$$
$$\text{Total area used for canola by all three farms} = X_{1C} + X_{2C} + X_{3C}$$

The objective is to maximize the total profit

$$\text{Maximize } Z = 450(X_{1A} + X_{2A} + X_{3A}) + 350(X_{1B} + X_{2B} + X_{3B}) + 150(X_{1C} + X_{2C} + X_{3C}) \quad (4.38)$$

Constraints:

1. Requirements of water in acre feet at each farm:

$$\text{Farm Lakefield: } 5X_{1A} + 4X_{1B} + 3X_{1C} \leq 1600 \quad (4.39)$$

$$\text{Farm Blackbull: } 5X_{2A} + 4X_{2B} + 3X_{2C} \leq 2000 \quad (4.40)$$

$$\text{Farm Buchanan: } 5X_{3A} + 4X_{3B} + 3X_{3C} \leq 1000 \quad (4.41)$$

2. Availability of usable acreage at each farm:

$$\text{Farm Lakefield: } X_{1A} + X_{1B} + X_{1C} \leq 500 \quad (4.42)$$

$$\text{Farm Blackbull: } X_{2A} + X_{2B} + X_{2C} \leq 600 \quad (4.43)$$

$$\text{Farm Buchanan: } X_{3A} + X_{3B} + X_{3C} \leq 400 \quad (4.44)$$

3. Usage of acreage for each crop due to equipment limitations:

$$\text{For wheat: } X_{1A} + X_{2A} + X_{3A} \leq 600 \quad (4.45)$$

$$\text{For potato: } X_{1B} + X_{2B} + X_{3B} \leq 700 \quad (4.46)$$

$$\text{For canola: } X_{1C} + X_{2C} + X_{3C} \leq 300 \quad (4.47)$$

4. To ensure the percentage of usable acreage is same at each farm:

$$\left(\frac{X_{1A} + X_{1B} + X_{1C}}{500}\right)100 = \left(\frac{X_{2A} + X_{2B} + X_{2C}}{600}\right)100 = \left(\frac{X_{3A} + X_{3B} + X_{3C}}{400}\right)100$$

or $\quad 6(X_{1A} + X_{1B} + X_{1C}) = 5(X_{2A} + X_{2B} + X_{2C})$

and $\quad 4(X_{2A} + X_{2B} + X_{2C}) = 6(X_{3A} + X_{3B} + X_{3C})$

or

$$6(X_{1A} + X_{1B} + X_{1C}) - 5(X_{2A} + X_{2B} + X_{2C}) = 0 \quad (4.48)$$

and

$$4(X_{2A} + X_{2B} + X_{2C}) - 6(X_{3A} + X_{3B} + X_{3C}) = 0 \quad (4.49)$$

So the final model becomes

$$\text{Maximize } Z = 450(X_{1A} + X_{2A} + X_{3A}) + 350(X_{1B} + X_{2B} + X_{3B}) + 150(X_{1C} + X_{2C} + X_{3C})$$

$\in$

$5X_{1A} + 4X_{1B} + 3X_{1C} \leq 1600$ Water availability in Lakefield

$5X_{2A} + 4X_{2B} + 3X_{2C} \leq 2000$ Water availability in Blackbull

$5X_{3A} + 4X_{3B} + 3X_{3C} \leq 1000$ Water availability in Buchanan

$X_{1A} + X_{1B} + X_{1C} \leq 500$ Land availability in Lakefield

$X_{2A} + X_{2B} + X_{2C} \leq 600$ Land availability in Blackbull

$X_{3A} + X_{3B} + X_{3C} \leq 400$ Land availability in Buchanan

$X_{1A} + X_{2A} + X_{3A} \leq 600$ Equipment restriction for wheat

$X_{1B} + X_{2B} + X_{3B} \leq 700$ Equipment restriction for potato

$X_{1C} + X_{2C} + X_{3C} \leq 300$ Equipment restriction for canola

$6\,(X_{1A} + X_{1B} + X_{1C}) - 5\,(X_{2A} + X_{2B} + X_{2C}) = 0$ Uniform loading

$4\,(X_{2A} + X_{2B} + X_{2C}) - 6\,(X_{3A} + X_{3B} + X_{3C}) = 0$ Uniform loading

$X_{ij} \geq 0, \quad i = 1, 2, 3;\ j = \text{A, B, C}$ Nonnegativity

Model (4.7)

Example 4.7: A manufacturing planning problem

A small machine shop has one drilling and five milling machines, which are to be used to produce a finished product consisting of two parts, P1 and P2. The productivity of each machine for the two parts is given below:

		Production Time (Minutes per Piece)	
Part	Profit ($ per Piece)	Drilling	Milling
P1	4	3	20
P2	5	5	15

It is desired to maintain a balanced loading on all machines such that no machine runs for more than 30 min per day longer than any other machine (assume that the milling load is split evenly among all five milling machines). Formulate a linear program to divide the work time of each machine to maximize the profit assuming an 8 h working day.

Decision variables:

Let X_i = number of part i ($i = 1$ and 2 for P1 and P2 respectively) produced per day

Objective function:

The objective is to maximize the profit.

$$\text{Maximize } Z = 4X_1 + 5X_2 \tag{4.50}$$

Constraints:

1. Loading constraint:

   The load on each milling machine (in minutes) $= (20X_1+15X_2)/5 = 4X_1 + 3X_2$.

   The load on the drilling machine (in minutes) $= 3X_1 + 5X_2$.

   Thus, the time restriction (in minutes) on each milling machine is

   $$4X_1 + 3X_2 \leq 8 \times 60 = 480 \text{ or}$$

   $$4X_1 + 3X_2 \leq 480 \tag{4.51}$$

   Similarly, for the drilling machine:

   $$3X_1 + 5X_2 \leq 480 \tag{4.52}$$

2. Machine balancing constraint:

   The machine balance constraint can be represented by

   $$|(4X_1 + 3X_2) - (3X_1 + 5X_2)| \leq 30 \text{ or } |X_1 - 2X_2| \leq 30$$

   This is a nonlinear constraint. The meaning of this constraint is if $X_1 \geq 2X_2$, then $(X_1 - 2X_2)$ would be a positive quantity that is expected to be less than or equal to 30, and if $2X_2 \geq X_1$, then $(2X_2 - X_1)$ would be positive that is expected to be less than or equal to 30.

   This constraint can be replaced by the following linear constraints:

   $$X_1 - 2X_2 \leq X_3 \tag{4.53}$$

   $$2X_2 - X_1 \leq X_3 \tag{4.54}$$

   where X_3 represents the positive difference.

   $$X_3 \leq 30 \tag{4.55}$$

   So the LP model is

   $$\text{Maximize } Z = 4X_1 + 5X_2$$

   $\in$

   $$\begin{aligned} 4X_1 + 3X_2 &\leq 480 && \text{Milling machine} \\ 3X_1 + 5X_2 &\leq 480 && \text{Drilling machine} \\ X_1 - 2X_2 - X_3 &\leq 0 && \text{Load balancing} \\ 2X_2 - X_1 - X_3 &\leq 0 && \text{Load balancing} \\ X_3 &\leq 30 && \text{Load balancing} \\ X_1, X_2, X_3 &\geq 0 && \text{Nonnegativity} \end{aligned} \qquad \text{Model (4.8)}$$

4.3.5 Maxi–Min or Mini–Max Objective Function

In many situations, one would like to minimize (or maximize) the maximum (or minimum) value of two or more variables or constraint functions. Two examples are provided below to show how to handle such conditions.

Example 4.8: A manufacturing planning problem
Consider Example 4.7 with a different objective. Formulate a linear program to divide the work time of each machine to obtain the maximum number of completed assemblies assuming an 8 h working day.

The number of completed assemblies cannot exceed the smaller value of part 1 and part 2 produced. Thus, the objective function is to maximize $Z = \text{minimum}\ (X_1, X_2)$. This is again a nonlinear function. However, another trick can be applied to represent it as a linear function.

Let $Y = \text{minimum of}\ (X_1, X_2)$, where Y represents the number of completed assemblies.

This means that

$$X_1 \geq Y \text{ or } X_1 - Y \geq 0 \tag{4.56}$$

$$X_2 \geq Y \text{ or } X_2 - Y \geq 0 \tag{4.57}$$

and the objective is to

$$\text{Maximize } Z = Y \tag{4.58}$$

Thus, the modified LP formulation becomes

$$\text{Maximize } Z = Y$$

$$\in$$

$$\left.\begin{array}{rl} 4X_1 + 3X_2 \leq 480 & \text{Milling machine} \\ 3X_1 + 5X_2 \leq 480 & \text{Drilling machine} \\ X_1 - 2X_2 - X_3 \leq 0 & \text{Load balancing} \\ 2X_2 - X_1 - X_3 \leq 0 & \text{Load balancing} \\ X_3 \leq 30 & \text{Load balancing} \\ X_1 - Y \geq 0 & \text{Minimum production} \\ X_2 - Y \geq 0 & \text{Minimum production} \\ X_1, X_2, X_3, Y \geq 0 & \text{Nonnegativity} \end{array}\right\} \quad \text{Model (4.9)}$$

Example 4.9: A production planning problem
Melbourne East Spares Manufacturer produces a specialized product. Each unit of the product consists of five units of component C1 and four

units of component C2. These two components (C1 and C2) require two different raw materials (R1 and R2) coming from outside sources. The availability of raw materials R1 and R2 is 120 and 240 units, respectively. The components can be produced using three different processes. Each process is capable of producing both components in a single production run (like a batch production mode). However, the number of components produced varies from process to process. The raw materials required per production run and the number of units for each component produced is given in the following table:

	Input per Production Run (Units)		Output per Production Run (Units)	
Process	Raw Material R1	Raw Material R2	Part C1	Part C2
1	7	5	6	4
2	4	7	5	8
3	2	9	7	3

Determine the number of production runs for each process so as to maximize the total number of completed units of the final product.

Decision variables:
Let X_i represent the number of production runs for process i ($i = 1, 2,$ and 3).

Objective function:
The objective is to maximize the total number of units of the final product.

The total number of units of component C1 produced by different methods is $6X_1 + 5X_2 + 7X_3$ and for component C2 is $4X_1 + 8X_2 + 3X_3$.

Since each unit of the final product requires five units of component C1 and four units of component C2, it is evident that the maximum number of units of the final product cannot exceed the smaller value of

$$\frac{6X_1 + 5X_2 + 7X_3}{5} \quad \text{and} \quad \frac{4X_1 + 8X_2 + 3X_3}{4}$$

Thus the objective is to maximize Z, where

$$Z = \text{Min}\left(\frac{6X_1 + 5X_2 + 7X_3}{5}, \frac{4X_1 + 8X_2 + 3X_3}{4}\right) \tag{4.59}$$

The above formulation violates the LP properties since the objective function is not linear. However, it can be easily reformulated into a linear function.

$$\text{Let } Y = \text{Min}\left(\frac{6X_1 + 5X_2 + 7X_3}{5}, \frac{4X_1 + 8X_2 + 3X_3}{4}\right)$$

It follows that $\frac{6X_1 + 5X_2 + 7X_3}{5} \geq Y$ and $\frac{4X_1 + 8X_2 + 3X_3}{4} \geq Y$

That is $6X_1 + 5X_2 + 7X_3 - 5Y \geq 0$

and $4X_1 + 8X_2 + 3X_3 - 4Y \geq 0$.

These two inequalities will be used as constraints and the objective will be to maximize Y.

Constraints:
The only other constraints required are for raw material availability:

$$\text{Raw material R1: } 7X_1 + 4X_2 + 2X_3 \leq 120 \tag{4.60}$$

$$\text{Raw material R2: } 5X_1 + 7X_2 + 9X_3 \leq 240 \tag{4.61}$$

Hence the final LP model becomes

Maximize Y

$\in$

$$\begin{aligned} 6X_1 + 5X_2 + 7X_3 - 5Y &\geq 0 && \text{Maxi–min constraint} \\ 4X_1 + 8X_2 + 3X_3 - 4Y &\geq 0 && \text{Maxi–min constraint} \\ 7X_1 + 4X_2 + 2X_3 &\leq 120 && \text{Raw material R1} \qquad \text{Model (4.10)} \\ 5X_1 + 7X_2 + 9X_3 &\leq 240 && \text{Raw material R2} \\ X_1, X_2, X_3 &\geq 0 && \text{Nonnegativity} \end{aligned}$$

4.3.6 Multi-Period Modelling

Many planning problems involve multi-period scenarios. Multi-period planning takes into account many time periods into the future. In such cases, the mathematical model is used to solve the planning process in the individual time periods as well as linking the planning process from one period to the next. Multi-period modelling is very useful when the model parameters such as demand, production level, price, and other factors fluctuate from one period to the other.

Example 4.10: A multi-period production planning problem
North Sydney Production Limited (NSPL) produces exhaust control valves to meet customer demands. The customers place their orders in advance and the demand per week is not constant. However, the company has a fixed production capacity of 1000 valves per week. The total demand of exhaust control valves over the next 4 weeks is given below:

Week (i)	Demand (D_i Units)
1	800
2	700
3	1200
4	1100

The unit production cost is \$50. The company is allowed to carry over any extra production, they make in a given week, to sell in a future week. However, it costs \$5 per unit per week for storage, handling, and insurance, commonly called inventory holding cost. The setup cost is assumed to be negligible. Formulate the problem to determine a production plan at minimum possible cost.

Assumptions:

- Products are either delivered to the customers or stored at the end of the week.
- There is no inventory at the beginning of the production and the final inventory must be zero.

Decision variables:

X_i = number of units produced in week i ($i = 1, 2, 3,$ and 4)

I_i = number of units in the inventory at the end of week i

Objective function:
The objective is to minimize the production plus inventory holding costs.

$$\text{Minimize } Z = 50(X_1 + X_2 + X_3 + X_4) + 5(I_1 + I_2 + I_3 + I_4) \quad (4.62)$$

Constraints:

1. Production capacity

   In any week, the production cannot exceed the available capacity.

$$X_i \leq 1000 \quad \text{for all } i \quad (4.63)$$

2. Meeting demand

   The production in a week plus the items (inventory) carried from the previous week must be equal to the demand in that week plus items stored for future weeks.

$$X_i + I_{i-1} = D_i + I_i \quad \text{for all } i$$

or

$$X_i + I_{i-1} - I_i = D_i \quad \text{for all } i$$

also

$$I_0 = I_4 = 0$$

Here, the inventory variables I_i link one week to the next. The model is flexible enough to allow inventory to be carried over from one week to any future week via the intermediate weeks. The model will suggest the minimum possible inventory to carry over from one week to future weeks since the objective is to minimize the total cost.

The final LP model is

$$\text{Minimize } Z = 50(X_1 + X_2 + X_3 + X_4) + 5(I_1 + I_2 + I_3 + I_4)$$

$\in$

$$\begin{aligned}
&X_1 \leq 1000 && \\
&X_2 \leq 1000 && \text{Capacity constraints} \\
&X_3 \leq 1000 && \\
&X_4 \leq 1000 && \qquad\qquad \text{Model (4.11)} \\
&X_1 - I_1 = 800 && \\
&X_2 + I_1 - I_2 = 700 && \text{Demand and linkage constraints} \\
&X_3 + I_2 - I_3 = 1200 && \\
&X_4 + I_3 = 1100 && \\
&X_i, I_i \geq 0 \quad \forall i && \text{Nonnegativity}
\end{aligned}$$

4.3.7 Transforming Infeasible Solutions to Satisfactory Solutions

A mathematical programming model can be infeasible for many reasons such as wrong formulation, wrong data, or inconsistent/infeasible constraints. By carefully checking or debugging the model, one can, in many cases, rectify the model and remove any inconsistencies existing in the constraints. However, if the constraints are infeasible, it is impossible to obtain a feasible solution to the problem. The problem of infeasibility is addressed below using an example.

Example 4.11: A product mix problem

Consider the following LP problem that maximizes the profit generated from producing four products subject to constraints involving available cash, working capital, and three technological constraints as follows.

Decision variables:
x_i = quantity produced of product i ($i = 1, 2, 3$, and 4).

Objective function:

$$\text{Maximize } Z = 112x_1 + 162x_2 + 192x_3 + 89x_4 \tag{4.64}$$

Subject to

Technological constraints:

$$x_1 + x_2 + x_3 + 0.333x_4 \leq 100 \quad \text{Technological 1} \tag{4.65}$$

$$x_2 + x_3 + 0.143x_4 \leq 100 \quad \text{Technological 2} \tag{4.66}$$

$$x_3 + x_4 \leq 150 \quad \text{Technological 3} \tag{4.67}$$

Other constraints:

$$168x_1 + 288x_2 + 288x_3 + 391x_4 \geq 67{,}000 \quad \text{Available cash} \tag{4.68}$$

$$150x_1 + 80x_2 + 86x_3 + 70x_4 \geq 16{,}000 \quad \text{Working capital} \tag{4.69}$$

$$x_i \geq 0 \quad \text{for all } i \tag{4.70}$$

The solution to the above LP model is

$$x_1 = 21.44,\ x_2 = 28.55,\ x_3 = 0,\ x_4 = 150, \text{ and } Z = \$20{,}948$$

Later, it is realized that the available cash requirement should have been increased from \$67,000 to \$72,000. That change makes the model infeasible as there is now no feasible solution space. So the LP approach is not appropriate any more for this problem. It is clear that the model has become infeasible because of the available cash constraint. Instead of reducing the RHS of that constraint just to obtain a feasible solution, a preferred option might be to develop a negotiated production plan to meet the financial plan. In such a situation, goal programming (GP) is an appropriate approach to find the satisfactory solutions.

Formulating the GP model:
The technological constraints are kept unchanged (due to other production limitations), so they are same as LP shown in Equations 4.65 through 4.67.

Goal constraints (changeable constraints):

1. Available cash (assuming cash is the most important or preferred goal):
   $168x_1 + 288x_2 + 288x_3 + 391x_4 \geq 67{,}000$ becomes

   $$168x_1 + 288x_2 + 288x_3 + 391x_4 + d_1^- - d_1^+ = 72{,}000 \tag{4.71}$$

   d_1^- and d_1^+ are the underachievement and overachievement factors (deviational variables) for available cash, respectively.

   Other goal constraints (in the order of importance)

2. Working capital:
$150x_1 + 80x_2 + 86x_3 + 70x_4 \geq 16{,}000$ becomes

$$150x_1 + 80x_2 + 86x_3 + 70x_4 + d_2^- - d_2^+ = 16{,}000 \quad (4.72)$$

d_2^- and d_2^+ are the underachievement and overachievement factors for working capital, respectively.

3. Profit—Objective function in LP:

$$112x_1 + 162x_2 + 192x_3 + 89x_4 + d_3^- - d_3^+ = 23{,}000 \quad (4.73)$$

d_3^- and d_3^+ are the underachievement and overachievement factors for the profit target, respectively. The profit target of \$23,000 is assumed because it is slightly higher than the LP solution.

GP objective function:
The objective is to minimize the weighted sum of all undesirable deviations:

$$\text{Minimize } Z = w_1 d_1^- + w_2 d_2^- + w_3 d_3^- \quad (4.74)$$

Here, d_1^-, d_2^-, and d_3^- are the undesirable deviations from available cash, working capital, and profit target, respectively. The terms w_1, w_2, and w_3 are the weights for first, second, and third goals, respectively, where $w_1 > w_2 > w_3$.

The final GP model is

$$\begin{aligned}
&\text{Minimize } Z = w_1 d_1^- + w_2 d_2^- + w_3 d_3^- \\
&\text{Subject to} \\
&x_1 + x_2 + x_3 + 0.333x_4 \leq 100 \\
&x_2 + x_3 + 0.143x_4 \leq 100 \\
&x_3 + x_4 \leq 150 \\
&168x_1 + 288x_2 + 288x_3 + 391x_4 + d_1^- - d_1^+ = 72{,}000 \\
&150x_1 + 80x_2 + 86x_3 + 70x_4 + d_2^- - d_2^+ = 16{,}000 \\
&112x_1 + 162x_2 + 192x_3 + 89x_4 + d_3^- - d_3^+ = 23{,}000 \\
&x_i, d_i^-, d_i^+ \geq 0 \quad \forall i
\end{aligned} \quad \text{Model (4.12)}$$

4.3.8 Single to Multiple Objectives

Goal programming forms a single objective optimization model where the objective is to minimize the sum of all undesirable deviations. In the above example (Example 4.11), a negotiated production plan was perceived in order to meet the financial plan—that meant a compromise between profit, available cash, and working capital. In other words, the objective was to indirectly maximize the profit, available cash, and working capital. Such a problem can be modelled as a multi-objective optimization model.

Example 4.12: A multi-objective product-mix problem
Consider Example 4.11. Formulate the problem as a multi-objective optimization model considering all the goals as objectives. Note that we expect the profit, available cash, and working capital as high as possible.

The following are the three objective functions (instead of goal constraints):

1. For available cash:

$$\text{Maximize } Z_1 = 168x_1 + 288x_2 + 288x_3 + 391x_4 \tag{4.75}$$

2. For working capital:

$$\text{Maximize } Z_2 = 150x_1 + 80x_2 + 86x_3 + 70x_4 \tag{4.76}$$

3. For profit, which was the original LP's objective:

$$\text{Maximize } Z_3 = 112x_1 + 162x_2 + 192x_3 + 89x_4 \tag{4.77}$$

The multi-objective LP model becomes

$$\begin{aligned}
&\text{Maximize } Z_1 = 168x_1 + 288x_2 + 288x_3 + 391x_4\\
&\text{Maximize } Z_2 = 150x_1 + 80x_2 + 86x_3 + 70x_4\\
&\text{Maximize } Z_3 = 112x_1 + 162x_2 + 192x_3 + 89x_4\\
&\text{Subject to}\\
&\qquad x_1 + x_2 + x_3 + 0.333x_4 \le 100\\
&\qquad x_2 + x_3 + 0.143x_4 \le 100\\
&\qquad x_3 + x_4 \le 150\\
&\qquad x_i \ge 0 \quad \forall i
\end{aligned} \qquad \text{Model (4.13)}$$

4.4 Special Types of Linear Programming

There are a number of problems that are recognized as special types of the LP problem such as transportation, assignment, transshipment, and other network flow problems. All network flow problems can be represented as a network—a collection of nodes connected by arcs. The nodes usually represent the locations and the arcs (lines) connecting the nodes, indicate the valid paths, routes, or connections. When the lines connecting the nodes in a network are arrows that indicate a direction, such arcs in the network are called *directed arcs*. In this section, we will introduce several simple network flow problems, as special types of LP, their structures and modelling approaches.

4.4.1 Transportation Problem

The transportation problem deals with the transportation of goods at minimum cost, from several sources to a number of different destinations. In this type of problem, the goods are only allowed to be directly transported from source to destination. That means, no transshipments are allowed between sources or between destinations. The capacities (always limited) of the sources, the demands of the destinations, and the unit transportation costs from each source to each and every destination are known. This is a specially structured problem and any variation to this description would not allow such a problem to be recognized as a transportation problem.

The problem, here, is to determine how many units should be shipped from each source to each destination so that all demands are satisfied at the minimum total shipping cost.

Example 4.13: A transportation problem

Austral Steel Company has two re-rolling plants, in locations P1 and P2 with supply capacities of 100 and 120 tonnes rod (of fixed diameter) per day, respectively. Austral has three distribution centers at locations DC1, DC2, and DC3. The distribution centers need, if possible, 80, 120, and 60 tonnes of rod each day, respectively, to meet their distribution demands. The shipping costs ($/tonne) from each plant to each distribution center are given below:

From/to	DC1	DC2	DC3
Plant-P1	1	2	3
Plant-P2	4	1	5

Formulate the problem as a mathematical model.

The above transportation problem can be explained by using a network as shown in Figure 4.1.

The arrows indicate the existing paths and directions of flow.

The problem can also be explained in a tabular form.

In Table 4.1

Rows 2 and 3 are associated with the sources of supply. Each source is represented by a row.

Columns 2, 3, and 4 are associated with the destinations. Each destination is represented by a column.

Column 5 (rows 2 and 3) represent the capacities of the sources.

Row 4 (columns 2, 3, and 4) represent the demands of the destination points.

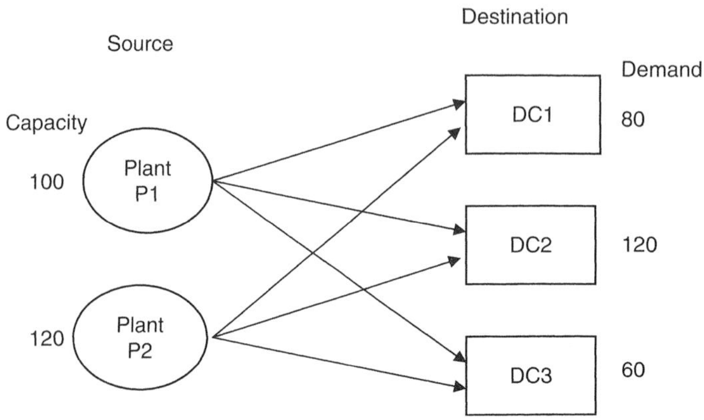

FIGURE 4.1
A transportation problem.

Each cell contains the transportation cost associated with the transport from source to destination (top right-hand corner), for example cell (P1–DC1) represents the transportation cost of \$1 for transporting one unit from source P1 to destination DC1.

Lower right-hand corner cell shows the total supply and total demand involved in this problem.

Defining variables:
X_{ij} = quantity to be shipped from source i ($i = 1$ for P1 and 2 for P2) to destination j ($j = 1$ for DC1, 2 for DC2, and 3 for DC3).

The transportation table with the decision variables is shown in Table 4.2.

Objective function:
The objective is to minimize the overall shipping costs. It is simply the sum of all variables (as written in the cells in Table 4.2) multiplied by their corresponding unit costs.

TABLE 4.1

Tabular Representation of Transportation Problem

From \ To	DC1	DC2	DC3	Supply
P1	1 Cell P1–DC1	2 Cell P1–DC2	3 Cell P1–DC3	100
P2	4 Cell P2–DC1	1 Cell P2–DC2	5 Cell P2–DC3	120
Demand	80	120	60	220 \ 260

TABLE 4.2

Tabular Representation with Decision Variables

From \ To	DC1	DC2	DC3	Supply
P1	1 X_{11}	2 X_{12}	3 X_{13}	100
P2	4 X_{21}	1 X_{22}	5 X_{23}	120
Demand	80	120	60	260 \ 220

$$\text{Minimize } Z = 1X_{11} + 2X_{12} + 3X_{13} + 4X_{21} + 1X_{22} + 5X_{23} \tag{4.78}$$

Constraints:
Only supply and demand constraints are involved in transportation problems. There are three possible relationships between total supply and total demand in transportation problems.

- Total supply < total demand: total demand cannot be satisfied by the available supply. Here, supply constraints are equalities (=) but demand constraints are less than or equal to (≤) inequalities.
- Total supply = total demand: total demand can be satisfied from the available supply. All the constraints are in the equalities (=) form.
- Total supply > total demand: there will be an excess supply. The demand constraints are equalities but the supply constraints are less than or equal to (≤) inequalities.

In our example, total demand (=260) is greater than total supply (=220). So the supply and demand constraints will be of equalities and less than or equal to inequalities, respectively.

Supply constraints: The total supply from a plant to all destinations must be equal to its capacity.

$$\text{Plant P1: } X_{11} + X_{12} + X_{13} = 100 \tag{4.79}$$

$$\text{Plant P2: } X_{21} + X_{22} + X_{23} = 120 \tag{4.80}$$

Demand constraints: The total supply received by a destination from all sources must be less than or equal to its requirements.

$$\text{Destination DC1: } X_{11} + X_{21} \leq 80 \tag{4.81}$$

$$\text{Destination DC2: } X_{12} + X_{22} \leq 120 \tag{4.82}$$

$$\text{Destination DC3: } X_{13} + X_{23} \leq 60 \tag{4.83}$$

The overall model is

Minimize $Z = 1X_{11} + 2X_{12} + 3X_{13} + 4X_{21} + 1X_{22} + 5X_{23}$
Subject to

$$
\begin{aligned}
X_{11} + X_{12} + X_{13} &= 100 && \text{Supply constraint for P1} \\
X_{21} + X_{22} + X_{23} &= 120 && \text{Supply constraint for P2} \qquad \text{Model (4.14)} \\
X_{11} + X_{21} &\leq 80 && \text{Demand constraint for DC1} \\
X_{12} + X_{22} &\leq 120 && \text{Demand constraint for DC2} \\
X_{13} + X_{23} &\leq 60 && \text{Demand constraint for DC3} \\
Xij &\geq 0 \quad \forall i, j
\end{aligned}
$$

Discussion question: What will happen if we put the "$\geq$" sign in place of the "$=$" in supply constraints?

In the model, the coefficient of each variable in the constraint is exactly one, and each variable exists exactly in two places—once in the supply constraints and once in the demand constraints. Although the above model can be solved as an LP model, the special structure of the model aids itself to the development of a more efficient solution approach for solving transportation problems.

4.4.2 Assignment Problem

This is a special class of transportation problem where the decision maker wants to assign a number of tasks to a number of individuals on a one-to-one basis so that no individual gets more than one task and no one task gets more than one individual assigned to it. The overall cost for an assignment problem requires to be minimized. The assignment problem can be represented as an equivalent transportation problem assuming that

- Each task is a source
- Each individual is a destination
- There is a supply of exactly one unit in each source
- There is a demand of exactly one unit in each destination
- The cost of transportation is equal to the cost of the assignment

Example 4.14: An assignment problem

A local government's public service unit wants to assign three health service teams to three remote rural locations, one team to each location. Each team is supposed to travel from the unit office to the assigned location on a daily basis. Because of isolation of these locations from the service unit office, the transportation costs and time windows for service differ from location to location. These differences are reflected as the cost per service hour as shown below. The problem is to find the assignment that minimizes the total cost for a given hours of service in each location.

Service Team	Location (L1)	Location (L2)	Location (L3)
S1	25	20	30
S2	20	15	35
S3	18	19	28

Table 4.3 gives the tabular format of the problem.

Defining variables:
X_{ij} = assignment of the ith ($i = 1, 2$, and 3) service team (or source) allocated to the jth ($j = 1, 2$, and 3) location (or destination).

Alternatively, we can define the binary variables as follows:

$$X_{ij} = \begin{cases} 1 & \text{if } i\text{th service team is assigned to } j\text{th location} \\ 0 & \text{otherwise} \end{cases} \quad \forall i, j$$

Objective function:
The objective is to minimize the overall assignment cost.

$$\text{Minimize } Z = 25X_{11} + 20X_{12} + 30X_{13} + 20X_{21} + 15X_{22} + 35X_{23} + 18X_{31} + 19X_{32} + 28X_{33} \quad (4.84)$$

Constraints:
There are two types of constraints similar to transportation problems.

Supply/service team constraint: One service team will be assigned to one of the three locations.

$$\text{Service team 1:} \quad X_{11} + X_{12} + X_{13} = 1 \quad (4.85)$$

$$\text{Service team 2:} \quad X_{21} + X_{22} + X_{23} = 1 \quad (4.86)$$

$$\text{Service team 3:} \quad X_{31} + X_{32} + X_{33} = 1 \quad (4.87)$$

Demand/location constraint: One location will receive only one of the three service teams.

TABLE 4.3

Tabular Representation of an Assignment Problem

Service Team \ Location	L1	L2	L3	Supply
S1	25	20	30	1
S2	20	15	35	1
S3	18	19	28	1
Demand	1	1	1	3 = 3

$$\text{Location 1:} \quad X_{11} + X_{21} + X_{31} = 1 \tag{4.88}$$

$$\text{Location 2:} \quad X_{12} + X_{22} + X_{32} = 1 \tag{4.89}$$

$$\text{Location 3:} \quad X_{13} + X_{23} + X_{33} = 1 \tag{4.90}$$

The overall model is

$$\text{Minimize } Z = 25X_{11} + 20X_{12} + 30X_{13} + 20X_{21} + 15X_{22} + 35X_{23} + 18X_{31} + 19X_{32} + 28X_{33}$$

Subject to

$$\begin{aligned}
X_{11} + X_{12} + X_{13} &= 1 \quad \text{Service team 1} \\
X_{21} + X_{22} + X_{23} &= 1 \quad \text{Service team 2} \\
X_{31} + X_{32} + X_{33} &= 1 \quad \text{Service team 3} \\
X_{11} + X_{21} + X_{31} &= 1 \quad \text{Location 1} \\
X_{12} + X_{22} + X_{32} &= 1 \quad \text{Location 2} \\
X_{13} + X_{23} + X_{33} &= 1 \quad \text{Location 3} \\
X_{ij} \geq 0 \ (\text{or } X_{ij} &\in 0, 1) \quad \forall i, j
\end{aligned} \qquad \text{Model (4.15)}$$

In addition to the special characteristics associated with transportation problems, all the right-hand sides are exactly 1 in assignment problems. The problem can be treated either as a linear program or an integer program. However, it is usually preferable to recognize assignment problems as linear programs because of the availability of efficient solution approaches for linear programs. Interestingly, there exists a very efficient algorithm for assignment problem, which is discussed later.

4.4.3 Transshipment Problem

Transshipment problems are known as generalized versions of transportation problems where shipments are allowed from a source to any destination via intermediate nodes if it saves on transportation costs. As you will see, most of the other types of network flow problems can all be viewed as simple variations to the transshipment problem. In fact, once you understand how to formulate the transshipment problem, the other problem types will be easy to model. In the transshipment problem, any node can act as one of the following:

- Supply node
- Demand node
- Transshipment (or intermediate) node
- Supply plus transshipment node
- Demand plus transshipment node

It is unusual for a node to act as both supply and demand node. If this is the case for a node, for some reason, the net supply/demand will be calculated separately and one then defines the type of node accordingly.

Example 4.15: A network flow problem

Consider a network flow problem illustrated in the Figure 4.2. There are six towns/cities in the mid-southern part of New South Wales: (1) Wagga Wagga, (2) Yass, (3) Cooma, (4) Canberra, (5) Goulburn, and (6) Batemans Bay. For a given product, the net supply or demand for each node in the network is indicated by a positive or negative number next to each node. Positive numbers represent the demand at a given node, and the negative numbers represent the supply available at the node. The transportation cost per unit is given on each arc. Develop a mathematical model that would minimize the overall transportation cost.

In this problem (Figure 4.2), Goulburn (node 5) is a supply node, Batemans Bay (node 6) is a demand node and all other nodes are transshipment nodes. Transshipment nodes can both send to and receive from other nodes in the network. In addition, a transshipment node may have its own supply or demand. For example, Wagga Wagga (node 1) has a demand of 100 units, would receive goods from node 5, and deliver goods to node 2.

Defining variables:
X_{ij} = the number of items shipped (or flowing) from node i to node j

Note that not all variables exist in this network. The variables that exist here are: X_{12}, X_{24}, X_{42}, X_{34}, X_{43}, X_{36}, X_{51}, X_{54}, and X_{56}. In network flow

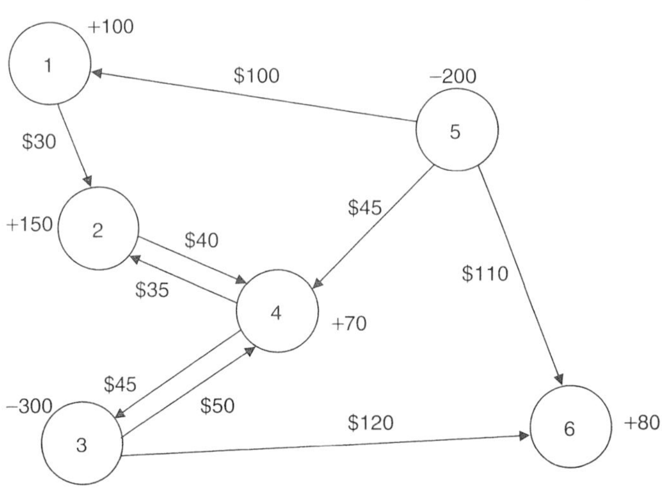

FIGURE 4.2
A transshipment problem.

problems, the number of variables in the LP model is equal to the number of arcs (one variable for each arc) in the network.

Objective function:
The objective is to minimize the overall transportation cost.

$$\text{Minimize } Z = 30X_{12} + 40X_{24} + 35X_{42} + 50X_{34} + 45X_{43} + 120X_{36} + 100X_{51} + 45X_{54} + 110X_{56} \quad (4.91)$$

Constraints:
The number of constraints in the LP model is equal to the number of nodes (one constraint for each node) in the network. In formulating the constraints for cost minimization network flow problems, the following *flow balance rules* are applied:

Condition	Flow Balance Rule at Each Node
Total supply = total demand	Inflow − outflow = supply or demand
Total supply > total demand	Inflow − outflow ≥ supply or demand
Total supply < total demand	Inflow − outflow ≤ supply or demand

It should be noted that if the total supply in a network flow problem is less than the total demand, then it will be impossible to satisfy all demands. In our problem, the total supply is (300 + 200) 500 and the total demand is (100 + 150 + 70 + 80) 400. That means, we will create a constraint at each node of the following form:

$$\text{Inflow} - \text{outflow} \geq \text{supply or demand}$$

For example, consider node 1 (Wagga Wagga). The inflow to this node is X_{51} and the outflow from the node is X_{12} and the demand is 100 units. According to the flow balance rule, the constraint for node 1 is

$$\text{Node 1: } X_{51} - X_{12} \geq +100 \quad (4.92)$$

The constraint for node 5 (Goulburn), where no arcs flow into this node, can be written as follows:

$$\text{Node 5: } -X_{51} - X_{54} - X_{56} \geq -200 \quad (4.93)$$

Constraints for the remaining nodes:

$$\text{Node 2: } X_{12} + X_{42} - X_{24} \geq +150 \quad (4.94)$$

$$\text{Node 3: } X_{43} - X_{34} - X_{36} \geq -300 \quad (4.95)$$

$$\text{Node 4: } X_{24} + X_{34} + X_{54} - X_{42} - X_{43} \geq +70 \quad (4.96)$$

$$\text{Node 6: } X_{56} + X_{36} \geq +80 \quad (4.97)$$

So the overall LP model is

$$\begin{aligned} \text{Minimize } Z = {} & 30X_{12} + 40X_{24} + 35X_{42} + 50X_{34} + 45X_{43} \\ & + 120X_{36} + 100X_{51} + 45X_{54} + 110X_{56} \end{aligned}$$

Subject to

$$\begin{array}{ll} X_{51} - X_{12} \geq +100 & \text{Node 1} \\ X_{12} + X_{42} - X_{24} \geq +150 & \text{Node 2} \\ X_{43} - X_{34} - X_{36} \geq -300 & \text{Node 3} \\ X_{24} + X_{34} + X_{54} - X_{42} - X_{43} \geq +70 & \text{Node 4} \\ -X_{51} - X_{54} - X_{56} \geq -200 & \text{Node 5} \\ X_{56} + X_{36} \geq +80 & \text{Node 6} \\ X_{ij} \geq 0 \quad \forall i, j & \end{array} \qquad \text{Model (4.16)}$$

As many optimization software products require the right-hand side of the constraint equations to be positive, we rewrite the form of the constraint equation for nodes 3 and 5 as follows. Note that this is not a problem for the latest version of a number of software products.

So the modified LP model is

$$\begin{aligned} \text{Minimize } Z = {} & 30X_{12} + 40X_{24} + 35X_{42} + 50X_{34} + 45X_{43} \\ & + 120X_{36} + 100X_{51} + 45X_{54} + 110X_{56} \end{aligned}$$

Subject to

$$\begin{array}{ll} X_{51} - X_{12} \geq 100 & \text{Node 1} \\ X_{12} + X_{42} - X_{24} \geq 150 & \text{Node 2} \\ -X_{43} + X_{34} + X_{36} \leq 300 & \text{Node 3} \\ X_{24} + X_{34} + X_{54} - X_{42} - X_{43} \geq 70 & \text{Node 4} \\ X_{51} + X_{54} + X_{56} \leq 200 & \text{Node 5} \\ X_{56} + X_{36} \geq 80 & \text{Node 6} \\ X_{ij} \geq 0 \quad \forall i, j & \end{array} \qquad \text{Model (4.17)}$$

4.4.4 Project Management Problem

A project is a well-defined sequence of events with a beginning and an end, directed toward achieving a defined goal, and is conducted by people satisfying time, cost, resource, and quality parameters. Managerial projects involving complex and interrelated activities can easily be modelled as

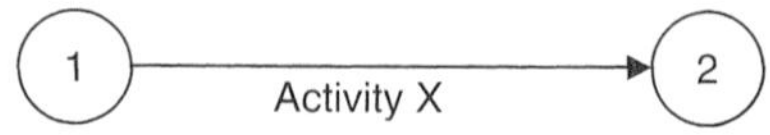

FIGURE 4.3
An activity-on-arc.

networks. A systematic analysis of project networks enables the manager to plan, monitor, and allocate resources so that objectives can be attained efficiently and on schedule. The project duration and other characteristic factors of projects are calculated using a network technique. Once the project has been represented by a network, the project management problem can be formulated as an LP model.

A project network is a logical and chronological set of activities and events, graphically illustrating the relationships among the various activities and events of the project. The project network can be drawn either as an activity-on-arc (AOA) or an activity-on-node (AON) network. In the AOA network, the arcs represent the activities, the arrow indicates the direction of flow thereby maintaining the precedence relationships, and the nodes represent the events (point in time). In the AON diagram, the nodes represent the activities and the arrows/arcs maintain the precedence relationships. In this section, we will discuss only the AOA networks for project management.

A single AOA is represented as shown in Figure 4.3. The line (arrow, arc) from node 1 to node 2 represents the activity X. Nodes 1 and 2, which represent events, must satisfy the time constraint that $t_2 > t_1$ (where t_1 and t_2 are the starting times for event 1 and 2, respectively). If the activity duration for X is X_d, then $t_1 + X_d \leq t_2$.

Let us consider a precedence requirement that activities B, C, and D can be conducted simultaneously, but none can start until activity A has been completed. This precedence requirement can be represented as Figure 4.4.

Assume that

T_{A} = starting time of activity 1–2 or A
T_{B} = starting time of activity 2–3 or B
T_{C} = starting time of activity 2–4 or C
T_{D} = starting time of activity 2–5 or D

and

X_{A} = activity time of activity A

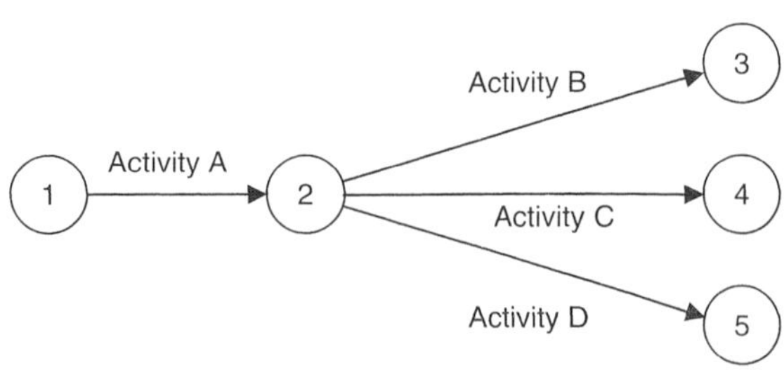

FIGURE 4.4
Example of precedence constraint 1.

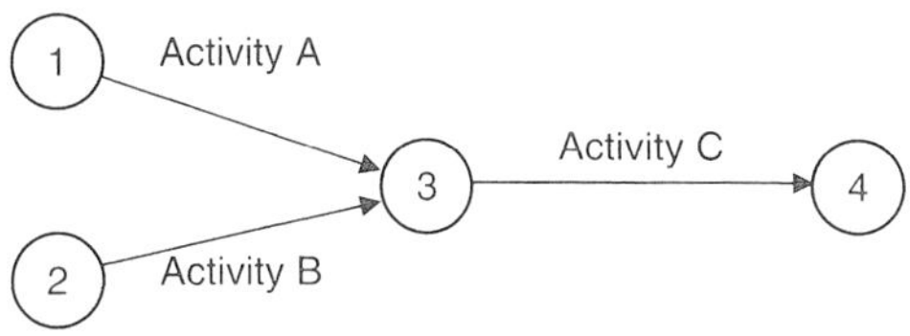

FIGURE 4.5
Example of precedence constraint 2.

The finishing time of activity A, which is $T_A + X_A$, must be less than or equal to the starting times of B, C, and D. That means

$$T_A + X_A \leq T_B \quad (4.98)$$

$$T_A + X_A \leq T_C \quad (4.99)$$

$$T_A + X_A \leq T_D \quad (4.100)$$

Another precedence constraint could be represented by having activities A and B starting simultaneously but C can only start after both A and B have finished. See the diagram in Figure 4.5 above.

Assume that

T_A = starting time of activity 1–3 or A
T_B = starting time of activity 2–3 or B
T_C = starting time of activity 3–4 or C

and

X_A = activity time of activity A
X_B = activity time of activity B

The finishing times for activities A and B are $T_A + X_A$ and $T_B + X_B$, respectively. These two completion times must be less than or equal to the starting time of C. That means

$$T_A + X_A \leq T_C \quad (4.101)$$

$$T_B + X_B \leq T_C \quad (4.102)$$

Consider four activities A, B, C, and D. A and B can start simultaneously, C can start after A is completed but D must wait for both A and B to finish. The corresponding network diagram is shown in Figure 4.6.

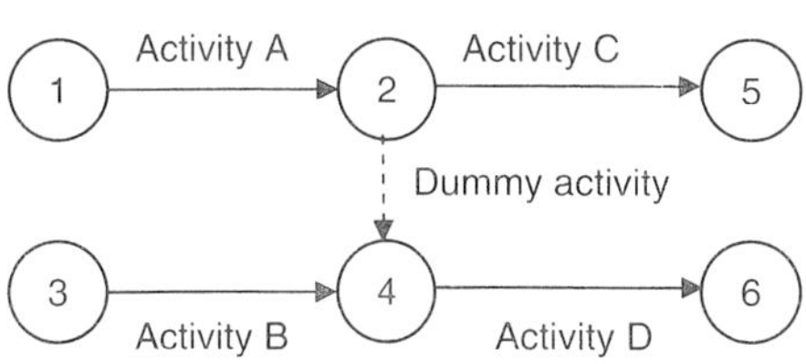

FIGURE 4.6
Example of precedence constraint 3.

The activity 2–3 is a dummy activity, which requires zero time and zero resources. The only function of the dummy activity is to designate a precedence relationship. Graphically, such activities are always shown as broken lines.

Assume that

T_A = starting time of activity 1–2 or A
T_B = starting time of activity 3–4 or B
T_C = starting time of activity 2–5 or C
T_D = starting time of activity 4–6 or D

and

X_A = activity time of activity A
X_B = activity time of activity B
X_{Dummy} = activity time of the dummy activity

The finishing times for activity A and B are $T_A + X_A$ and $T_B + X_B$. The corresponding constraints are

$$\text{For activity C: } T_A + X_A \leq T_C \quad (4.103)$$

$$\text{For activity D: } T_A + X_A + X_{Dummy} \leq T_D \quad (4.104)$$

$$\text{and } T_B + X_B \leq T_D \quad (4.105)$$

Now consider six activities A, B, C, D, E, and F. E can start after A and C are completed but F must wait until A, B, C, and D are completed. The corresponding network diagram is as Figure 4.7.

Assume that

T_A = starting time of activity 1–3 or A
T_B = starting time of activity 1–4 or B
T_C = starting time of activity 2–3 or C
T_D = starting time of activity 2–4 or D
T_E = starting time of activity 3–5 or E
T_F = starting time of activity 4–5 or F

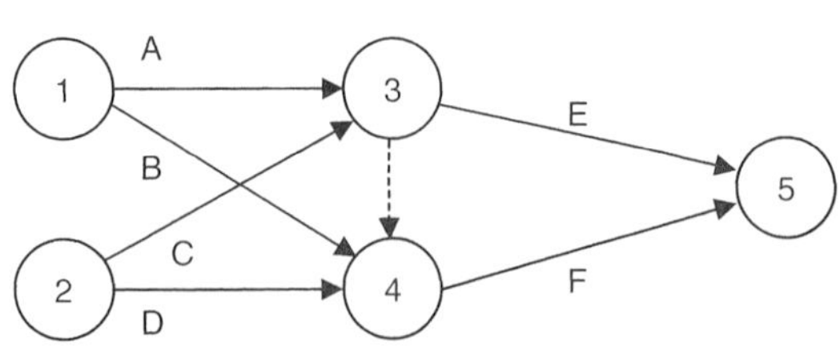

FIGURE 4.7
Example of precedence constraint 4.

and

X_A = activity time of activity A
X_B = activity time of activity B
X_C = activity time of activity C
X_D = activity time of activity D
X_{Dummy} = activity time of dummy activity

The corresponding constraints are

$$\text{For activity E: } T_A + X_A \leq T_E \tag{4.106}$$

and

$$T_C + X_C \leq T_E \tag{4.107}$$

$$\text{For activity F: } T_A + X_A + X_{Dummy} \leq T_F \tag{4.108}$$

$$T_C + X_C + X_{Dummy} \leq T_F \tag{4.109}$$

$$T_B + X_B \leq T_F \tag{4.110}$$

and

$$T_D + X_D \leq T_F \tag{4.111}$$

The project completion time in any project is defined as the time required to complete all the activities of the project. If the project network has several alternative paths from the first activity (or node) to the last activity (or node), the completion time for the longest path is the project duration. The longest path is also known as the critical path as any delay in this path will increase the completion time of the whole project.

Example 4.16: A simple project management problem
Consider a small renovation project with the following activities:

Activity	Description	Duration	Preceding Activities
a	Paper work and drafting	3	None
b	Manpower planning	4	a
b	Material planning	4	a
d	Transporting materials	2	c
e	Site preparation	4	a
f	Work Planning	6	c
g	Hiring equipment	3	d, b
h	Plan evaluation	1	e
i	Renovation work	12	f, g
j	Inspection and certification	2	i, h

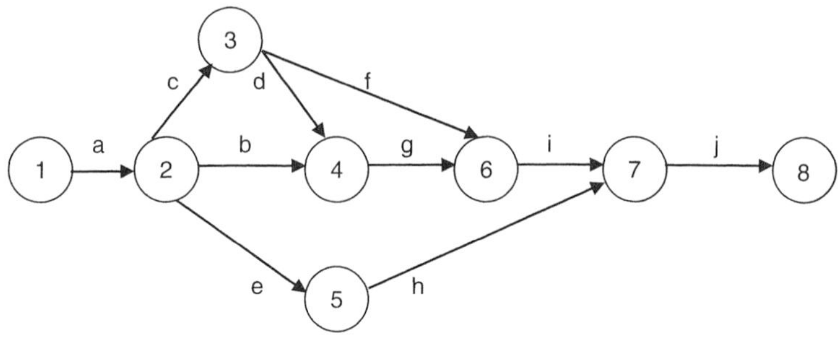

FIGURE 4.8
A project network.

The network of the above project can be drawn as Figure 4.8.

The activities other than on the longest path have flexibility in starting times as they can be started either at their earliest possible or latest possible time. By definition, the earliest possible start time for an activity to occur is immediately after all the preceding activities have been completed. The latest start time is the time that allows an activity to occur without causing a delay in the project-completion time. The LP model, for determining project duration, is usually developed considering one of the above two start times.

Defining variables:
Considering the earliest start times

T_k = earliest start time of activity k (k = a, b, ..., j, where a, b, ..., j represents all the activities).

Defining project data:
t_k = activity time of activity k (k = a, b, ..., j).

Objective function:
The objective is to minimize the overall project completion time. It can be achieved by minimizing the sum of the earliest start times of all activities.

$$\text{Minimize } Z = T_a + T_b + T_c + \cdots + T_j \tag{4.112}$$

Constraints:
The constraints represent only the precedence constraints.

At node 2:

$$T_b \geq T_a + t_a \tag{4.113}$$

$$T_c \geq T_a + t_a \tag{4.114}$$

$$T_e \geq T_a + t_a \tag{4.115}$$

At node 3:

$$T_d \geq T_c + t_c \tag{4.116}$$

$$T_f \geq T_c + t_c \tag{4.117}$$

At node 4:

$$T_g \geq T_b + t_b \tag{4.118}$$

$$T_g \geq T_d + t_d \tag{4.119}$$

At node 5:

$$T_e \geq T_h + t_h \tag{4.120}$$

At node 6:

$$T_i \geq T_f + t_f \tag{4.121}$$

$$T_i \geq T_g + t_g \tag{4.122}$$

At node 7:

$$T_j \geq T_i + t_i \tag{4.123}$$

$$T_j \geq T_h + t_h \tag{4.124}$$

$$T_i \geq 0 \quad \text{for all activities} \tag{4.125}$$

After organizing the variables in the left-hand side of the constraints and the constants in the right-hand side, the final model becomes

$$\text{Minimize } Z = T_a + T_b + T_c + \cdots + T_j$$

Subject to

$$\begin{aligned}
T_b - T_a &\geq t_a \\
T_c - T_a &\geq t_a \\
T_e - T_a &\geq t_a \\
T_d - T_c &\geq t_c \\
T_f - T_c &\geq t_c \\
T_g - T_b &\geq t_b \\
T_g - T_d &\geq t_d \\
T_e - T_h &\geq t_h \\
T_i - T_f &\geq t_f \\
T_i - T_g &\geq t_g \\
T_j - T_i &\geq t_i \\
T_j - T_h &\geq t_h \\
T_i &\geq 0 \quad \text{for all activities}
\end{aligned} \qquad \text{Model (4.18)}$$

After solving the model, the value of $T_j + t_j$ would provide the project duration.

4.5 Summary

In this chapter, we have introduced how one can use subscripts in variables to make the mathematical model formulation more concise. A number of examples have been provided to demonstrate the simple modelling techniques as well as the use of subscripts. The interested readers may find more examples in the books listed in the bibliography section of the chapter. In the next chapter, we will provide more modelling techniques involving integer and nonlinear variables. We will continue to present more examples using integer and nonlinear variables.

Exercises

1. The commanding officer (CO) plans to move a part of his battalion to another location. There are two types of vehicle available to him, a Ford vehicle that can carry 25 m^3 plus 5 personnel, and a Holden vehicle that can carry 15 m^3 plus 10 personnel. For security reason, all vehicles must move together and they will be used for a single trip. The required materials to be taken to the new location are organized in unit pallet load of 2 m^3. The CO requires transporting of a total of 60 pallet loads and 40 personnel. There are a maximum of 30 Ford and 40 Holden vehicles available. Each Ford vehicle is estimated to use 50 L of fuel per trip, whereas the Holden vehicle will only use 30 L. If the CO wants to move all the required materials and personnel at minimum fuel use, what mix of Ford and Holden vehicles should the CO choose? Formulate the problem as an LP model.
2. CPS Investment Services must develop an investment portfolio for a new client. CPS is considering six investment plans of three different types for this client. The investment plans with their expected yield are given below:

Investment Plan	Estimated Annual Return (%)
A1	6
A2	5
B1	7
B2	8
C1	10
C2	9

The client is interested to invest $80,000. To minimize the risk in investment, it is the CPS's policy that at least 40% of the entire amount be invested in units of type A and not more than 35% in type B or type

C plan. Formulate the problem as an LP model in order to maximize the total return.

3. A transport company requires the following number of drivers for its trucks during 24 h.

Time Period	Number Required
00–04 h	6
04–08 h	10
08–12 h	20
12–16 h	14
16–20 h	20
20–24 h	10

A driver may join for duty at midnight, 04, 08, 12, 16, and 20 h, and work for either 4 or 8 h long shift. To ensure smooth operation, at least half the drivers must be employed for 8 h shift in a 24 h cycle. Formulate the problem to find the number of drivers start at the beginning of each time period for both 4 and 8 h shift by minimizing the total number of driver–hours to be employed.

4. A paint company produces three grades of paints—inner paint, ceiling paint, and wall paint. The production process is run on a three shift basis. The limited resources with their requirements in each paint production and availability are shown below:

Resource Requirement	Inner Paint	Ceiling Paint	Wall Paint	Availability per Month
Special additive (kg/L)	0.25	0.15	0.65	550 tonnes
Processing (kL/shift)	3.0	4.0	6.0	90 shifts
Packaging (kL/shift)	10.0	10.0	10.0	75 shifts

There are no limitations on other resources. The demands and profits of these products are estimated as follows:

	Inner Paint	Ceiling Paint	Wall Paint
Demand per month (kL)	150	350	550
Profit ($/kL)	5000	4000	3000

As per the contract signed, a minimum of 250 kL per month of wall paint must be delivered to Home Mart. If the company is unable to meet the market demand from its monthly production, it is the company policy that the percentage of demand unmet must be same for all paints. Formulate an LP model for determining the monthly production plan in order to maximize the overall profit.

5. FFB fresh fruit business in Young, NSW, mixes three fruits—apples, peaches, and nectarines to make three different types of baskets for local market. Each basket contains approximately 5 kg fruits. The content of three types of baskets is specified as follows:

Basket Type	Apple	Peach	Nectarine
1	At least 30%	At most 20%	Rest
2	Rest	At most 40%	At least 20%
3	At least 20%	Rest	At most 30%

   FFB purchases apple at a cost of \$1.00/kg, peach \$1.50/kg, and nectarine \$1.80/kg, and sells type-1 basket at \$2.25/kg, type-2 basket \$3.00/kg, and type-3 basket \$2.60/kg. The daily supply of fruit is limited to 60 kg of apples, 70 kg of peaches, and 50 kg of nectarines. FFB is able to sell all the fruit baskets they prepare for a given day. Formulate an LP model to determine how the fruit be mixed in order to maximize the profit.

6. A popular product is produced at three plants and is required to ship to three warehouses. The supply from each plant, the delivery required in each warehouse, and the unit transportation cost from each plant to each warehouse is shown below:

	Warehouse			
Plant	WH1	WH2	WH3	Plant Capacity
PL1	20	15	25	400
PL2	10	12	8	500
PL3	14	20	12	200
Warehouse demand	300	200	600	

   Develop an LP model for minimizing the transportation cost while not violating the supply and demand constraints.

7. A supply chain network of Gordon Company consists of three plants, two warehouses, and four customer centers. Plant capacities and shipping costs (\$) from each plant to each warehouse are given below:

	Warehouse		
Plant	WH1	WH2	Capacity
PL1	5	8	500
PL2	9	6	400
PL3	6	7	300

Customer demand and shipping costs ($) per unit from each warehouse to each customer centers are shown below:

	Customer Center			
Warehouse	**CC1**	**CC2**	**CC3**	**CC4**
WH1	7	5	9	5
WH2	4	7	8	9
Demand	200	300	400	300

Develop an LP model for minimizing the transportation cost while not violating the supply, demand, and flow balance constraints.

8. Refer to Problem 7 in Exercises. Suppose that shipments between two warehouses are permitted at $3 per unit and direct shipment can be made from plant 2 to customer center 3 at a cost of $10 per unit. Formulate an LP model for this problem.
9. Refer to Problem 8 in Exercises. Suppose that shipment between warehouse 1 and customer center 4 cannot be made due to poor road condition. Revise the LP model developed for Problem 7 in Exercises to reflect this change.
10. A manufacturing firm has two drilling and four milling machines, which are to be used to produce a finish product consisting of two parts, P1 and P2. These parts can also be sold separately. The productivity of each machine for the two parts is given below:

		Production Time (Minutes per Piece)	
Part	**Profit ($ per Piece)**	**Drilling**	**Milling**
P1	10	8	15
P2	12	5	12

It is desired to maintain a balanced loading on all machines such that no machine (by type) runs for more than 30 min per day longer than any other machine (assume that the load is split evenly among all machines of a given type). Formulate an LP model to divide the work time of each machine to maximize the profit assuming an 8 h working day.

11. Consider Problem 10 in Exercises with a different objective. Formulate an LP model to divide the work time of each machine to obtain the maximum number of completed assemblies assuming an 8 h working day.

12. SunTech Electronics produces DVD recorders in their city plant. The estimated demand for the product for next four periods is 3,000, 8,000, 5,000, and 2,000. At the beginning of period 1, SunTech has 50 workers. SunTech spends $1,500 to hire a worker and $5,000 to fire a worker. Workers are paid $12,000 per period. A newly hired worker can make up to 50 recorders per period, whereas a previously hired worker can make up to 80 recorders per period. Each DVD recorder is sold for $295. It costs $15 to hold a DVD recorder in inventory for a period. Assume that workers are hired and fired at the beginning of each period and the products produced in any period are available to meet demand of that period. Inventory at the beginning of period 1 is 500. Assume that demand is lost if it cannot be met from either current stock or production of current period. Develop a mathematical model for determining the worker level in each period in order to maximize SunTech's profit.

Bibliography

Bazaraa, M., Jarvis, J., and Sherali, H., *Linear Programming and Network Flows,* John Wiley & Sons, New York, 1990.

Bradley, S.P., Hax, A.C., and Magnanti, T.L., *Applied Mathematical Programming,* Addison-Wesley Publishing Company, Readings, MA, 1977.

Daellenbach, H.G., George, J.A., and McNickle, D.C., *Introduction to Operations Research Techniques,* Allyn & Bacon, Boston, MA, 1983.

Eiselt, H.A., Pederzoli, G., and Sandblom, C.-L., *Continuous Optimization Models,* Walter deGruyter, Berlin, 1987.

Hadley, G., *Linear Programming,* Narosa Publishing House, New Delhi, 1990 (reprinted).

Hillier, F. and Lieberman, G., *Introduction to Operations Research*, 8th ed., McGraw-Hill, Boston, 2005.

Martin, R.K., *Large Scale Linear and Integer Optimization,* Kluwer Academic Publishers, Boston, MA, 1999.

Morse, P.M. and Kimball, G.E., *Methods of Operations Research,* John Wiley & Sons, New York, 1951.

Papadimitriou, C.H. and Steiglitz, K., *Combinatorial Optimization: Algorithms and Complexity,* 2nd ed., Dover Publications, New York, 1998.

Render, B. and Stair, R., *Quantitative Analysis for Management,* Prentice Hall, New Jersey, 1994.

Sierksma, G., *Linear and Integer Programming: Theory and Practice,* Marcel Dekker, New York, 1996.

Taha, H.A., *Operations Research: An Introduction,* 4th ed., MacMillan Publishing Company, New York, 1989.

Turban, E. and Meredith, J., *Fundamentals of Management Science,* 6th ed., Irwin McGraw-Hill, Boston, MA, 1994.

5

Simple Modelling Techniques II

5.1 Introduction

In Chapter 4, we have introduced linear programming–based basic modelling techniques. In practice, many problems require integer (includes binary) and nonlinear variables for appropriately modelling the problems. In this chapter, we would like to discuss simple integer and nonlinear modelling techniques frequently used by the modellers. In addition, few other modelling techniques are presented. A number of examples will be provided to demonstrate the use of integer and nonlinear variables in modelling practical problems.

This chapter is organized as follows. After introduction of the chapter, we discuss a number of simple modelling techniques such as precedence constraints, either–or constraints, K out of N constraints, yes/no decision, functions with N possible values, mutually exclusive alternatives and contingent decision, linking constraints, piecewise linear function, approximating nonlinear functions, deterministic models with probability terms, alternative objective functions, constrained to unconstrained problem, simplifying cross product of binary variables, fractional programming, unrestricted variables, regression model, stochastic programming, and constraint programming.

5.2 Precedence Constraints

In manufacturing planning and in project planning, the jobs, tasks, or activities must be performed in a certain order or sequence. The order of the jobs may be either prefixed or needs to be determined. For the case of a prefixed order, which may be due to technical reasons, it is important to find the starting and finishing times of each job. For the latter case, one is required to determine not only the starting and finishing times but also the order (/sequence) of the jobs.

Example 5.1: A job sequencing problem
Suppose, there are two tasks A and B for a given job, and task B cannot be started until task A is finished. In technical terms, task A is to precede task B. Formulate a precedence constraint for these two tasks.

Note that the precedence constraint concept has been used in project management in Section 4.4.4. We present another example here for convenience of explaining other modelling techniques in job sequencing.

Variables and parameters:
Let X_{A} and X_{B} be the starting times of tasks A and B, respectively, and d_{A} is the duration of performing task A.

Constraint:
The resulting precedence constraint is

$$X_{\mathrm{A}} + d_{\mathrm{A}} \leq X_{\mathrm{B}} \tag{5.1}$$

5.3 Either–or Constraints

Assume d_{B} is the duration of performing task B mentioned in Example 5.1. In that example, if one was asked to determine the sequence of tasks, then either $X_{\mathrm{A}} + d_{\mathrm{A}} \leq X_{\mathrm{B}}$ or $X_{\mathrm{B}} + d_{\mathrm{B}} \leq X_{\mathrm{A}}$ would exist (not both of them) depending, respectively, on whether A precedes B or B precedes A in the optimal solution. That means, one of the two constraints must hold.

Example 5.2: Sequencing with an either–or decision
Formulate the sequencing constraints for the above conditions.

Variables:
The presence of either–or constraints poses a problem since the model is no longer of the linear programming (LP) format (i.e., the either–or constraint creates a nonconvex solution space). This difficulty is overcome by introducing the binary variables Y_{AB} defined by

$$Y_{\mathrm{AB}} = \begin{cases} 1 & \text{if task A preceeds task B} \\ 0 & \text{if task B preceeds task A} \end{cases}$$

Constraint:
A transformation of the either–or constraints, discussed above, is required as follows:

$$X_{\mathrm{A}} + d_{\mathrm{A}} \leq X_{\mathrm{B}} + M(1 - Y_{\mathrm{AB}}) \tag{5.2}$$

$$X_{\mathrm{B}} + d_{\mathrm{B}} \leq X_{\mathrm{A}} + MY_{\mathrm{AB}} \tag{5.3}$$

Here, M is a sufficiently large number. If $Y_{AB} = 1$ (i.e., task A precedes task B) in the optimal solution, the above two constraints can be written as follows:

$$X_A + d_A \leq X_B \quad \text{Active constraint} \tag{5.4}$$

$$X_B + d_B \leq X_A + M \quad \text{Redundant} \tag{5.5}$$

The second constraint is redundant because the right-hand side is very big compared to the left-hand side. The redundant constraint has no effect on the optimum solution.

If $Y_{AB} = 0$ (i.e., task B precedes task A) in the optimal solution, then the second constraint becomes active and the first constraint redundant.

5.4 *K* out of *N* Constraints Must Hold

Consider the case where a model includes a set of N constraints such that only K of these constraints must hold (where $K < N$). In the optimization process, it is to choose the combination of K constraints that permits the objective function to reach its best possible value. The (N–K) constraints not chosen are, practically, eliminated from the problem, although feasible solutions might coincidentally still satisfy some of them. This is a general case for either–or constraints where we consider $K = 1$ and $N = 2$.

Example 5.3: A problem with three mutually exclusive constraints
Consider the following problem with three mutually exclusive constraints where $K = 1$ and $N = 3$.

$$\begin{aligned}
&\text{Maximize } Z = 4x_1 + 7x_2 \\
&\text{Subject to only one of the following constraint:} \\
&\qquad 2x_1 + x_2 \leq 6{,}000 \\
&\text{or} \qquad 3x_1 + 7x_2 \leq 13{,}000 \\
&\text{or} \qquad 5x_1 + 6x_2 \leq 12{,}500 \\
&\qquad x_1, x_2 \geq 0
\end{aligned} \qquad \text{Model (5.1)}$$

In order to incorporate the condition of only one of the above constraints, one defines an auxiliary variable

$$y_i = \begin{cases} 1 & \text{if constraint } i \text{ is chosen} \\ 0 & \text{otherwise} \end{cases}$$

Hence the equivalent integer programming model becomes

$$\begin{aligned}
&\text{Maximize } Z = 4x_1 + 7x_2 \\
&\text{Subject to} \\
&\qquad 2x_1 + x_2 \le 6{,}000 + M(1 - y_1) \\
&\qquad 3x_1 + 7x_2 \le 13{,}000 + M(1 - y_2) \\
&\qquad 5x_1 + 6x_2 \le 12{,}500 + M(1 - y_3) \\
&\qquad y_1 + y_2 + y_3 = 1 \\
&\qquad x_1, x_2 \ge 0 \\
&\qquad \text{and } y_i = 0, 1 \quad \text{for all } i
\end{aligned} \qquad \text{Model (5.2)}$$

M is a very large number to ensure that the other two constraints become redundant. The constraint ($y_1 + y_2 + y_3 = 1$) ensures that only one constraint will be active. If $y_1 = 1$, then constraint 1 is active and constraints 2 and 3 are redundant because $y_2 = y_3 = 0$.

This model can be formulated in a slightly different way by reversing the definition of the auxiliary variable:

$$y_i = \begin{cases} 0 & \text{if constraint } i \text{ is chosen} \\ 1 & \text{otherwise} \end{cases}$$

The alternative integer programming model becomes

$$\begin{aligned}
&\text{Maximize } Z = 4x_1 + 7x_2 \\
&\text{Subject to} \\
&\qquad 2x_1 + x_2 \le 6{,}000 + My_1 \\
&\qquad 3x_1 + 7x_2 \le 13{,}000 + My_2 \\
&\qquad 5x_1 + 6x_2 \le 12{,}500 + My_3 \\
&\qquad y_1 + y_2 + y_3 = 2 \\
&\qquad x_1, x_2 \ge 0 \\
&\qquad \text{and } y_i = 0, 1 \quad \text{for all } i
\end{aligned} \qquad \text{Model (5.3)}$$

M is a very large number to ensure that the other two constraints become redundant. The constraint ($y_1 + y_2 + y_3 = 2$) ensures that two out of three constraints will be redundant. If $y_1 = 0$, then constraint 1 is active and constraints 2 and 3 are redundant as $y_2 = y_3 = 1$.

5.5 Yes-or-No Decisions

There exist many decision-making problems that involve a number of interrelated yes-or-no decisions. In such decisions, the only two possible choices are either yes or no. For example, should we undertake a particular project? Should we hire a particular person? Should we make a particular investment? Should we purchase a particular aircraft? Should we locate a facility in a given site? We can represent such qualitative decisions (yes/no) by decision variables that are restricted to just two values, for example zero and one. Such variables are known as *binary variables* (or 0–1 variables). These variables are defined as follows:

$$x_i = \begin{cases} 1 & \text{if decision } i \text{ is yes} \\ 0 & \text{if decision } i \text{ is no} \end{cases} \quad \forall i$$

Although examples of such decision variables have already been given in Examples 3.6 and 3.7, we will provide yet another example that involves not only yes-or-no decisions but also a secondary decision such as the size of a capacity expansion plan.

Example 5.4: A capacity planning problem

A retail chain is considering whether to expand a distribution center or not. If so, the level of the expansion should be at least 3000 units but not more than 8000 units. The expansion incurs a high initial cost and the cost of expansion follows a nonlinear cost function. Disregarding all other information, formulate the expansion decision constraints only.

Decision variables:

$X =$ the amount of expansion to be made

$$y = \begin{cases} 1 & \text{if the decision for expansion is yes} \\ 0 & \text{if the decision for expansion is no} \end{cases}$$

We will ignore the objective function for the time being and will discuss it later.

Constraints:
If the decision for the expansion is yes, then it should be at least 3000 units but not more than 8000, that is

$$X \geq 3000$$
$$X \leq 8000$$

To incorporate the yes/no decision with the above constraints, we rewrite them as follows:

$$X \geq 3000y \tag{5.6}$$

$$X \leq 8000y \tag{5.7}$$

or

$$X - 3000y \geq 0 \tag{5.8}$$

$$X - 8000y \leq 0 \tag{5.9}$$

where y = either 0 or 1.

If $y = 1$, as per the above two inequalities X can be any value between 3000 and 8000. However, if $y = 0$, then X must be exactly equal to zero.

Note that this constraint may not work unless it is properly linked with other constraints and/or the objective function.

5.6 Functions with N Possible Values

Consider the situation where a given constraint or function is required to choose one of N different discrete values. For example,

$$f(x_1, x_2, \ldots, x_n) = b_1 \quad \text{or} \quad b_2 \cdots b_n \tag{5.10}$$

The equivalent form of this function is

$$f(x_1, x_2, \ldots, x_n) = b_1y_1 + b_2y_2 + b_3y_3 + \cdots + \mathrm{b}_ny_n \tag{5.11}$$

$$\text{where } y_1 + y_2 + y_3 + \cdots + y_n = 1 \tag{5.12}$$

and y_i is binary for $i = 1, 2, \ldots, n$.

Note that this constraint may not work unless it is properly linked with other constraints and/or the objective function.

Example 5.5: A production planning problem

Consider Example 4.8. The available time on the milling machines is 480 min. In order to share the milling machine with another product, management now wants to impose the restriction that the milling time used must be either 250 or 300 or 350 min.

So the milling machine constraint is

$$4X_1 + 3X_2 = 250 \text{ or } 300 \text{ or } 350 \tag{5.13}$$

Let us define three binary variables y_1, y_2, and y_3 for the right-hand side values of 250, 300, and 350, respectively, where

$$y_i = \begin{cases} 1 & \text{if right-hand side } i \text{ is chosen} \\ 0 & \text{otherwise} \end{cases} \qquad (i = 1, 2, 3)$$

So the equivalent constraints are

$$4X_1 + 3X_2 = 250y_1 + 300y_2 + 350y_3$$

$$\text{or } 4X_1 + 3X_2 - 250y_1 - 300y_2 - 350y_3 = 0 \qquad (5.14)$$

$$y_1 + y_2 + y_3 = 1 \qquad (5.15)$$

and y_1, y_2, and y_3 are binary.

Note that this constraint may not work unless it is properly linked with other constraints and/or the objective function.

5.7 Mutually Exclusive Alternatives and Contingent Decisions

In mathematical modelling, there are situations where the mutually exclusive alternatives and contingent decisions are considered. The mutually exclusive alternatives could be either at the variable level or at the constraint level. The Examples 5.2 and 5.3, in fact, represent the examples of mutually exclusive alternative constraints. In this section, one example on mutually exclusive decision variables along with contingent decisions is provided.

Example 5.6: A location problem

Southern Production International has decided to expand their production capacity by building a new plant in one of three major cities in Australia. It is also considering of building a new distribution center in whichever city selected for the new plant. However, a new distribution center may not be built if it is not profitable. The return (in net present value term) and the investment required for each of these alternatives are shown in the following table:

Decision Number	Facility and Location	Return (million \$)	Investment Required (million \$)
1	Plant in Melbourne	8	18
2	Plant in Adelaide	6	16
3	Plant in Sydney	9	14
4	Distribution center in Melbourne	5	11
5	Distribution center in Adelaide	4	8
6	Distribution center in Sydney	6	7

The company has \$36 million for the proposed expansion plan. The objective is to find the feasible combination of alternatives that maximizes the total net present value.

Decision variables:

$$x_i = \begin{cases} 1 & \text{if decision } i \text{ is yes} \\ 0 & \text{if decision } i \text{ is no} \end{cases} \quad (i = 1, 2, 3, 4, 5, 6).$$

Objective function:
The objective is to maximize the total return

$$\text{Maximize } Z = 8x_1 + 6x_2 + 9x_3 + 5x_4 + 4x_5 + 6x_6 \tag{5.16}$$

Constraints:

1. Investment constraint: The company has only $36 million to invest

$$18x_1 + 16x_2 + 14x_3 + 11x_4 + 8x_5 + 7x_6 \leq 36 \tag{5.17}$$

2. Mutually exclusive alternative 1: The company wants to build only one new plant

$$x_1 + x_2 + x_3 = 1 \tag{5.18}$$

3. Mutually exclusive alternative 2: The company wants to build at best one new distribution center

$$x_4 + x_5 + x_6 \leq 1 \tag{5.19}$$

4. Contingent decision constraint: The company would consider building a distribution center in a city only if the new plant is also built there. This contingency is taken into account by the following constraints:

$$x_4 - x_1 \leq 0 \quad \text{Melbourne} \tag{5.20}$$

$$x_5 - x_2 \leq 0 \quad \text{Adelaide} \tag{5.21}$$

$$x_6 - x_3 \leq 0 \quad \text{Sydney} \tag{5.22}$$

In the optimal solution, if $x_4 = 0$ then x_1 is either equal to zero or one. That indicates a plant may or may not be built in Melbourne; however, no distribution center will be built there. If $x_4 = 1$ then x_1 must be equal to one, which means both plant and distribution center would be built in Melbourne. The similar condition is true for x_5 and x_2, and x_6 and x_3.

The complete binary integer programming model is

Maximize $Z = 8x_1 + 6x_2 + 9x_3 + 5x_4 + 4x_5 + 6x_6$

$\in$

$$\begin{aligned}
&18x_1 + 16x_2 + 14x_3 + 11x_4 + 8x_5 + 7x_6 \leq 36 \\
&x_1 + x_2 + x_3 = 1 \\
&x_4 + x_5 + x_6 \leq 1 \\
&x_4 - x_1 \leq 0 \\
&x_5 - x_2 \leq 0 \\
&x_6 - x_3 \leq 0 \\
&x_i \text{ either 0 or 1} \quad \text{for all } i
\end{aligned} \qquad \text{Model (5.4)}$$

5.8 Linking Constraints with the Objective Function

The use of binary variables has made mathematical modelling simpler in many practical situations and has also made the development of models for some difficult cases possible. A fixed-charge problem (which involves a combination of fixed and variable costs) is one of the well-known examples where binary variables play a key role in modelling. However, it requires careful linkage of the constraints with the objective function so that the logic of introducing binary variables works properly. As you may recall, we ignored the development of the objective function of Example 5.4 as such complexity had not been addressed earlier.

Example 5.7: A fixed-charge problem

PQR Corporation is planning to produce at least 800 exhaust control valves, for a particular model of vehicle, using their three production lines. The production lines with the setup costs, unit processing costs, and capacities are given below. Determine the production lines to use in order to minimize the total cost.

Production Line	Setup Cost ($)	Unit Processing Cost ($)	Maximum Capacity
PL-A	750	20	400
PL-B	150	55	700
PL-C	420	35	600

The setup cost is one-off fixed cost per line and it is an expense only when the line is used. The unit processing cost is a variable cost where the total variable cost for a line is equal to the number of product produced in that line multiplied by the unit processing cost.

Decision variables:

x_i = the quantities to be produced on production line i (i = A, B, and C)

$$y_i = \begin{cases} 1 & \text{if line } i \text{ is used} \\ 0 & \text{otherwise} \end{cases} \quad (i = \text{A}, \text{B}, \text{C})$$

Constraints:

1. Demand constraint: The company must produce at least 800 exhaust control valves.

$$x_\text{A} + x_\text{B} + x_\text{C} \geq 800 \tag{5.23}$$

2. Capacity constraints: Each line should produce less than its capacity.

$$\text{Line A:} \quad x_\text{A} \leq 400 \tag{5.24}$$

$$\text{Line B:} \quad x_\text{B} \leq 700 \tag{5.25}$$

$$\text{Line C:} \quad x_\text{C} \leq 600 \tag{5.26}$$

Objective function:
There are two types of costs: fixed and variable. The fixed cost is incurred if the line is used and it is independent of the quantity produced. However, the variable cost is charged per unit production.

$$\text{Total variable cost} = 20x_\text{A} + 55x_\text{B} + 35x_\text{C} \tag{5.27}$$

$$\text{Total fixed cost} = 750y_\text{A} + 150y_\text{B} + 420y_\text{C} \tag{5.28}$$

So the objective function is

$$\text{Minimize } Z = 750y_\text{A} + 150y_\text{B} + 420y_\text{C} + 20x_\text{A} + 55x_\text{B} + 35x_\text{C} \tag{5.29}$$

Something is missing in the model. What is it?

The binary variables are only present in the objective function. All the constraints are technically correct. As per the objective function, the fixed cost of a line selected for production could be zero as our objective is to minimize the total cost. However, there is no condition in the above equations that stipulates that if a production line is selected for production then the fixed cost must be added along with the variable cost. How does one impose such a linkage? One simple way is to modify the capacity constraints as follows:

$$\text{Line A:} \quad x_\text{A} \leq 400y_\text{A} \tag{5.30}$$

$$\text{Line B:} \quad x_\text{B} \leq 700y_\text{B} \tag{5.31}$$

$$\text{Line C:} \quad x_\text{C} \leq 600y_\text{C} \tag{5.32}$$

One can now check to see whether this works logically or not. If $y_A = 1$ in the optimal solution (i.e., we select production line A for production), the fixed plus variable cost components in the objective function for line A are $(750 + 20x_A)$. The capacity constraint for line A, in effect, is $(x_A \leq 400)$. If $y_A = 0$ in the optimal solution (i.e., line A is not selected for production), the fixed plus variable cost components in the objective function for line A is only $(20x_A)$. However, the capacity constraint for line A ensures that $x_A = 0$ which follows that the total cost component for line A is zero. So the logic is working. Now you can check for the production lines B and C.

So the final mixed integer LP model becomes

$$\begin{aligned}
\text{Minimize } Z = 750y_A + 150y_B + 420y_C + 20x_A + 55x_B + 35x_C \\
x_A + x_B + x_C \geq 800 \quad & \text{Requirements} \\
x_A - 400y_A \leq 0 \quad & \text{Production capacity of line A} \\
x_B - 700y_B \leq 0 \quad & \text{Production capacity of line B} \\
x_C - 600y_C \leq 0 \quad & \text{Production capacity of line C} \\
x_i \geq 0 \quad \text{for all } i & \\
y_i = \text{either 0 or 1} \quad \text{for all } i &
\end{aligned}$$

Model (5.5)

5.9 Piecewise Linear Functions

Many real-world problems involve continuous functions that are piecewise linear. Examples of piecewise linear functions occur when there are increasing or decreasing returns to scale, marginal costs, etc. Consider a piecewise profit curve with three line segments as shown in the Figure 5.1. Binary variables are usually used to model such a piecewise linear curve.

Example 5.8: A piecewise linear function

An objective function can be expressed as

$$f(y) = \begin{cases} 5y, & \text{if } 0 \leq y \leq 4 \\ y, & \text{if } 4 \leq y \leq 10 \\ 3y, & \text{if } 10 \leq y \leq 15 \end{cases} \tag{5.33}$$

Disregarding all other information of the problem, formulate the above function as an LP model.

From the above function, the slopes of three line segments can be written as $S_1 = 5$, $S_2 = 1$, and $S_3 = 3$. Other data: $u_1 = 4$, $u_1 + u_2 = 10$, and $u_1 + u_2 + u_3 = 15$.

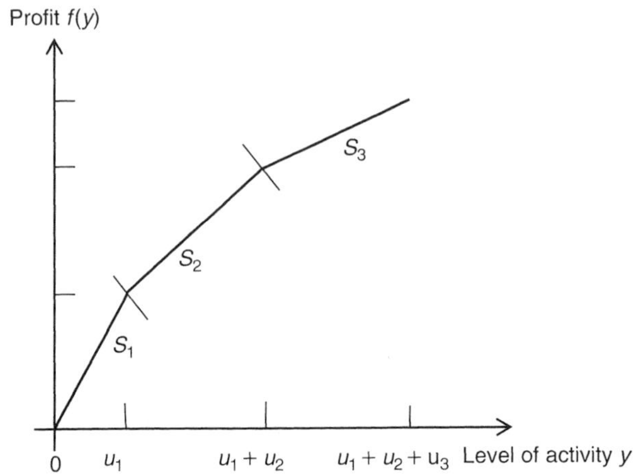

FIGURE 5.1
A piecewise linear function.

Let us introduce three variables x_1, x_2, and x_3 so that each of these variables is linear and set (see Figure 5.2)

$$y = x_1 + x_2 + x_3 \tag{5.34}$$

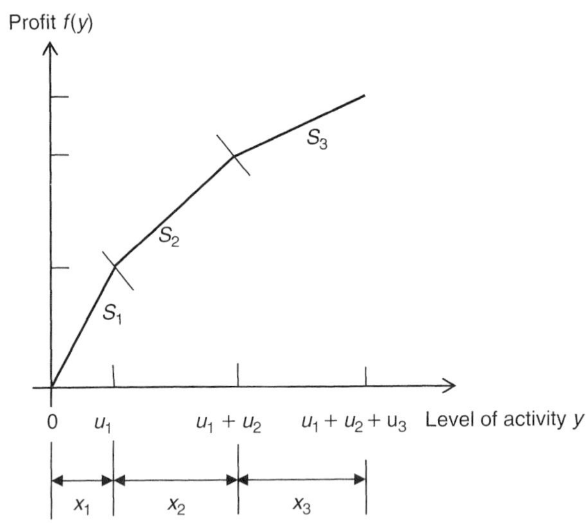

FIGURE 5.2
A piecewise linear function with defined variables.

where

$$0 \leq x_1 \leq 4,$$
$$0 \leq x_2 \leq 6, \text{ and}$$
$$0 \leq x_3 \leq 5.$$

Then using the slopes $S_1 = 5$, $S_2 = 1$, and $S_3 = 3$, the objective function $f(y)$ can be written as: $f(y) = 5x_1 + x_2 + 3x_3$ with the special restriction that

$$x_2 = 0, \quad \text{if } x_1 < 4$$
$$x_3 = 0, \quad \text{if } x_2 < 6$$

Let us now introduce two binary variables to represent the above two restrictions in the mathematical model:

$$z_1 = \begin{cases} 0 & \text{if } x_1 < 4 \\ 1 & \text{if } x_1 = 4 \end{cases}$$

$$z_2 = \begin{cases} 0 & \text{if } x_2 < 6 \\ 1 & \text{if } x_2 = 6 \end{cases}$$

To ensure that the proper conditional constraints hold, we write the constraints as follows:

$$y = x_1 + x_2 + x_3 \tag{5.35}$$

$$4z_1 \leq x_1 \leq 4 \tag{5.36}$$

$$6z_2 \leq x_2 \leq 6z_1 \tag{5.37}$$

$$0 \leq x_3 \leq 5z_2 \tag{5.38}$$

$$z_1, z_2 \in 0, 1$$

If $z_1 = 0$, then $z_2 = 0$, to maintain the feasibility of the constraint imposed upon x_2, the three boundary constraints reduce to $0 \leq x_1 \leq 4$, $x_2 = 0$, and $x_3 = 0$.

If $z_1 = 1$ and $z_2 = 0$, then $x_1 = 4$, $0 \leq x_2 \leq 6$, and $x_3 = 0$.

Finally, if $z_1 = 1$ and $z_2 = 1$, then $x_1 = 4$, $x_2 = 6$, and $0 \leq x_3 \leq 5$.

The same general technique can be applied to piecewise linear curves with any number of segments. The general constraint imposed upon the variable x_i for the ith segment will read: $L_i z_i \leq x_i \leq L_i z_{i-1}$, where L_i is the length of the segment.

5.10 Nonlinear to Approximate Functions

One of the most useful applications of the piecewise linear representation is the approximation of nonlinear functions.

Consider a concave (nonlinear) profit function $f(y)$, as shown in Figure 5.3, which occurs quite frequently in practice. It is convenient to approximate such a function by using a sequence of connected piecewise linear functions.

Suppose the function is divided into three segments to apply linear approximation. One introduces three variables x_1, x_2, and x_3 as shown in the Figure 5.4 and set

$$y = x_1 + x_2 + x_3 \tag{5.39}$$

where

$$0 \leq x_1 \leq u_1,$$
$$0 \leq x_2 \leq u_2, \text{ and}$$
$$0 \leq x_3 \leq u_3.$$

Note that the accuracy of the approximation can be improved by increasing the number of line segments, which indeed increases the number of variables and constraints.

Using the slopes S_1, S_2, and S_3, the objective function $f(y)$ can be written as $f(y) = S_1x_1 + S_2x_2 + S_3x_3$ with the special restrictions that

$$\begin{aligned} x_2 &= 0 \quad \text{if } x_1 < u_1 \\ x_3 &= 0 \quad \text{if } x_2 < u_2 \end{aligned}$$

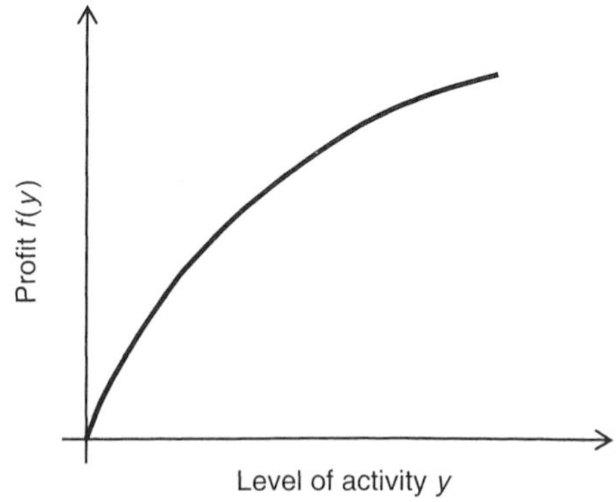

FIGURE 5.3
A nonlinear function.

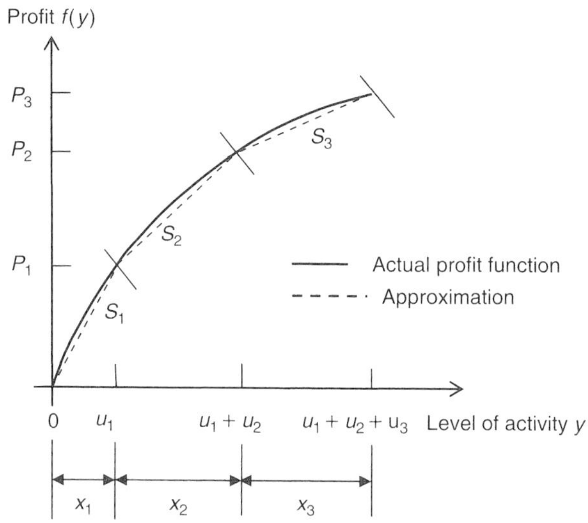

FIGURE 5.4
Piecewise linear approximation.

Now one can introduce two binary variables to represent the above two restrictions in the mathematical model:

$$Z_1 = \begin{cases} 0 & \text{if } x_1 < u_1 \\ 1 & \text{if } x_1 = u_1 \end{cases}$$

$$Z_2 = \begin{cases} 0 & \text{if } x_2 < u_2 \\ 1 & \text{if } x_2 = u_2 \end{cases}$$

To ensure that the proper conditional constraints hold, we write the constraints as follows:

$$y = x_1 + x_2 + x_3 \tag{5.40}$$

$$u_1 z_1 \leq x_1 \leq u_1 \tag{5.41}$$

$$u_2 z_2 \leq x_2 \leq u_2 z_1 \tag{5.42}$$

$$0 \leq x_3 \leq u_3 z_2 \tag{5.43}$$

$$z_1, z_2 \in 0, 1$$

Example 5.9: Approximating a nonlinear function
A nonlinear profit function can be divided into three parts to approximate three line segments. The points of the proposed intersections of the line

segments are given as: $(f(y), y) = (20, 4)$, $(26, 10)$, and $(41, 15)$. Formulate an equivalent but approximate LP model of the function.

From the data, we can write that $P_1 = 20$, $P_2 = 26$, $P_3 = 41$, $u_1 = 4$, $u_1 + u_2 = 10$, and $u_1 + u_2 + u_3 = 15$.

The slopes of the proposed line segments can be calculated as follows:

$$S_1 = \frac{P_1 - P_0}{y_1 - y_0} = \frac{20 - 0}{4 - 0} = 5 \tag{5.44}$$

$$S_2 = \frac{P_2 - P_1}{y_2 - y_1} = \frac{26 - 20}{10 - 4} = 1 \tag{5.45}$$

$$S_3 = \frac{P_3 - P_2}{y_3 - y_2} = \frac{41 - 26}{15 - 10} = 3 \tag{5.46}$$

That means, $S_1 = 5$, $S_2 = 1$, and $S_3 = 3$. Now one can proceed modelling the profit function using the procedure discussed above.

We introduce two binary variables to represent the above two restrictions in the mathematical model:

$$z_1 = \begin{cases} 0 & \text{if } x_1 < 4 \\ 1 & \text{if } x_1 = 4 \end{cases}$$

$$z_2 = \begin{cases} 0 & \text{if } x_2 < 6 \\ 1 & \text{if } x_2 = 6 \end{cases}$$

So the overall model will be

$$\begin{aligned} &\text{Maximize } f(y) = 5x_1 + x_2 + 3x_3 \\ &\text{Subject to} \\ &\qquad y = x_1 + x_2 + x_3 \\ &\qquad 4z_1 \leq x_1 \leq 4 \\ &\qquad 6z_2 \leq x_2 \leq 6z_1 \\ &\qquad 0 \leq x_3 \leq 5z_2 \\ &\qquad x_1, x_2, x_3 \geq 0 \\ &\qquad z_1, z_2 \in 0, 1 \end{aligned} \qquad \text{Model (5.6)}$$

5.11 Deterministic Models with Probability Terms

In modelling stochastic situations, probability and probability distributions play an important role. However, it is not a common practice to use probability or probability distributions in conjunction with deterministic modelling. In deterministic mathematical modelling, probabilities of occurrence usually take the form of a parameter. In such cases, the model is designed either (1) to maximize (or minimize) the probability or expected value of an occurrence or to (2) satisfy certain conditions using probability or expected-value terms. An example of such a problem together with its mathematical model is discussed below.

Example 5.10: A war planning problem

The Eastern Command receives orders to interrupt the enemy long-range missile production. The enemy has four key factories located in different cities, and certain successful interdictions could effectively halt the production of deadly missiles. The fuel supply is limited to 30,000 L for this particular mission. Any bomber aircraft sent to any particular city must have at least enough fuel for the round trip plus 100 L for safety reason.

The number of bomber aircraft available to the commander and their specific details are as follows:

Bomber Type	Description	km/L	Number Available
B	B2 Precision	2	30
F	F-117 Regular	3	25

Information about the location of the factories and their probability of hitting by a B2 or an F-117 bomber aircraft is given below:

Factory	Distance from Base (km)	Probability of Destruction by	
		B2 Bomber	F-117 Bomber
1	350	0.15	0.10
2	400	0.30	0.15
3	450	0.25	0.12
4	550	0.35	0.20

How many of each type of bomber aircraft should be dispatched and how should they be allocated across the four targets to maximize the probability of success?

Decision variables:

X_{ij} = the number of bomber type i (i = B and F) that will be sent to location/factory j (j = 1, 2, 3, and 4).

Objective function:
The objective is to maximize the probability of success in destroying the factories, or expressed in other words to minimize the probability of failing to destroy the factories.

The probability of succeeding to destroy plant 1 by one B2 bomber is 0.15. The probability, therefore, of not succeeding (failing) to destroy plant 1 by one B2 bomber is (1 − 0.15 =) 0.85. The probability of not succeeding to destroy plant 1 by X_{B1} number of B2 bombers is $(0.85)^{X_{B1}}$. Therefore, the probability of succeeding to destroy plant 1 by X_{B1} B2 bombers is $1 - (0.85)^{X_{B1}}$. The probability of failure and success for all other assignments can be found in a similar way. The objective then is to maximize the success of destroying all factories or in other words to minimize the failure of the overall mission. This can be expressed as an objective function where the failure of destroying all factories is to be minimized.

$$\text{Minimize } Z = (0.85)^{X_{B1}} \times (0.70)^{X_{B2}} \times (0.75)^{X_{B3}} \times (0.65)^{X_{B4}} \times (0.90)^{X_{F1}} \times (0.85)^{X_{F2}} \times (0.88)^{X_{F3}} \times (0.80)^{X_{F4}} \quad (5.47)$$

So the objective function is nonlinear.

Constraints:

1. Fuel supply limitation
   Fuel required (liters) for each trip of $X_{B1} = 2\frac{350}{2} + 100 = 450$
   Fuel required (liters) for each trip of $X_{B2} = 500$
   Fuel required (liters) for each trip of $X_{B3} = 550$
   Fuel required (liters) for each trip of $X_{B4} = 650$
   Fuel required (liters) for each trip of $X_{F1} = 2\frac{350}{2} + 100 = 333.33$
   Fuel required (liters) for each trip of $X_{F2} = 366.67$
   Fuel required (liters) for each trip of $X_{F3} = 400$
   Fuel required (liters) for each trip of $X_{F4} = 466.67$
   The fuel supply constraint is

$$450X_{B1} + 500X_{B2} + 550X_{B3} + 650X_{B4} + 333.33X_{F1} + 366.67X_{F2} + 400X_{F3} + 466.67X_{F4} \leq 30{,}000 \quad (5.48)$$

2. Constraint for the number of aircraft

$$\text{Type B: } X_{B1} + X_{B2} + X_{B3} + X_{B4} \leq 30 \quad (5.49)$$

$$\text{Type F: } X_{F1} + X_{F2} + X_{F3} + X_{F4} \leq 25 \quad (5.50)$$

The final nonlinear model is therefore

$$\text{Minimize } Z = (0.85)^{X_{B1}} \times (0.70)^{X_{B2}} \times (0.75)^{X_{B3}} \times (0.65)^{X_{B4}} \times (0.90)^{X_{F1}} \times (0.85)^{X_{F2}} \times (0.88)^{X_{F3}} \times (0.80)^{X_{F4}}$$

$\in$ Model (5.7)

$$450X_{B1} + 500X_{B2} + 550X_{B3} + 650X_{B4} + 333.33X_{F1} + 366.67X_{F2} + 400X_{F3} + 466.67X_{F4} \leq 30{,}000$$
$$X_{B1} + X_{B2} + X_{B3} + X_{B4} \leq 30$$
$$X_{F1} + X_{F2} + X_{F3} + X_{F4} \leq 25$$
$$X_{ij} \geq 0 \quad \text{for all } i \text{ and } j.$$

5.12 Alternate Objective Functions

We have previously seen how a nonlinear objective function can be approximated by a number of piecewise linear functions to solve a model using the integer LP approach. There are cases, however, where nonlinear objective functions can be converted into equivalent linear forms. Such conversions make models easier in terms of solution approaches.

Example 5.11: A war planning problem with a nonlinear objective
Consider the problem in Example 5.10. As formulated, the objective function of the problem is nonlinear. Convert the nonlinear objective function into an equivalent linear form.

As per Example 5.10, the objective function is

$$\text{Minimize } Z = (0.85)^{X_{B1}} \times (0.70)^{X_{B2}} \times (0.75)^{X_{B3}} \times (0.65)^{X_{B4}} \times (0.90)^{X_{F1}} \times (0.85)^{X_{F2}} \times (0.88)^{X_{F3}} \times (0.80)^{X_{F4}} \tag{5.51}$$

Minimizing Z is equivalent to minimizing $\log Z$, and $\log Z$ is linear.

$$\text{Minimize } \log Z = X_{B1}(\log 0.85) + X_{B2}(\log 0.70) + X_{B3}(\log 0.75) + X_{B4}(\log 0.65) + X_{F1}(\log 0.90) + X_{F2}(\log 0.85) + X_{F3}(\log 0.88) + X_{F4}(\log 0.80) \tag{5.52}$$

or

$$\text{Minimize } \log Z = -(0.071X_{B1} + 0.155X_{B2} + 0.125X_{B3} + 0.187X_{B4} + 0.046X_{F1} + 0.071X_{F2} + 0.056X_{F3} + 0.097X_{F4}) \tag{5.53}$$

This is clearly a linear function.

We can transpose the function into a maximization problem as follows:

$$\text{Maximize} - \log Z\left(= \log\frac{1}{Z}\right) = (0.071X_{B1} + 0.155X_{B2} + 0.125X_{B3} + 0.187X_{B4} + 0.046X_{F1} + 0.071X_{F2} + 0.056X_{F3} + 0.097X_{F4}) \tag{5.54}$$

Example 5.12: A joint inventory batch-sizing problem
In a joint inventory optimum batch-sizing problem, the following nonlinear function had to be minimized (Sarker and Khan, 2001):

$$Z = \sqrt{2D_p\left(A_p + \frac{A_i}{X_i}\right)\left[\frac{D_p}{P_p}(H_p + r_i H_i) + H_i r_i (X_i - 1)\right]} \tag{5.55}$$

X_i's are the only variables in the function. Find an alternative function that would be easier to solve.

In this problem, Z can be minimized through minimization of Z^2. That is,

$$\begin{aligned}\text{Minimize } Z^2 &= 2D_p\left(A_p + \frac{A_i}{X_i}\right)\left[\frac{D_p}{P_p}(H_p + r_i H_i) + H_i r_i (X_i - 1)\right] \\ &= 2D_p\left\{A_p\left[\frac{D_p}{P_p}(H_p + r_i H_i) - H_i r_i\right] + A_i H_i r_i\right\} \\ &\quad + 2D_p\left\{\frac{A_i}{X_i}\left[\frac{D_p}{P_p}(H_p + r_i H_i) - H_i r_i\right] + A_p H_i r_i X_i\right\}\end{aligned} \tag{5.56}$$

The constant components have no affect on the optimization of this function. Removing the constant components of the function, we can rewrite the Equation 5.56 so that it becomes an equivalently unconstrained optimization model as follows:

$$\text{Minimize } Y = \frac{A_i}{X_i}\left[\frac{D_p}{P_p}(H_p + r_i H_i) - H_i r_i\right] + A_p H_i r_i X_i \tag{5.57}$$

Although the above function is still nonlinear, it is not difficult to solve such simple nonlinear function.

5.13 Constrained to Unconstrained Problem

Nonlinear constrained problems are frequently converted to equivalent unconstrained problems to make solving and analyzing problems more convenient. One of the popular methods of conversion is the penalty method. The penalty method is briefly described below.

Consider a constrained mathematical model as follows:

$$\begin{aligned}&\text{Minimize } f(\mathbf{x})\\&\text{Subject to}\\&\qquad g(\mathbf{x}) \le 0\\&\qquad h(\mathbf{x}) = 0\\&\qquad \mathbf{x} \in X\end{aligned} \qquad \text{Model (5.8)}$$

where $\mathbf{g}$ (or $g_i(\mathbf{x})$) is a vector function with components $g_1, \ldots, g_m$, and $\mathbf{h}$ (or $h_i(\mathbf{x})$) is a vector function with components $h_1, \ldots, h_k$. Here f, $g_1, \ldots, g_m$, $h_1, \ldots, h_k$ are functions on R^n (n-dimensional real-number space) and X is a nonempty set in R^n.

The constrained model can be transformed into an unconstrained model as follows:

$$\begin{aligned}&\text{Minimize } f_p(\mathbf{x}) = f(\mathbf{x}) + \mu\alpha(\mathbf{x})\\&\text{Subject to } \mathbf{x} \in X\end{aligned} \qquad \text{Model (5.9)}$$

where

$\mu > 0$ is a large number

$\alpha(\mathbf{x})$ is the penalty function

For a minimization problem, the new unconstrained function, $f_p(\mathbf{x})$, is the simple sum of the usual (un-penalized) objective function, $f(\mathbf{x})$, and a penalty function, $\mu\alpha(\mathbf{x})$. Here, μ is a given penalty parameter. Higher value of μ means higher weight for the penalty. A suitable penalty function incurs a positive penalty for infeasible points and does not incur a penalty for feasible points. The penalty function α is usually of the form

$$\alpha(\mathbf{x}) = \sum_{i=1}^{m} \{\text{maximum } [0, g_i(\mathbf{x})]\}^p + \sum_{i=1}^{l} |h_i(\mathbf{x})|^p \qquad (5.58)$$

where p is a positive integer.

Example 5.13: A constrained nonlinear model

Transform the following constrained model into an unconstrained model:

$$\begin{aligned}&\text{Minimize } Z = (x_1 - 3)^4 + (x_1 - 3x_2)^2\\&\text{Subject to}\\&\qquad 3x_1^2 - x_2 \le 30\\&\qquad x_1^2 - 3x_2 = 0\end{aligned} \qquad \text{Model (5.10)}$$

The expression for $\alpha_1(\mathbf{x})$ and $\alpha_2(\mathbf{x})$ are derived as follows:

For constraint 1 (a less than equal to constraint) in Model 5.10, a positive value of the expression $(3x_1^2 - x_2 - 30)$ indicates a constraint violation. Therefore, the expression for $\alpha_1(\mathbf{x})$ with p = 2 is

$$\alpha_1(\mathbf{x}) = \left\{\text{maximum of } \left[0, \left(3x_1^2 - x_2 - 30\right)\right]\right\}^2 \tag{5.59}$$

For constraint 2 (an equal to constraint), any value (positive or negative) other than zero for the left-hand side of the constraint, $(x_1^2 - 3x_2)$, indicates a constraint violation. To derive the expression for $\alpha_2(\mathbf{x})$, we need to deal with the absolute value of such constraint. So the expression is

$$\alpha_2(\mathbf{x}) = \left\{\text{maximum of } \left[0, \left(|x_1^2 - 3x_2|\right)\right]\right\}^2 \tag{5.60}$$

The equivalent unconstrained model is

$$\begin{aligned}
&\text{Minimize } Z_p = (x_1 - 3)^4 + (x_1 - 3x_2)^2 + \mu_1\alpha_1(\mathbf{x}) + \mu_2\alpha_2(\mathbf{x}) \\
&\text{where} \quad \alpha_1(\mathbf{x}) = \left\{\text{maximum of } \left[0, \left(3x_1^2 - x_2 - 30\right)\right]\right\}^2 \\
&\qquad\quad\ \ \alpha_2(\mathbf{x}) = \left\{\text{maximum of } \left[0, \left(|x_1^2 - 3x_2|\right)\right]\right\}^2
\end{aligned} \qquad \text{Model (5.11)}$$

In the above Model 5.11, μ_1 and μ_2 are the penalty parameters (set by the modeller) for constraints 1 and 2, respectively, and the value of p is assumed to be 2.

5.14 Simplifying Cross Product of Binary Variables

In mathematical modelling, the cross product of two or more binary variables introduces a polynomial term. To make the model simpler, such a polynomial term can be replaced by a continuous, rather than integer, linear variable with few additional constraints. To demonstrate the methodology, we consider the following example:

$$\begin{aligned}
&\text{Maximize } Z = X_1X_2 \\
&\text{Subject to some constraints} \\
&\text{Both } X_1 \text{ and } X_2 \text{ are binary}
\end{aligned} \qquad \text{Model (5.12)}$$

As shown by Glover and Woolsey (1974), we replace the product term, X_1X_2, by a single continuous variable Y. This substitution requires the following additional constraints:

$$Y - X_1 \leq 0 \tag{5.61}$$

$$Y - X_2 \leq 0 \tag{5.62}$$

$$X_1 + X_2 - Y \leq 1 \tag{5.63}$$

$$0 \leq Y \leq 1 \tag{5.64}$$

If we assume Y is a binary, the result will be same.

The revised objective function is

$$\text{Maximize } Z = Y \tag{5.65}$$

We now check whether the above method works or not.

	Value of			
Case	X_1	X_2	Y	$X_1\,X_2$
1	0	0	0	0
2	1	1	1	1
3	1	0	0	0
4	0	1	0	0

As per the first two constraints given by Equations 5.61 and 5.62, the possible value for Y is shown in column 4, which also satisfies the last two constraints (Equations 5.63 and 5.64). In all cases, the value of Y is equal to the value of (X_1X_2). An example with application of this method will be provided in a later chapter.

Let us now consider another problem with higher degree function as the following model:

Maximize $Z = (X_1X_2)^p$
Subject to some constraints
Both X_1 and X_2 are binary
Model (5.13)

Since both X_1 and X_2 are binary, the function $(X_1X_2)^p$ can be replaced by (X_1X_2), for any positive value of p, without affecting the value of the function. One can now simplify the cross product, (X_1X_2), in a similar manner discussed above. The cross product of more than two variables can also be handled using the same concept (for more details see Watters, 1967; Glover and Woolsey, 1974).

5.15 Fractional Programming

In some practical problems, the objective function is expressed as the ratio of two functions. This is the case when we like to maximize the performance or productivity of a system where the performance or productivity is determined as the ratio of output to input. Examples of such problems include machine utilization (output to capacity), productivity of organization (output to human-hours), and rate of return on investment (profit to capital used).

Assume for a given system, we have the following two functions:

$$\text{Input function: } f_1(\mathbf{x}) = a_0 + \mathbf{ax} \tag{5.66}$$

$$\text{Output function: } f_2(\mathbf{x}) = b_0 + \mathbf{bx} \tag{5.67}$$

Here, $\mathbf{a}$ and $\mathbf{b}$ are row vectors, $\mathbf{x}$ is a column vector, and a_0 and b_0 are scalars.

So the productivity can be expressed as the ratio of these two functions as follow:

$$\text{Productivity: } f(\mathbf{x}) = \frac{f_2(\mathbf{x})}{f_1(\mathbf{x})} = \frac{b_0 + \mathbf{bx}}{a_0 + \mathbf{ax}} \tag{5.68}$$

A mathematical programming model with an objective function of this type (as Equation 5.68) is known as fractional programming. For ease of explanation, let us have a set of constraints: $\mathbf{Cx} \leq \mathbf{d}$ and $\mathbf{x} \geq 0$ for the above fractional program. As both $f_1(\mathbf{x})$ and $f_2(\mathbf{x})$ are linear in Equation 5.68, the objective function can be transformed to an equivalent linear form. To do so, we assume

$$\mathbf{y} = \frac{\mathbf{x}}{a_0 + \mathbf{ax}} \tag{5.69}$$

and

$$p = \frac{1}{a_0 + \mathbf{ax}} \tag{5.70}$$

From Equations 5.69 and 5.70, $\mathbf{x} = \mathbf{y}/p$ or $p\mathbf{x} = \mathbf{y}$. Similarly, from Equation 5.70, $pa_0 + p\mathbf{ax} = 1$ or $pa_0 + \mathbf{ay} = 1$. Substituting $\mathbf{y}$ and p in Equation 5.68 and in the constraints, we get

$$\begin{aligned}
&\text{Maximize } Z = b_0 p + \mathbf{by} \\
&\text{Subject to} \\
&\qquad \mathbf{Cy} - \mathbf{d}p \leq 0 \\
&\qquad \mathbf{ay} + a_0 p = 1 \\
&\qquad \mathbf{y} \geq 0,\ p \geq 0
\end{aligned} \qquad \text{Model (5.14)}$$

This is clearly an LP model. The fractional programming technique is demonstrated with an example below.

Example 5.14: Productivity maximization

A manufacturing industry produces two products: product-1 and product-2. The total sales revenue and total cost of the products can be expressed by the following functions:

$$\text{Cost function: } f_1(\mathbf{x}) = 40 + 10x_1 + 15x_2 \tag{5.71}$$

$$\text{Revenue function: } f_2(\mathbf{x}) = 20 + 16x_1 + 28x_2 \tag{5.72}$$

Where x_1 and x_2 are the number of products produced for product-1 and product-2, respectively.

The industry has the following technological constraints related to the products to be produced:

$$\text{Constraint 1: } 2x_1 + 5x_2 \leq 180 \tag{5.73}$$

$$\text{Constraint 2: } 4x_1 + 3x_2 \leq 150 \tag{5.74}$$

and

$$x_1, x_2 \geq 0$$

The objective of the problem is to maximize the productivity of producing two products while satisfying the technological constraints. Transform the problem to an equivalent LP problem.

LP formulation: The objective of the problem can be written as

$$\text{Objective function: } f(\mathbf{x}) = \frac{f_2(\mathbf{x})}{f_1(\mathbf{x})} = \frac{20 + 16x_1 + 28x_2}{40 + 10x_1 + 15x_2} \tag{5.75}$$

We now assume

$$y_1 = \frac{x_1}{40 + 10x_1 + 15x_2} \tag{5.76}$$

$$y_2 = \frac{x_2}{40 + 10x_1 + 15x_2} \tag{5.77}$$

$$p = \frac{1}{40 + 10x_1 + 15x_2} \tag{5.78}$$

From Equations 5.76 through 5.78 we can write

$$x_1 = y_1/p, \tag{5.79}$$

$$x_2 = y_2/p, \text{ and} \tag{5.80}$$

$$40p + 10x_1p + 15x_2p = 1 \tag{5.81}$$

Substituting x_1 and x_2, as of Equations 5.79 and 5.80, in Equation 5.81 we can write

$$40p + 10y_1 + 15y_2 = 1 \tag{5.82}$$

$$\begin{aligned}\text{Objective function: } f(\mathbf{x}) &= 20p + 16x_1p + 28x_2p \\ &= 20p + 16y_1 + 28y_2\end{aligned} \tag{5.83}$$

The technological constraints:

$$\begin{aligned}\text{Constraint 1: } 2x_1 + 5x_2 &\le 180 \text{ or} \\ 2y_1 + 5y_2 &\le 180p \text{ or} \\ 2y_1 + 5y_2 - 180p &\le 0\end{aligned} \tag{5.84}$$

$$\begin{aligned}\text{Constraint 2: } 4x_1 + 3x_2 &\le 150 \text{ or} \\ 4y_1 + 3y_2 &\le 150p \text{ or} \\ 4y_1 + 3y_2 - 150p &\le 0\end{aligned} \tag{5.85}$$

So the equivalent LP model will be

$$\begin{aligned}&\text{Maximize } Z = 16y_1 + 28y_2 + 20p \\ &\text{Subject to} \\ &\quad 2y_1 + 5y_2 - 180p \le 0 \\ &\quad 4y_1 + 3y_2 - 150p \le 0 \\ &\quad 10y_1 + 15y_2 + 40p = 1 \\ &\quad y_1, y_2, p \ge 0\end{aligned} \qquad \text{Model (5.15)}$$

5.16 Unrestricted Variables

Most traditional optimization techniques assume that all variables are non-negative. However, there are situations in practice where the values of variables can be either positive, zero, or negative. For example, the temperature is a variable for certain decisions. The money can be expressed as either positive, zero, or negative value where positive value indicates profit and negative value means loss. These variables are usually defined as variables with unrestricted in sign or free variables. To handle such variables in mathematical modelling, each variable is usually replaced by two new variables where the values of the new variables are greater than or equal to zero. To demonstrate the procedure, let us consider the following example:

$$\text{Minimize } Z = 3X_1 + 2X_2 + 2X_3 + X_4$$
Subject to
$$X_1 + 4X_2 + X_3 \geq 7 \qquad \text{Model (5.16)}$$
$$2X_1 + X_2 + X_4 \geq 10$$
$$X_1, X_2, X_3 \geq 0 \text{ and } X_4 \text{ unrestricted in sign}$$

Since X_4 is unrestricted in sign, we replace it by $X_4' - X_4''$ where both X_4' and $X_4'' \geq 0$. So the revised formulation is as follows:

$$\text{Minimize } Z = 3X_1 + 2X_2 + 2X_3 + X_4' - X_4''$$
Subject to
$$X_1 + 4X_2 + X_3 \geq 7 \qquad \text{Model (5.17)}$$
$$2X_1 + X_2 + X_4' - X_4'' \geq 10$$
$$X_1, X_2, X_3, X_4', X_4'' \geq 0$$

After solving the model, the value of X_4 will be calculated from $X_4 = X_4' - X_4''$.

5.17 Changing Constraint and Objective Type

As seen in the last section, any unrestricted variable can be replaced by two nonnegative variables. If necessary, inequality constraints can be converted to equality constraints and equality constraints to inequality constraints. The maximization (/minimization) objective function can be changed to minimization (/maximization) objective function by applying simple modelling tricks. In this section, we will provide examples of changing constraint and objective type.

5.17.1 From $\leq$ to $=$ Constraints

Consider a less than equal to constraint as follows:

$$\sum_i a_{1i} x_i \leq b_1 \tag{5.86}$$

This constraint can be converted to equal to constraint by simply adding a variable in the left-hand side of the constraint. The new variable is known as slack variable.

$$\sum_i a_{1i} x_i + S_1 = b_1 \tag{5.87}$$

5.17.2 From $\geq$ to $=$ Constraints

Now consider a greater than equal to constraint.

$$\sum_i a_{2i}x_i \geq b_2 \tag{5.88}$$

This constraint can be converted to equal to constraint by simply subtracting a variable from the left-hand side of the constraint. The new variable is known as surplus variable.

$$\sum_i a_{2i}x_i - S_2 = b_2 \tag{5.89}$$

5.17.3 From $\geq$ to $\leq$ Constraints

Consider the greater than equal to constraint (Equation 5.88). It can be written as an equivalent less than equal to constraint by simply multiplying both sides with a minus sign. As some software and methodology do not allow negative right-hand sides, we move them to the left-hand side.

$$-\sum_i a_{2i}x_i \leq -b_2 \text{ or} \tag{5.90}$$

$$-\sum_i a_{2i}x_i + b_2 \leq 0 \tag{5.91}$$

5.17.4 From $\leq$ to $\geq$ Constraints

The less than equal to constraint can also be converted to an equivalent greater than equal to constraint. See the example with constraint (Equation 5.86) below:

$$-\sum_i a_{1i}x_i \geq -b_1 \text{ or} \tag{5.92}$$

$$-\sum_i a_{1i}x_i + b_1 \geq 0 \tag{5.93}$$

5.17.5 From $=$ Constraint to $\geq$ and $\leq$ Constraints

Any equality constraint can be converted to two inequality constraints where one is less than equal to constraint and the other is greater than equal to constraint. Consider the following equal to constraint:

$$\sum_i a_{3i}x_i = b_3 \tag{5.94}$$

This can be written as

$$\sum_i a_{3i}x_i \le b_3 \quad \text{and} \quad \sum_i a_{3i}x_i \ge b_3 \tag{5.95}$$

The above Equations are only true when Equation 5.94 is true.

5.17.6 Changing Objective Type

A maximization (/minimization) objective function can be converted to an equivalent minimization (/maximization) function by changing the sign of the function. For example, the maximization objective function (Equation 5.96) is equivalent to minimization objective function (Equation 5.97).

$$\text{Maximize } Z_1 = \sum_i r_i x_i \tag{5.96}$$

$$\text{Minimize } Z_2 = -\sum_i r_i x_i \tag{5.97}$$

Example 5.15: Changing constraints and objective type
Change the following model to maximization objective with less than equal to constraints form.

$$\begin{aligned}
&\text{Minimize } Z_1 = 5x_1 + 4x_2 - 2x_3\\
&\text{Subject to}\\
&\qquad 2x_1 + 2x_2 + 3x_3 = 16\\
&\qquad 4x_1 + 3x_2 + 2x_3 \ge 15\\
&\qquad x_1, x_2, x_3 \ge 0
\end{aligned} \qquad \text{Model (5.18)}$$

The model can be written as

$$\begin{aligned}
&\text{Maximize } Z_2 = -5x_1 - 4x_2 + 2x_3\\
&\text{Subject to}\\
&\qquad 2x_1 + 2x_2 + 3x_3 \le 16 \qquad\qquad 2x_1 + 2x_2 + 3x_3 \le 16\\
&\qquad 2x_1 + 2x_2 + 3x_3 \ge 16 \quad \text{or} \quad -2x_1 - 2x_2 - 3x_3 \le -16\\
&\qquad 4x_1 + 3x_2 + 2x_3 \ge 15 \qquad\quad -4x_1 - 3x_2 - 2x_3 \le -15\\
&\qquad x_1, x_2, x_3 \ge 0 \qquad\qquad\qquad x_1, x_2, x_3 \ge 0
\end{aligned} \qquad \text{Model (5.19)}$$

The final form of the model is

$$\begin{aligned}
&\text{Maximize } Z_2 = -5x_1 - 4x_2 + 2x_3 \\
&\text{Subject to} \\
&\qquad 2x_1 + 2x_2 + 3x_3 \le 16 \\
&\qquad -2x_1 - 2x_2 - 3x_3 + 16 \le 0 \\
&\qquad -4x_1 - 3x_2 - 2x_3 + 15 \le 0 \\
&\qquad x_1, x_2, x_3 \ge 0
\end{aligned} \qquad \text{Model (5.20)}$$

5.18 Conditional Constraints

In some situations, the conditional relationships may exist between two or more constraints. For example, the conditional relationship in the following equation:

$$\sum_i a_{1i}x_i > b_1 \qquad (5.98)$$

implies that

$$\sum_i a_{2i}x_i \le b_2$$

As shown by Castillo et al. (2002), this conditional constraint is not satisfied only when

$$\sum_i a_{1i}x_i > b_1 \quad \text{and} \quad \sum_i a_{2i}x_i > b_2 \qquad (5.99)$$

So the equivalent alternative set of constraints for Equation 5.98 can be written as following:

$$\sum_i a_{1i}x_i \le b_1 \quad \text{and} \quad \sum_i a_{2i}x_i \le b_2 \qquad (5.100)$$

These constraints (Equation 5.100) can be expressed in the model as follows:

$$\begin{aligned}
&\sum_i a_{1i}x_i \le b_1 + M(1 - y_1) \\
&\sum_i a_{2i}x_i \le b_2 + M(1 - y_2) \\
&\qquad y_1 + y_2 \ge 1 \\
&\qquad y_1, y_2 \in \{0, 1\} \text{ and M is a big number}
\end{aligned} \qquad (5.101)$$

5.19 Dual Formulation

For every maximization LP problem, we can formulate an associated minimization problem, and vice versa. In fact, these LP problems exist in pairs. In this case, the original problem is called the *primal*, and the complementary problem is termed as *dual*. The dual problem is considered as an important part of LP problem analysis. In this section, we will show how the dual problem is formulated from the primal version.

A primal (/dual) model is suitable for dual (/primal) formulation, if it contains one of the two forms: (1) maximization with $\leq$ constraints or (2) minimization with $\geq$ constraints. If a primal model deviates from such structures, it must be transformed into one of these standard formats before dual formulation. In formulating a dual model, we define one new variable (known as dual variable) for each constraint of the primal model. Then the dual model corresponds to the primal model as follows:

Primal Model		Dual Model	
Maximize $Z = \mathbf{cx}$		Minimize $W = \mathbf{ub}$	
Subject to		Subject to	
	$A\mathbf{x} \leq \mathbf{b}$		$\mathbf{u}A \geq \mathbf{c}$
and	$\mathbf{x} \geq 0$	and	$\mathbf{u} \geq 0$

Here, $\mathbf{x}$ is a decision variable vector in the primal problem and $\mathbf{u}$ is a dual variable vector in the dual problem. A, $\mathbf{c}$, and $\mathbf{b}$ are the constraint coefficient matrix, objective function coefficient row vector, and constraint right-hand side column vector of the primal problem, respectively.

Example 5.16: Dual formulation-1

Consider a maximization problem with less than equal to constraints. We recognize this problem as primal.

Maximize $Z = 5x_1 + 4x_2 + 2x_3$

Subject to

$$2x_1 + 2x_2 + 3x_3 \leq 16 \qquad \text{Model (5.21)}$$

$$4x_1 + 3x_2 + 2x_3 \leq 15$$

$$x_1, x_2, x_3 \geq 0$$

For each constraint in the primal, we assume one decision variable in the dual as shown below:

	Dual Variables
Max $Z = 5x_1 + 4x_2 + 2x_3$	
Subject to	
$2x_1 + 2x_2 + 3x_3 \leq 16$	u_1
$4x_1 + 3x_2 + 2x_3 \leq 15u_2$	u_2
$x_1, x_2, x_3 \geq 0$	

The corresponding dual formulation is

$$\begin{aligned} &\text{Minimize } W = 16u_1 + 15u_2 \\ &\text{Subject to} \\ &\qquad 2u_1 + 4u_2 \geq 5 \\ &\qquad 2u_1 + 3u_2 \geq 4 \\ &\qquad 3u_1 + 2u_2 \geq 2 \\ &\qquad u_1, u_2 \geq 0 \end{aligned} \qquad \text{Model (5.22)}$$

If we develop a dual of the above dual model by assuming the dual variables x_1, x_2, and x_3 for Model 5.22, it will produce Model 5.21. That indicates dual of the dual is a primal.

Example 5.17: Dual formulation-2

Find the dual of the problem in Example 5.15. For convenience, we reproduce the model here as Model 5.23.

$$\begin{aligned} &\text{Min } Z = 5x_1 + 4x_2 - 2x_3 \\ &\text{Subject to} \\ &\qquad 2x_1 + 2x_2 + 3x_3 = 16 \\ &\qquad 4x_1 + 3x_2 + 2x_3 \geq 15 \\ &\qquad x_1, x_2, x_3 \geq 0 \end{aligned} \qquad \text{Model (5.23)}$$

This model can be transformed to dual using two different approaches. The first approach is to convert the equality constraint to two inequality constraints and then formulate the dual model. The equality constraint is converted as follows:

$$2x_1 + 2x_2 + 3x_3 = 16 \tag{5.102}$$

$$2x_1 + 2x_2 + 3x_3 \leq 16 \quad \text{and} \quad 2x_1 + 2x_2 + 3x_3 \geq 16 \tag{5.103}$$

$$-2x_1 - 2x_2 - 3x_3 \geq -16 \quad \text{and} \quad 2x_1 + 2x_2 + 3x_3 \geq 16 \tag{5.104}$$

The equivalent model for Model 5.23 is

$$\begin{aligned} &\text{Minimize } Z = 5x_1 + 4x_2 - 2x_3 \\ &\text{Subject to} \\ &\qquad -2x_1 - 2x_2 - 3x_3 \geq -16 \\ &\qquad 2x_1 + 2x_2 + 3x_3 \geq 16 \\ &\qquad 4x_1 + 3x_2 + 2x_3 \geq 15 \\ &\qquad x_1, x_2, x_3 \geq 0 \end{aligned} \qquad \text{Model (5.24)}$$

We assume the dual variables u_1', u_1'', and u_2 for the constraints 1, 2, and 3 of Model 5.24, respectively. Now the corresponding dual formulation is

$$\begin{aligned} &\text{Maximize } W = -16u_1' + 16u_1'' + 15u_2, \\ &\text{Subject to} \\ &\qquad -2u_1' + 2u_1'' + 4u_2 \leq 5 \\ &\qquad -2u_1' + 2u_1'' + 3u_2 \leq 4 \\ &\qquad -3u_1' + 3u_1'' + 2u_2 \leq -2 \\ &\qquad u_1', u_1'', u_2 \geq 0 \end{aligned} \qquad \text{Model (5.25)}$$

The second approach is to keep the equality constraint as it is and to define its corresponding dual variable as unrestricted (free) variable. We assume u_1 and u_2 to be the dual variables for the Model 5.23. So the corresponding dual formulation is

$$\begin{aligned} &\text{Maximize } W = 16u_1 + 15u_2 \\ &\text{Subject to} \\ &\qquad 2u_1 + 4u_2 \leq 5 \\ &\qquad 2u_1 + 3u_2 \leq 4 \\ &\qquad 3u_1 + 2u_2 \leq -4 \\ &\qquad u_1 \text{ unrestriced in sign and } u_2 \geq 0 \end{aligned} \qquad \text{Model (5.26)}$$

Both the Models 5.25 and 5.26 are dual formulation of Model 5.23. So these two models should be either same or equivalent in some way. If we assume $u_1 = (u_1'' - u_1')$, then the two models are equivalent but with different variable definitions.

Note that the constraint in a dual will be with equal to sign if a primal variable is unrestricted in sign.

5.20 Regression Model

In many occasions, the decision makers require to establish a functional relationship between dependent variable and one or more independent variables. The process starts with a set of known data points. Assume we have n independent variables (x_i) and one dependent variable (y). To develop a relationship between y and x_i, suppose we have m data points, which can be represented as y_j, and x_{ij}, where $i=1-n$ and $j=1-m$. A simple relationship that can be developed is

$$y_j = b_0 + b_1x_{1j} + b_2x_{2j} + \cdots + b_nx_{nj} + \varepsilon_j \tag{5.105}$$

Here $b_0, b_1, b_2, \ldots, b_n$ are the model parameters and ε_j is the error term. In the above function, the parameters ($b_0, b_1, b_2, \ldots, b_n$) are unknown. Let us now rewrite the function (Equation 5.105) as the following, where $\hat{y}_j$ is the estimated value of the right-hand side of Equation 5.105.

$$\hat{y}_j = b_0 + b_1x_{1j} + b_2x_{2j} + \cdots + b_nx_{nj} \tag{5.106}$$

For a given set of the parameter values, we can find the value of $\hat{y}_j$ and the error for each data item, which is $(y_j - \hat{y}_j)$. This term can be written as

$$(y_j - \hat{y}_j) = y_j - (b_0 + b_1x_{1j} + b_2x_{2j} + \cdots + b_nx_{nj}) \tag{5.107}$$

In finding the appropriate values of the parameters, we treat the parameters as variables (unknowns) and use the data points for y_j and x_{ij} as parameters. We then apply optimization process to find the values of $b_0, b_1, b_2, \ldots,$ and b_n by minimizing the sum of either absolute errors or square errors. The corresponding mathematical model with sum of square errors is as follows:

$$\text{Minimize } Z = \sum_j \left[y_j - (b_0 + b_1x_{1j} + b_2x_{2j} + \cdots + b_nx_{nj})\right]^2 \qquad \text{Model (5.27)}$$

The variables here are unrestricted in sign as they can be either positive, zero, or negative.

This is an unconstrained nonlinear optimization model. However, the resulting relationship will be linear as shown in Equation 5.106.

Example 5.18: Sales prediction

A retailer chain is interested to develop a relationship between its sales and the amount of money it spends on advertisement for a given product. The data for six previous instances are given below. Develop a mathematical model that will help to determine the relationship between sales and advertising costs.

Observation	Advertising Costs ($1000)	Sales Volume ($1000)
1	45	105
2	48	108
3	60	122
4	55	120
5	68	135
6	62	121

The mathematical model for the prediction problem is

$$\text{Minimize } Z = \sum_j \left[y_j - \left(b_0 + b_1 x_j\right)\right]^2 \text{or}$$

$$\begin{aligned}\text{Minimize } Z = {} & (105 - b_0 - 45b_1)^2 + (108 - b_0 - 48b_1)^2 \\ & + (122 - b_0 - 60b_1)^2 + (120 - b_0 - 55b_1)^2 \\ & + (135 - b_0 - 68b_1)^2 + (121 - b_0 - 62b_1)^2\end{aligned} \qquad \text{Model (5.28)}$$

Both b_0 and b_1 are unrestricted in sign.

5.21 Stochastic Programming

The models discussed so far are deterministic in nature. The parameters in such models are known with certainty. However, real-world problems almost invariably include some unknown or uncertain parameters. The mathematical models of such problems are known as stochastic programming models. Stochastic programming models are similar to deterministic models in style but they require probability distributions to generate/estimate data. The purpose here is to find solution that is feasible for all (or almost all) the possible data instances and maximizes the expectation of some functions of the decision and random variables. As our focus in this book is mainly on deterministic modelling, we would not provide further details on stochastic programming. However, interested readers may find details of stochastic programming in the books by Birge and Louveaux (1997), Kall and Mayer (2006), and Kall and Wallace (1995).

5.22 Constraint Programming

The constraint programming is a special case of mathematical programming models where the purpose is to satisfy a set of constraints only. That means the model does not require an objective function like other mathematical programming models discussed earlier. To demonstrate the concept of constraint programming, let us revisit some of the examples of this book.

- Example 3.2: Finding the quantity of foodstuff A, B, and C that will satisfy the constraints (Equations 3.12 through 3.14) only. This is the case where a certain mix of foodstuff is crucial but the cost of food is not a concern. In other words, we are looking for a feasible solution only.
- Example 3.6: Identify the projects that would satisfy the budget constraint. Note that it can be done in a number of different ways. We are interested for one of such feasible solutions.
- Example 3.10: Finding a location for the new tower satisfying the distance constraints.

The practical constraint programming problems are not as simple as the above examples (see Section 7.5.1). In some problems, a feasible solution may not exist. In such cases, our purpose is to find a solution with a minimum constraint violation. Note that these cases can be formulated as traditional mathematical programming models where the objective function represents the constraint violation. Interested readers can find more on constraint programming in Apt (2003) and Abdennadher and Frühwirth (2003).

5.23 Summary

In this chapter, we have introduced a number of simple modelling techniques and demonstrated their use using numerical examples. The interested readers may find more examples in the books listed in the bibliography section of the chapter. In the next chapter, we further refine the mathematical models into a more concise form by using the summation sign and the subset sign to separate certain variables. We will continue to present a number of applications to demonstrate the generalization of specific models.

Exercises

1. Royal Australian Navy (RAN) is considering three types of attack aircraft to equip its carriers: a supersonic type, a subsonic type, and a boost glide type. The effectiveness of any aircraft to the fleet is determined by the expected military value of targets the aircraft can destroy during military engagements of a certain length. These have been estimated for the three types mentioned as follows:

Type	Expected Value of Target Destroyed
Supersonic	32
Subsonic	26
Boost glide	27

The numbers of aircraft that could be accommodated, if the entire deck is allocated to one specific type, are 60 supersonic, 120 subsonic, or 160 boost glide, or any other linear combination of these type. The personnel requirements and monthly maintenance costs are given below:

Type	Personnel per Aircraft	Maintenance Cost per Aircraft ($)
Supersonic	16	6000
Subsonic	14	4000
Boost glide	18	7000

A carrier has facilities for 1,400 personnel, and the navy's monthly maintenance budget for aircraft is $600,000 per carrier.

The problem is to find how many of each of the three types of aircraft should be purchased per carrier in order to maximize the value of the attack capability of a carrier. Formulate the problem as an integer programming model.

2. Consider a budget allocation problem where $30 million is available for a number of projects listed below. The investment required in each project along with the expected return in terms of utility is also given.

Project Number	Project	Cost ($ million)	Expected Utility
1	After-school program	6	18
2	Road security	18	16
3	Crime reduction	10	12
4	Road extension	9	25
5	Child care facility	4	14

(i) The problem is to find which projects should be financed in order to maximize the total expected utility not exceeding the budget limitation.

(ii) Now suppose there are additional conditions in selecting the projects. Write the constraints for these conditions.

- Any two of the first four projects must be undertaken.
- Projects 1 and 3 must be taken simultaneously or not taken at all.

- Project-1 will be undertaken only if project-3 is undertaken but project-3 is not conditional on project-1 (that means, you can have project-3 without project-1, but you cannot have project-1 unless project-3 is undertaken).

3. A certain paint can be produced using four different production processes. The processing cost of each liter in any of the four available processes, the maximum capacity of each process, and the setup costs are given below:

Process	Setup Cost ($)	Processing Cost ($/L)	Capacity (L)
P1	5,000	0.60	20,000
P2	6,000	0.50	15,000
P3	10,000	0.40	40,000
P4	6,000	0.30	25,000

Assume that a daily demand of 45,000 L must be fulfilled. Formulate the problem as an integer programming model to determine the daily production schedule in order to minimize total costs.

4. Consider the following mathematical programming model:

$$\text{Maximize } Z = 3x_1 - 2x_2 + 4x_3$$

Subject to

$$x_1 + 2x_2 + 4x_3 \leq 22$$
$$2x_1 + 3x_2 - x_3 \leq 20$$
$$3x_1 + x_2 + 3x_3 \geq 18$$
$$2x_1 - x_2 + 3x_3 \geq 17$$

Use integer programming techniques to express the following:

- At least three of the constraints must hold
- No more than any single constraint must hold
- No more than any two constraints must hold

5. A local power company is considering increasing its generating capacity to meet expected demand over the next 5 years. Currently, the company has a generating capacity of 600 MW. As per the demand forecast, the minimum generating capacities required over the next 5 years are as follows:

	Year				
	1	2	3	4	5
Minimum capacity (MW)	670	760	890	1020	1180

The company can increase its transmission capacity by purchasing five different types of generators: 20, 50, 80, 100, and 120 MW. The costs of acquiring, installing, and operating each of the five types of generators are summarized below:

Generator Type (MW)	Acquiring and Installation Cost/Unit[a]	Operating Cost/Unit/Year[a]
20	5.0	2.0
50	10.0	4.0
80	13.0	6.0
100	17.0	8.0
120	22.0	10.0

[a]In million dollars.

Any number of generators (of any type) can be added in any year. However, once a generator is added, it must be operated in the future years as required. Formulate a mathematical programming model to determine the least cost expansion plan, for 5 years, while fulfilling all yearly demand.

6. The manager of City Carpet has just received an order for carpets for a new office block. The order is for 1500 m of carpet 2 m wide, 6000 m of carpet 3 m wide, and 3000 m of carpet 4 m wide. The manager can manage two kinds of carpet rolls, which he needs to cut to fill this order. One type of roll is 6 m wide and 30 m long, and the second is 8 m wide and 30 m long. It costs $1500 per roll for the first type and $2000 per roll for the second type. City Carpet needs to determine how many of the two types of carpet rolls to buy and how they should be cut. Formulate the problem as an integer programming model in order to minimize the cost of carpets.
7. The city council is interested in locating two ambulances, in a newly developed housing zone, to maximize the number of residents that can be reached within 5 min in emergency situations. The housing zone is divided into six regions, and the average times required (in minutes) to travel from one region to the next are summarized as follows:

	To Region					
From Region	1	2	3	4	5	6
1	0	2	3	6	8	10
2	2	0	6	4	7	6
3	3	6	0	3	4	8
4	6	4	3	0	2	2
5	8	7	3	3	0	3
6	10	6	8	2	3	0

The population in regions 1, 2, 3, 4, 5, and 6 are estimated to be 20,000, 60,000, 45,000, 55,000, 36,000, and 48,000, respectively. The question is which two regions should be chosen for the ambulances. Formulate an integer linear programming model for the problem.

8. An engineering company produces two products A and B. Each unit of product A requires 1 h of engineering service and 5 h of machining time. One unit of B requires 2 h of engineering service and 8 h of machining time. There are 110 h of engineering and 400 h of machining time available. The cost of production is a step function as follows:

Product	Production (Units)	Unit Cost ($)
A	0–50	12
	50–100	9
B	0–40	8
	40–100	5

The selling price for product A and B is $14 and $15, respectively. Formulate an integer programming model to maximize the total profit.

Bibliography

Bazaraa, M., Jarvis, J., and Sherali, H., *Linear Programming and Network Flows*, John Wiley & Sons, New York, 1990.

Bradley, S.P., Hax, A.C., and Magnanti, T.L., *Applied Mathematical Programming*, Addison-Wesley Publishing Company, Readings, MA, 1977.

Daellenbach, H.G., George, J.A., and McNickle, D.C., *Introduction to Operations Research Techniques*, Allyn & Bacon, Boston, MA, 1983.

Eiselt, H.A., Pederzoli, G., and Sandblom, C.-L., *Continuous Optimization Models*, Walter deGruyter, Berlin, 1987.

Hadley, G., *Linear Programming*, Narosa Publishing House, New Delhi, 1990 (reprinted).

Hillier, F. and Lieberman, G., *Introduction to Operations Research*, 8th ed., McGraw-Hill, Boston, MA, 2005.

Martin, R.K., *Large Scale Linear and Integer Optimization*, Kluwer Academic Publishers, Boston, MA, 1999.

Morse, P.M. and Kimball, G.E., *Methods of Operations Research*, John Wiley & Sons, New York, 1951.

Papadimitriou, C.H. and Steiglitz, K., *Combinatorial Optimization: Algorithms and Complexity*, 2nd ed., Dover Publications, New York, 1998.

Render, B. and Stair, R., *Quantitative Analysis for Management*, Prentice Hall, Englewood Cliffs, NJ, 1994.

Sierksma, G., *Linear and Integer Programming: Theory and Practice*, Marcel Dekker, New York, 1996.

Taha, H.A., *Operations Research: An Introduction*, 4th ed., MacMillan Publishing Company, New York, 1989.

Turban, E. and Meredith, J., *Fundamentals of Management Science*, 6th ed., Irwin McGraw-Hill, Boston, MA, 1994.

References

Abdennadher, S. and Frühwirth, T., *Essentials of Constraint Programming*, Springer, Berlin, 2003.

Apt, K., *Principles of Constraint Programming*, Cambridge University Press, Cambridge, 2003.

Birge, J.R. and Louveaux, F., *Introduction to Stochastic Programming*, Springer, Berlin, 1997.

Castillo, E., Conejo, A.J., Pedregal, P., Garcia, R., and Alguacil, N., *Building and Solving Mathematical Programming Models in Engineering and Science*, John Wiley & Sons, Canada, 2002.

Glover, F. and Woolsey, E., Converting 0–1 polynomial programming problem to a 0–1 linear program, *Operations Research*, 22, 180, 1974.

Kall, P. and Mayer, J., *Stochastic Linear Programming: Properties, Solution Methods, and Applications*, Springer, 2006.

Kall, P. and Wallace, S.W., *Stochastic Programming*, John Wiley & Sons, New York, 1995.

Sarker, R. and Khan, L., Optimum batch size under periodic delivery policy, *International Journal of Systems Science*, 32, 1089, 2001.

Watters, L., Reduction of integer polynomial programming problems to zero–one linear programming problems, *Operations Research*, 15, 1171, 1967.

6

Modelling Large-Scale and Well-Known Problems I

6.1 Introduction

In Chapter 4, we introduced the use of subscripts in variable notation to make the model more mathematically compact. In this chapter, we introduce the use of the summation sign when expressing a number of variables together and the use of the subset sign to separate certain variables, which should make the model even more compact specifically when we are dealing with large-scale modelling. We also present a number of well-known problems frequently mentioned in the literature and that have arisen in practice. A number of practical applications of these problems are briefly discussed with reference to the literature.

The chapter is organized as follows. After introducing summation sign and subset sign, we have introduced models of different problems such as shortest path, maximum flow, multi-commodity flow, capital budgeting, bin packing, cutting stock, facility location, facility layout, production planning and scheduling, and airlift.

6.2 Use of the Summation (Σ) Sign

To demonstrate the use of the summation sign, let us reproduce the transportation model as formulated in Example 4.13.

Defining variables:
X_{ij} = quantity to be shipped from source i ($i = 1$ for P1 and 2 for P2) to destination j ($j = 1$ for DC1, 2 for DC2, and 3 for DC3)

Objective function:
The objective is to minimize the overall shipping costs. It is simply the sum of all variables multiplied by their corresponding unit costs.

$$\text{Minimize } Z = 1X_{11} + 2X_{12} + 3X_{13} + 4X_{21} + 1X_{22} + 5X_{23} \qquad (6.1)$$

Constraints:
Supply constraints: The total supply from a plant to all destinations must be equal to its capacity.

$$\text{Plant P1:} \quad X_{11} + X_{12} + X_{13} = 100 \tag{6.2}$$

$$\text{Plant P2:} \quad X_{21} + X_{22} + X_{23} = 120 \tag{6.3}$$

Demand constraints: The total supply received by a destination from all sources must be less than or equal to its requirements.

$$\text{Destination DC1:} \quad X_{11} + X_{21} \leq 80 \tag{6.4}$$

$$\text{Destination DC2:} \quad X_{12} + X_{22} \leq 120 \tag{6.5}$$

$$\text{Destination DC3:} \quad X_{13} + X_{23} \leq 60 \tag{6.6}$$

Let us now assume the parameters are as follows:

C_{ij} = cost ($) of shipping a unit from source i to destination j
S_i = supply (in units) at source i
D_j = demand (in units) at destination j

Using summation sign, the objective function and the constraints can now be written as

$$\text{Minimize } Z = \sum_i \sum_j C_{ij} X_{ij} \quad \forall i, j \tag{6.7}$$

Supply constraints: The total supply from a plant to all destinations must be equal to its capacity.

$$\sum_j X_{ij} = S_i \quad \forall i \tag{6.8}$$

Demand constraints: The total supply received by a destination from all sources must be less than or equal to its requirements.

$$\sum_i X_{ij} \leq D_j \quad \forall j \tag{6.9}$$

The parameters required as input for the model of Example 4.13 are as follows:

$C_{11} = 1, C_{12} = 2, C_{13} = 3, C_{21} = 4, C_{22} = 1, C_{23} = 5$
$S_1 = 100, S_3 = 120$
$D_1 = 80, D_2 = 120, D_3 = 60.$

So the overall transportation model can be formulated as follows:

$$\text{Minimize } Z = \sum_i \sum_j C_{ij} X_{ij}$$

Subject to

$$\sum_j X_{ij} = S_i \quad \forall i \qquad \text{Model (6.1)}$$

$$\sum_i X_{ij} \leq D_i \quad \forall i$$

$$X_{ij} \geq 0 \quad \forall i, j$$

The above formulation is valid whether there are a few sources and destinations or there are a large number of sources and destinations as long as one can represent the data matrices separately.

During the model formulation, one does not need to use all the data as the data requirements can be finalized after the model development. However, sometimes data may dictate the model structure, which is discussed in a later chapter. At the modelling stage, one should concentrate on the logic and practical issues of modelling. In addition, for a large-scale problem, writing the model and incorporating all the data will be tedious. Interestingly, most large-scale software packages allow the data to be entered separately, which makes the overall task easier.

If the number of sources and destinations are known, one can easily calculate the number of variables and constraints from the model. For example, if there are I sources and J destinations, there will be $I \times J$ variables and $I + J$ constraints (I supply constraints and J demand constraints) in the model.

Using the above notation, the assignment model (discussed in Chapter 4) can be written as

$$\text{Minimize } Z = \sum_i \sum_j C_{ij} X_{ij}$$

Subject to

$$\sum_j X_{ij} = 1 \quad \forall i \qquad \text{Model (6.2)}$$

$$\sum_i X_{ij} \leq 1 \quad \forall j$$

$$X_{ij} \geq 0 \text{ (or } X_{ij} \in 0, 1) \quad \forall i, j$$

6.3 Use of the Subset (∈) Sign

In the above transportation model, we assumed that there were connections from all the sources to all the destinations. In many practical situations, this is not the case. For example, if there is no connection from source 1 to destination 3, the model can be written in a simpler form as follows:

$$\begin{aligned}
&\text{Minimize } Z = X_{11} + 2X_{12} + 3X_{13} + 4X_{21} + X_{22} + 5X_{23} \\
&\text{Subject to} \\
&\quad X_{11} + X_{12} = S_1 \quad \text{Supply constraint 1} \\
&\quad X_{21} + X_{22} + X_{23} = S_2 \quad \text{Supply constraint 2} \\
&\quad X_{11} + X_{21} \le D_1 \quad \text{Demand constraint 1} \\
&\quad X_{12} + X_{22} \le D_2 \quad \text{Demand constraint 2} \\
&\quad X_{23} \le D_3 \quad \text{Demand constraint 3} \\
&\quad X_{ij} \ge 0 \quad \forall i, j
\end{aligned} \qquad \text{Model (6.3)}$$

However, how would one handle this situation using the compact form of the model? At the modelling stage, one can easily define a subset of destinations (or sources) for each source (or destination) where the connections exist.

Assume that A is a set of directed arcs (i, j) that exist in the transportation network. Involving the directed arcs, the transportation model can be revised as follows:

$$\begin{aligned}
&\text{Minimize } Z = \sum_{(i,j)\in A} C_{ij}X_{ij} \\
&\text{Subject to} \\
&\quad \sum_{\{j:(i,j)\in A\}} X_{ij} = S_i \quad \forall i \quad \text{Supply constraints} \\
&\quad \sum_{\{i:(i,j)\in A\}} X_{ij} \le D_j \quad \forall j \quad \text{Demand constraints} \\
&\quad X_{ij} \ge 0 \quad \forall i, j
\end{aligned} \qquad \text{Model (6.4)}$$

In the objective function, one sums over all existing arcs A. In the supply constraint, one sums over j only for the existing arcs (i, j). Similarly, in the supply constraint, one sums over i but only for the existing arcs (i, j). In our list, the existing arcs (i, j) are $(1, 1)$, $(1, 2)$, $(2, 1)$, $(2, 2)$, and $(2, 3)$.

It would be a tedious job to write down all the constraints for a large-scale general network flow problem as illustrated in Example 4.15. The modelling of such problems could be made more convenient by using the subset concept.

The flow balance constraint at each node is

Inflow to node i – outflow from node $i \ge$ supply or demand at i (b_i).

or

$$\sum_{\{j:(j,i)\in A\}} X_{ji} - \sum_{\{j:(i,j)\in A\}} X_{ij} \ge b_i \quad \forall i \in N \qquad (6.10)$$

N is the set of nodes.

The revised model then becomes

$$\text{Minimize } Z = \sum_{(i,j)\in A} C_{ij}X_{ij}$$

Subject to

$$\sum_{\{j:(j,i)\in A\}} X_{ji} - \sum_{\{j:(i,j)\in A\}} X_{ij} \geq b_i \quad \forall i \in N$$

$$X_{ij} \geq 0 \quad \forall (i,j) \in A$$

Model (6.5)

The above mathematical model will always be the same, irrespective of the number of nodes and arcs in a network flow problem, unless additional or side constraints are imposed.

There are a number of well-known practical optimization problems that are frequently referred to in both the theoretical and application domains. Some of these problems with their general mathematical models are discussed in the next few sections.

6.4 Network Flow Problems

Network flow problems are well known for their application to a variety of problem areas. A number of network flow problems are discussed and their formulations are presented below.

6.4.1 Shortest Path Problem

The shortest path problem is considered to be the simplest type of network flow problem. In this type of problem, one seeks to find a path of minimum cost (or length) from a specified source node to another specified sink (or destination) node, assuming that each arc on the network has an associated cost (or length). In this problem, we assume there is only one source node and only one sink node.

Parameters:

N = number of nodes

A = set of existing arcs (i, j)

C_{ij} = arc length (or arc cost) associated with each arc (i, j)

i = s for source node

i = d for sink or destination node

Variables:

x_{ij} = the flow from node i to node j (it is assumed that the arcs are directed)

Objective function:
The objective is to minimize the flow cost (or length) from the source to the sink.

$$\text{Minimize } Z = \sum_{(i,\,j)\in A} C_{ij}x_{ij} \tag{6.11}$$

Constraint:
The total inflow to a node i minus the total outflow from that node must be equal to its supply or demand. It is assumed that there will be only one unit available at the source for supply and only one unit of demand at the sink node.

$$\sum_{\{j:(j,\,i)\in A\}} x_{ji} - \sum_{\{j:(i,\,j)\in A\}} x_{ij} = \begin{cases} -1 & \text{if } i = s \\ 0 & \text{if } i \neq s \text{ or } d \quad \forall i \in N \\ 1 & \text{if } i = d \end{cases} \tag{6.12}$$

Note that, as discussed in Chapter 4, positive numbers represent the demand at the destination/sink node, zero represent the transshipment nodes, and the negative numbers represent the supply available at the supply/source node.

The final linear programming (LP) model is therefore

$$\text{Minimize } Z = \sum_{(i,\,j)\in A} C_{ij}x_{ij}$$

Subject to

$$\sum_{\{j:(j,\,i)\in A\}} x_{ji} - \sum_{\{j,\,(i,\,j)\in A\}} x_{ij} = \begin{cases} -1 & \text{if } i = s \\ 0 & \text{if } i \neq s \text{ or } d \quad \forall i \in N \\ 1 & \text{if } i = d \end{cases} \qquad \text{Model (6.6)}$$

$$x_{ij} \geq 0 \quad \forall (i,\,j) \in A$$

The shortest path problems arise frequently in practice. The problem can be viewed as finding the shortest paths (1) from one node to another given node, (2) one node to all other nodes, or (3) from every node to every other node. The model provided above is for case (1). In addition to the basic problem types described in (2) and (3), the shortest path problem may include additional constraints such as constraints on path reliability, capacity, and resources required. A number of direct applications of shortest path problems can be found in Ahuja et al. (1993). Erkut and Ingolfsson (2005) developed a shortest-path-based model for hazardous materials transportation. Pattanamekar et al. (2003) considered travel time uncertainty in modelling shortest path problems. Sung et al. (2000) formulated a mathematical model of the shortest path problem on time-dependent networks, where the travel time of each link depends on the time interval.

6.4.2 Maximum Flow Problem

The maximum flow problem seeks a solution that identifies the maximum amount of flow that can occur from a specified source node to another specified sink node assuming each arc on the network has a limited flow capacity.

Parameters:

N = number of nodes
A = set of existing arcs (i, j)
U_{ij} = upper bound of flow (arc capacity) in each arc (i, j)
i = s source node (only one)
i = d sink or destination node (only one)

Variables:

x_{ij} is the flow from node i to node j (assuming that the arcs are directed)
f is the total flow from the source to the sink

Objective function:
The objective is to maximize the total flow from the source to the sink through the network.

$$\text{Maximize } Z = f \tag{6.13}$$

Constraints:

1. Flow balance: The total inflow to a node i minus the total outflow from that node must be equal to its supply or demand. It is assumed that there will be a minimum of f units available at the source for supply. f is a variable here.

$$\sum_{\{j:(j,\,i)\in A\}} x_{ji} - \sum_{\{j:(i,\,j)\in A\}} x_{ij} = \begin{cases} -f & \text{if } i = s \\ 0 & \text{if } i \neq s \text{ or } d \\ f & \text{if } i = d \end{cases} \quad \forall i \in N \tag{6.14}$$

   Note that, as discussed in Chapter 4, positive numbers represent the demand at the destination/sink node, zero represents the transshipment nodes, and negative numbers represent the supply available at the supply/source node.

2. Arc capacity: The total flow through any arc should not exceed its capacity.

$$x_{ij} \leq U_{ij} \quad \forall (i, j) \in A \tag{6.15}$$

The final LP model for the maximal flow problem becomes

Maximize $Z = f$
Subject to

$$\sum_{\{j:(j,\,i)\in A\}} x_{ji} - \sum_{\{j:(i,\,j)\in A\}} x_{ij} = \begin{cases} -f & \text{if } i = s \\ 0 & \text{if } i \neq s \text{ or } d \\ f & \text{if } i = d \end{cases} \quad \forall i \in N$$

$$x_{ij} \leq U_{ij} \quad \forall (i, j) \in A \qquad \text{Model (6.7)}$$

$$x_{ij} \geq 0 \quad \forall (i, j) \in A$$

$$f \geq 0$$

6.4.3 Multi-Commodity Flow Problem

Multi-commodity flow problems arise when several commodities use the same underlying network. Different commodities may have different origins and destinations, and the commodities have separate mass balance constraints at each node. However, the sharing of the common arc capacities binds the different commodities together. The problem is to allocate the capacity of each arc to the individual commodities in a way that minimizes overall flow costs.

Parameters:

N = number of nodes
A = set of existing arcs (i,j)
K = number of commodities
C_{ijk} = unit cost for commodity k in arc (i,j)
U_{ij} = capacity of arc (i,j)
B_{ik} = the demand or supply for commodity k in node i

Variables:
x_{ijk} is the quantity transported of commodity k from node i to node j.

Objective function:
The objective is to minimize the sum of all commodity flow costs.

$$\text{Minimize } Z = \sum_{k} \sum_{(i,\,j)\in A} C_{ijk} x_{ijk} \tag{6.16}$$

Constraints:
Flow balance: For each commodity, the total inflow to a node i minus the total outflow from that node must be equal to its supply or demand. It is assumed that the total supply for each commodity is equal to its demand.

$$\sum_{\{j:(j,\,i)\in A\}} x_{jik} - \sum_{\{j:(i,\,j)\in A\}} x_{ijk} = B_{ik} \quad \forall i \in N, \quad k \in K \tag{6.17}$$

Note that, as discussed in Chapter 4, positive numbers represent the demand at the destination/sink node, zero represent the transshipment nodes, and the negative numbers represent the supply available at the supply/source node. So the right hand side B_{ik} can be positive, zero, or negative.

Arc capacity:
The total flow of all commodities through any arc should not exceed its capacity.

$$\sum_{k} x_{ijk} \leq U_{ij} \quad \forall (i, j) \in A \tag{6.18}$$

The final LP model for multi-commodity flow problem becomes

$$\text{Minimize } Z = \sum_{k} \sum_{(i,j)\in A} C_{ijk} x_{ijk}$$

Subject to

$$\sum_{\{j:(j,i)\in A\}} x_{jik} - \sum_{\{j:(i,j)\in A\}} x_{ijk} = B_{ik} \quad \forall i \in N, \quad k \in K \qquad \text{Model (6.8)}$$

$$\sum_{k} x_{ijk} \leq U_{ij} \quad \forall (i, j) \in A$$

$$x_{ijk} \geq 0 \quad \forall (i, j) \in A, \quad k \in K$$

For more on multi-commodity flow problems, see Gabrel et al. (1999), Leighton et al. (1995), Ahuja et al. (1993), and Bazaraa et al. (1990). Gabrel et al. (1999) studied multi-commodity network optimization problems using general discontinuous step-increasing cost functions. This class of problems includes the so-called single-facility and multiple-facility capacitated network loading problems as special cases. They reported computational experiments for networks containing up to 20 nodes, 37 links, and cost functions with an average six steps per link. Hadjiat et al. (2000) studied the minimum linear cost multi-commodity flow problem where the given traffic demand is satisfied through routes having less than a given maximum number of edges. Their model has been tested on some real-world instances given by the national French telecommunication operator (France Telecom).

Castro and Nabona (1996) presented an approach for solving the multi-commodity network flow problem with a linear or nonlinear objective function by considering additional linear side constraints that link arcs of the same or different commodities. Several tests were reported using random problems obtained from different network generators and real problems arising from the fields of long- and short-term hydrothermal scheduling of electricity generation and traffic assignment, with sizes of up to 150,000 variables and 45,000 constraints.

6.5 Knapsack Problem

The knapsack problem is to decide what should be put in a knapsack given a weight limitation on how much can be carried. The term "knapsack problem" invokes the image of the backpacker who is constrained by a fixed-size knapsack and so must fill it only with the most useful or essential items. However, any problem that matches a similar analogy from any other problem area is also recognized as a knapsack problem, for example, the capital budgeting problem. The knapsack problems are represented by integer programming models. A simple formulation for a capital budgeting problem is discussed below as an example of general knapsack model. Two examples on knapsack problems are already provided in Examples 3.6 and 3.7.

6.5.1 Capital Budgeting Problem

The problem is to decide which projects to fund given a strict limitation on capital availability. The return for each project is calculated in terms of net present value and the objective of the problem is to maximize the sum of returns from the selected projects subject to capital constraint.

Parameters:

N = the number of projects under consideration

NP_i = the net present value of project i

c_i = the capital required to fund project i

B = the capital (or budget) available for all projects

Variables:

$$x_i = \begin{cases} 1 & \text{if project } i \text{ is funded} \\ 0 & \text{otherwise} \end{cases}$$

Objective function:
The objective is to maximize the sum of the net present value of the selected projects.

$$\text{Maximize } Z = \sum_{i=1}^{N} \mathrm{NP}_i x_i \tag{6.19}$$

Constraint:
The projects must be funded from within the available capital.

$$\sum_{i=1}^{N} c_i x_i \leq B \tag{6.20}$$

The final capital budgeting model becomes

$$\begin{aligned} &\text{Maximize } Z = \sum_{i=1}^{N} \text{NP}_i x_i \\ &\text{Subject to} \\ &\qquad \sum_{i=1}^{N} c_i x_i \leq B \\ &\qquad x_i \in \{0, 1\} \quad \forall i \end{aligned} \qquad \text{Model (6.9)}$$

For more on the modelling of capital budgeting problems see Weingartner (1963). Hall et al. (1992) developed a mathematical model for a project funding decision facing a U.S. Cancer Institute and Chan et al. (2005) studied a defense modernization acquisition decision problem using a multi-criteria optimization model. This model explicitly considers the diverse functions of the organization. In particular, the synergism among the functions is modelled as a multiplicative value function. The model highlights how technology acquisitions can be affected as the priorities of each organizational function changes.

6.5.2 Bin Packing Problem

When there are multiple knapsacks, the problem is probably better thought of as a bin packing problem. In this type of problem, the aim is to determine how to put the most objects in the least number of fixed-space bins. There are many variants of bin packing problems, such as one-, two-, and three-dimensional, linear, pack by volume, pack by weight, minimize volume, maximize value, and fixed-shape objects (Martello and Toth, 1990). Practical examples include how best to fit boxes into trucks (or shipping containers or storage bins) to minimize the number of trucks (or containers or bins) needed to ship everything. A mathematical model for a multiple container packing problem is presented below.

Parameters:

- M = number of containers (index i)
- N = number of items (index j)
- W_j = weight of item j
- V_j = value of item j
- C_i = capacity of container i (weight)

Variables:

$$x_{ij} = \begin{cases} 1 & \text{if item } j \text{ is assigned to container } i \\ 0 & \text{otherwise} \end{cases}$$

Objective function:
The objective is to maximize the total value of all assigned items.

$$\text{Maximize } Z = \sum_{i=1}^{M} \sum_{j=1}^{N} V_j x_{ij} \tag{6.21}$$

Constraints:
Item assignment: Each item is assigned to no more than one container.

$$\sum_{i=1}^{M} x_{ij} \leq 1 \quad \forall j \tag{6.22}$$

Container capacity: The total weight of all items assigned to a container must not exceed its capacity.

$$\sum_{j=1}^{N} W_j x_{ij} \leq C_i \quad \forall i \tag{6.23}$$

The final mathematical model becomes

$$\begin{aligned} &\text{Maximize } Z = \sum_{i=1}^{M} \sum_{j=1}^{N} V_j x_{ij} \\ &\text{Subject to} \\ &\quad \sum_{i=1}^{M} x_{ij} \leq 1 \quad \forall j \\ &\quad \sum_{j=1}^{N} W_j x_{ij} \leq C_i \quad \forall i \\ &\quad x_{ij} \in \{0, 1\} \quad \forall i, j \end{aligned} \qquad \text{Model (6.10)}$$

A slightly different approach to the one-dimensional bin packing problem, which is similar to the cutting stock model presented in Section 6.5.3, can be found in Elhedhli (2005). Lodi et al. (2002) studied two-dimensional packing problems. They considered problems in which the aim was to allocate a set of rectangular items to larger rectangular standardized units by minimizing the waste. In two-dimensional bin packing problems these units are finite rectangles, and the objective is to pack all the items into the minimum number of units, while in two-dimensional strip packing problems a single standardized unit of given width is used, and the objective is to pack all the items within the minimum height. For three-dimensional problems, see Chien and Wu (1998, 1999). They presented new procedures for loading different-sized boxes into a freight container of known dimensions so as to minimize the waste of container space.

6.5.3 Cutting Stock Problem

The cutting stock problem is very much similar to bin packing problems from the optimization modelling point of view. For example, find the best arrangement of different shapes out of a larger rectangular entity to minimize waste or the number of rectangles. This problem arises often in manufacturing. For instance, deciding how to cut pieces of cloth for pants from a roll of material or punching shapes from sheet metal. A mathematical model for a simple two-dimensional cutting stock problem, to minimize the number of rolls to be used to cut out all the items, is presented below as of Valerio de Carvalho (2002).

Parameters:

K = the known upper bound on the number of rolls needed (index k)
N = the number of clients (index i)
B_i = the number of items required by client i
W_i = the width of items required by client i
TW = the total width of each roll

Variables:

x_{ik} the number of times item i is cut out of roll k

$$y_k = \begin{cases} 1 & \text{if roll } k \text{ is used} \\ 0 & \text{otherwise} \end{cases}$$

Objective function:
The objective is to minimize the number of rolls to be used to cut out all the items.

$$\text{Minimize } Z = \sum_{k=1}^{K} y_k \tag{6.24}$$

Constraints:
Requirements: The number of items cut of a given width must be greater than or equal to the demand.

$$\sum_{k=1}^{K} x_{ik} \geq B_i \quad \forall i \tag{6.25}$$

Width limitation: The total width of the items cut from a roll must be less than or equal to the width of the roll.

$$\sum_{i=1}^{N} W_i x_{ik} \leq \text{TW} y_k \quad \forall k \tag{6.26}$$

So the final cutting stock model becomes

$$\text{Minimize } Z = \sum_{k=1}^{K} y_k$$

Subject to

$$\sum_{k=1}^{K} x_{ik} \geq B_i \quad \forall i$$

$$\sum_{i=1}^{N} W_i x_{ik} - \text{TW} y_k \leq 0 \quad \forall k \qquad \text{Model (6.11)}$$

$$x_{ik} \geq 0 \text{ and integer} \quad \forall i, k$$

$$y_k \in \{0, 1\} \quad \forall k$$

The relevant literature on cutting stock problems is briefly outlined here. Valerio de Carvalho (2002) has reviewed several LP formulations for the one-dimensional cutting stock and bin packing problems. Correia et al. (2004) describe a real-world industrial problem of production planning and cutting optimization of reels and sheets, occurring at a Portuguese paper mill. It focuses on a particular module of the global problem, which is concerned with the determination of the width combinations of the items involved in the planning process. The main goal consists of satisfying an order set of reels and sheets that must be cut from master reels. The width combination process will determine the quantity/weight of the master reels to be produced and their cutting patterns, in order to minimize waste, while satisfying production orders.

Johnston and Sadinlija (2004) developed a mathematical model that resolves the nonlinearity in the one-dimensional cutting stock problem, between pattern variables and pattern run lengths, by using 0–1 variables. The model does not require prespecification of cutting patterns. It includes practical constraints such as pattern cardinality, minimum run length, and sequence-related constraints.

Degraeve et al. (2002) developed an integer programming model for solving the layout problem in the fashion industry. Before cutting, several layers of cloth are put on a cutting table and several templates, indicating how to cut out all material for a specific size, are fixed on top of the stack. The problem consists of finding good combinations of templates and the associated height of the stack of cloth to satisfy demand while minimizing total excess production.

Schilling and Georgiadis (2002) presented a mathematical programming model for the problem of determining the optimal manner in which several product rolls of given sizes are to be cut out of raw rolls of one or more standard types. The objective is to perform this task so as to maximize the profit, taking into account the revenue from the sales, the costs of the original rolls, the costs of changing the cutting pattern, and the costs of disposal of the trim.

Morabito and Garcia (1998) proposed a two-phase integer programming model for the cutting stock problems of a hardboard industry. Rönnqvist (1995) reported a problem of how to cut wood products with different quality requirements from wooden boards, which in turn consist of a number of different qualities. A major difficulty was the existence of distinct time limits since the cross cutting of the boards was performed as a real-time process. The mathematical model derived for the problem included a set packing problem that accurately described the real-world problem of a Swedish company.

6.6 Facility Location and Layout

Facility location and layout are very important factors for efficient operation of manufacturing systems, chemical plants, and many other service organizations. In this section, we present simple but representative formulations and discuss the relevant literature.

6.6.1 Facility Location Problem

The mathematical modelling of facility siting has attracted the interest of many researchers for over four decades. Investigators have modelled location problems for both the private sector (e.g., industrial plants, banks, retail facilities, warehouses, shopping centers, recreation centers, etc.) and the public sector (e.g., ambulances, clinics, schools, etc.). Each model has differences and similarities relative to each other, but the peculiarities of each problem provide the basis for the hundreds of investigations.

These problems concern the location of facilities to serve clients economically. We are given a set N of potential facility locations and a set of clients i. A facility placed at location j costs F_j. Each client has a demand for a certain type of goods, and the total cost of satisfying the demand of client i from a facility at j is C_{ij}. The problem is to choose a subset of locations at which to place facilities and then to assign the clients to these facilities so as to minimize the total cost.

Parameters:

N = the number of customers

M = the number of potential sites

C_{ij} = the total cost of satisfying demand for customer i from site j

F_j = the cost of installation of facility j

Variables:

y_{ij} the fraction of demand of client i that is satisfied from a facility at j

$$x_j = \begin{cases} 1 & \text{if a facility is placed at } j \\ 0 & \text{otherwise} \end{cases}$$

Objective function:
The objective is to minimize the sum of costs for the installation and the satisfaction of demand.

$$\text{Minimize } Z = \sum_{j=1}^{M} F_j x_j + \sum_{i=1}^{N} \sum_{j=1}^{M} C_{ij} y_{ij} \tag{6.27}$$

Constraints:
Demand constraint: The customers' demands must be fulfilled.

$$\sum_{j=1}^{M} y_{ij} = 1 \quad \forall i \tag{6.28}$$

Facility constraint: A client i cannot be served from site j unless a facility is placed at j.

$$y_{ij} - x_j \leq 0 \quad \forall i, j \tag{6.29}$$

The above model is for a uncapacitated problem as it considers that a facility can serve any number of clients. Now, let us assume that a facility located at j has a capacity of U_j (capacitated problem) and the ith client has a demand of D_i. For this situation, we need to change the definition of C_{ij} and y_{ij} as follows:

C_{ij} unit shipment cost for customer i supplying from site j

y_{ij} the quantity of goods for client i that would be satisfied from a facility at j

The corresponding constraints for the capacitated problem are
Demand constraint: The customers' demands must be fulfilled.

$$\sum_{j=1}^{M} y_{ij} = D_i \quad \forall i \tag{6.30}$$

Facility constraint: A client i cannot be served from site j unless a facility is placed at j and the total supply from a site j to all customers must be less than or equal to its capacity.

$$\sum_{i=1}^{N} y_{ij} - U_j x_j \leq 0 \quad \forall j \tag{6.31}$$

The final capacitated facility location model becomes

$$\text{Minimize } Z = \sum_{j=1}^{M} F_j x_j + \sum_{i=1}^{N} \sum_{j=1}^{M} C_{ij} y_{ij}$$

Subject to

$$\sum_{j=1}^{M} y_{ij} = D_i \quad \forall i$$

$$\sum_{i=1}^{N} y_{ij} - U_j x_j \leq 0 \quad \forall j$$

$$y_{ij} \geq 0 \quad \forall i, j$$

$$x_j \in \{0, 1\} \quad \forall j$$

Model (6.12)

In the above model, if we remove the binary variables x_j (variables for identifying facility locations) and facility installation cost F_j, it is similar to the transportation model discussed in Chapter 4.

ReVelle and Eiselt (2005) have reviewed many facets of the facility location field through reference to both seminal works and current reviews. Other specific models are discussed below.

Eben-Chaime et al. (2002) have studied a capacitated location–allocation problem and developed an appropriate mathematical model. McGarvey and Cavalier (2005) have examined a competitive facility location problem occurring in two dimensions. They have developed a new gravity-based utility model in which the capacity of a facility serves as its measure of attractiveness. Gue (2003) has developed a multi-period, facility location and material flow model, and has shown how to configure the land-based distribution system over time to support a given battle plan with minimum inventory. Drezner et al. (2002) and Drezner (1998) have studied a multiple competitive facilities location problem in which a franchise of several facilities is to be located in a trade area where competing facilities already exist. The objective of their model is to maximize the market share captured by the franchise as a whole.

6.6.2 Facility Layout Problem

Consider a warehouse layout problem. In a warehouse, the operating staff must decide where to locate the different items of goods they receive and later where to deliver the items of goods to. Suppose that a company is using a warehouse for storing n items. The warehouse has only one loading and unloading dock. The warehouse stores items on pallets. For any item i, the cost of movement per year, between the dock and the storage region, is dependent on the distance traveled by a forklift and the average number of pallet loads received and delivered per year. The warehouse layout problem is to determine the regions for storing each of the n items that will minimize the total transportation cost between the items and the dock.

For the convenience of modelling, the warehouse floor area is divided into m square grids of equal size, numbered from 1 to m. Each grid-square can accommodate only one pallet.

Parameters:

G_i = the total number of grid-squares required to store item i (as an item may require more than one grid-square)

F_i = the average number of pallet loads, for item i, received and delivered in a year

D_j = the distance between the dock and the center of grid-square j

P_i = the cost per pallet per unit distance incurred in transporting item i between the dock and its storage region

We assume that the total number of grid-squares required by all items is less than or equal to m, that is,

$$\sum_{i=1}^{n} F_i \leq m \tag{6.32}$$

The transportation cost per pallet load for item i between the dock and the grid-square location j is $D_j \times P_i$. The average number of pallet loads per grid for item $i = F_i/G_i$. So the average cost of locating a pallet of item i in a given year $= C_{ij} = (F_i/G_i) \times D_j \times P_i$.

Decision variables:

$$x_{ij} = \begin{cases} 1 & \text{if item } i \text{ is stored in grid-square } j \\ 0 & \text{otherwise} \end{cases}$$

Objective function:
The objective is to minimize the total cost locating all the items in the warehouse.

$$\text{Minimize } Z = \sum_{i=1}^{n} \sum_{j=1}^{m} C_{ij} x_{ij} \tag{6.33}$$

Constraints:

1. The number of placed pallets for item i must be equal to the number of grid-squares required.

$$\sum_{j=1}^{m} x_{ij} = G_i \quad \forall i \tag{6.34}$$

2. Each grid-square can, at most, contain one pallet.

$$\sum_{i=1}^{n} x_{ij} = 1 \quad \forall j \tag{6.35}$$

So the overall model becomes

$$\text{Minimize } Z = \sum_{i=1}^{n} \sum_{j=1}^{m} C_{ij} x_{ij}$$

Subject to

$$\sum_{j=1}^{m} x_{ij} = G_i \quad \forall i \qquad \text{Model (6.13)}$$

$$\sum_{i=1}^{n} x_{ij} = 1 \quad \forall j$$

$$x_{ij} \in \{0, 1\} \quad \forall i, j$$

Facility layout is considered an important planning and design issue for many facilities and organizations, such as conventional machine-shops, chemical processing plants, flexible manufacturing systems (FMSs), warehouses, hospitals, and shopping malls. Malmborg and Krishnakumar (1989) determined the optimal warehouse layout based on a mathematical model for the case of aisle-capture and multi-command order picking cycles.

Potts and Whitehead (2001) solved combined scheduling and machine layout problems in an FMS using the mathematical modelling approach. In the model, for a known set of products, the primary objective was to maximize the throughput and the secondary objective was to minimize the movement of work between machines.

Mak et al. (1998) developed a mathematical model to examine the machine layout and the pattern of material flow for the typical job shop and flow shop manufacturing environments. The analysis also considered various practical aspects, such as the constraints of restricted areas and reserved machinery locations, and also the irregularity of the shapes of manufacturing plants.

Georgiadis et al. (1999) presented a mathematical model to address the problem of allocating items of equipment in a given two- or three-dimensional space. The problem was formulated as a mixed-integer linear programming (MILP) model where equipment of various sizes and geometries were taken into account.

The dynamic plant layout problem (DPLP) deals with the design of multi-period layout plans. The dynamic problem involves selecting a static layout for each period and then deciding on whether to change to a different layout in the next period. For more detail on DPLP, see Balakrishnan et al. (2003) and Baykasoglu and Gindy (2001).

6.7 Production Planning and Scheduling

The problem is to determine a production schedule for N products over T periods. This problem is also known as the *multi-product lot sizing/scheduling problem*. Demand for each product must be met in each period and no stockouts are allowed. There is a marginal cost of production and there is a marginal inventory holding cost for items held at the end of each period. There is also a fixed setup cost for each product if the production level in a period is greater than zero. All the products share a common but limited machine capacity and other resources.

Parameters:

T = number of periods
N = number of products
D_{it} = demand for product i in period t
F_{it} = fixed cost associated with the production of product i in period t
H_{it} = unit inventory holding cost of product i at the end of period t
C_{it} = unit production cost of product i in period t
P_t = production capacity available in period t
PC_i = production capacity consumed by each unit of product i

Variables:

x_{it} units of product i produced in period t

$$y_{it} = \begin{cases} 1 & \text{for nonzero production of product } i \text{ in period } t \\ 0 & \text{otherwise} \end{cases}$$

I_{it} units of product i held as inventory at the end of period t

Objective function:
The objective is to minimize the sum of production, setup, and inventory holding costs.

$$\text{Minimize } Z = \sum_{i=1}^{N} \sum_{t=1}^{T} (C_{it}x_{it} + F_{it}y_{it} + H_{it}I_{it}) \tag{6.36}$$

Constraints:
Capacity constraint: The total production in each period t must not exceed the available production capacity.

$$\sum_{i=1}^{N} \text{PC}_i x_{it} \leq P_t \quad \forall t \tag{6.37}$$

Demand constraint: The demand of each product i in each period t must be met from the production of the current period plus the inventory carried over from the previous period minus the inventory to be carried over for the next period.

$$I_{i,\,t-1} + x_{it} - I_{it} = D_{it} \quad \forall i, t \tag{6.38}$$

Fixed charge or setup cost forcing constraint: It ensures y_{it} to be nonzero if x_{it} is nonzero. In other words, the constraint links production with setup that is x_{it} with y_{it}.

$$x_{it} \leq M_{it} y_{it} \quad \forall i, t \tag{6.39}$$

where M_{it}s are chosen as very big numbers—also known as the "Big M" in operations research and optimization literature. Alternatively, we can use P_t in place of M_{it}.

The final production scheduling model is

$$\text{Minimize } Z = \sum_{i=1}^{N} \sum_{t=1}^{T} (C_{it} x_{it} + F_{it} y_{it} + H_{it} I_{it})$$

Subject to

$$\begin{aligned} & \sum_{i=1}^{N} \text{PC}_i x_{it} \leq P_t \quad \forall t \\ & I_{i,\,t-1} + x_{it} - I_{it} = D_{it} \quad \forall i, t \\ & x_{it} \leq M_{it} y_{it} \quad \forall i, t \\ & x_{it}, I_{it} \geq 0 \quad \forall i, t \\ & y_{it} \in \{0, 1\} \quad \forall i, t \end{aligned} \qquad \text{Model (6.14)}$$

6.7.1 Relevant Literature

Lot sizing is one of the most important and also one of the most difficult problems in production planning. Karimi et al. (2003) have studied single-level lot sizing problems and their variants. They have introduced factors affecting the formulation and the complexity of production planning problems, and discussed different variants of lot sizing and scheduling problems

(LSSPs). Gnoni et al. (2003) have dealt with LSSPs of a multi-site manufacturing system with capacity constraints and uncertain multi-product and multi-period demand. Wolsey (1997) has presented a survey that covers a number of topics such as handling of changeovers, both sequence independent and sequence dependent, in production planning and machine sequencing, within the realm of the question of how to model time.

de Matta and Miller (2004) have studied a problem of coordinating the short-term production and inter-facility transportation scheduling decisions between a plant that produces intermediate products and a finishing plant, which processes the intermediate products into finished goods. The goal was to develop a better understanding of the general relationships between production and transportation scheduling decisions, in particular, how changes in plant capacity and costs affect the coordination of scheduling decisions as well as the choice of transportation modes and carriers. They have formulated the problem as a mixed-integer programming model.

Wu and Chang (2004) have attempted to assess an optimal production-planning program in response to varying environmental costs in an uncertain environment. The optimal production strategy concerning the numerous screenings of possible production alternatives of dyeing cloth in a textile-dyeing firm in terms of market demand, resource availability, and impact of environmental costs was treated as an integral part of the multi-criteria decision-making framework based on the gray compromise programming approach. It covered not only the regular part of production costs and the direct income from product sales but also the emission/effluent charges and water resource fees reflecting part of the goals for internalization of external costs in a sustainable society.

Yan et al. (2004) have explored the hierarchical production planning (HPP) problem of flexible automated workshops (FAWs), each of which had a number of FMSs. The objective was to decompose medium-term production plans into short-term production plans (to be executed by FMSs in the FAW) so as to minimize cost on the condition that demands have just been met.

Göthe-Lundgren et al. (2002) have developed a production-planning and scheduling model for an oil refinery company. The production process consists of one distillation unit and two hydrotreatment units. In the process, crude oil is transformed into bitumen and naphthenic special oils. The aim of the scheduling is to decide the mode of operation to use in each processing unit at each point in time, in order to satisfy the demand while minimizing the production cost and taking storage capacities into account. Ortíz-Gómez et al. (2002) have described three mixed-integer multi-period optimization models of varying complexity for the oil production planning of wells of an oil reservoir.

Tang et al. (2000) presented a model for hot rolling production scheduling. The project was part of a large-scale effort to upgrade production and operations management systems of major iron and steel companies in China. Hot rolling production involves sequence-dependent setup costs. They proposed a parallel strategy to model the scheduling problem and solve it using a new modified genetic algorithm. Combining the model and the human–machine interactive method, a scheduling system was developed. The result of one year's running in Shanghai Baoshan Iron & Steel Complex showed 20% improvement over the previous manual-based system.

6.8 Logistics and Transportation

The logistics and transportation problems involve efficient transportation of goods under different conditions and environments. The basic transportation problem has been discussed in Chapter 4. The mathematical model of an airlift problem and the relevant literature on logistics are discussed in this section.

6.8.1 Airlift Problem

Suppose a number of transport aircraft, each of limited capacity, have been assigned a task of airlifting materials, from a number of locations, and delivering them to different zones in a prespecified region. For technical reasons and for a better distribution of the airlifted loads among the aircraft, a maximum allowable load from a location for each aircraft is given. The problem is to determine the airlift operation plan satisfying the demand at various zones at minimum cost.

Parameters:

N = number of aircraft available (index k)
M = number of source locations (index i)
P = number of zones receiving airlifted materials (index j)
L_k = load capacity of aircraft k
S_i = quantity available for transportation from location i
D_j = the tonnage of materials required by zone j
C_{ijk} = unit cost of transporting from location i to zone j by aircraft k
U_{ik} = maximum allowable load that can be transported from location i by aircraft k in a given period.

Variables:
x_{ijk} = the number of trips required by aircraft k from location i to zone j

Objective function:
The objective is to minimize the overall airlift cost.

$$\text{Minimize } Z = \sum_{i=1}^{M} \sum_{j=1}^{P} \sum_{k=1}^{N} C_{ijk} x_{ijk} \tag{6.40}$$

Constraints:
Demand constraint: The demand of each zone must be satisfied.

$$\sum_{i=1}^{M}\sum_{k=1}^{N} x_{ijk}L_k \geq D_j \quad \forall j \tag{6.41}$$

Supply constraint: The transportation from a location to all zones must be less than or equal to its available supply.

$$\sum_{j=1}^{P}\sum_{k=1}^{N} x_{ijk}L_k \leq S_i \quad \forall i \tag{6.42}$$

Load constraint: The total load to be transported from a location, by an aircraft, to all zones must be less than or equal to the allowable load limit. This limit can also be set based on environmental and operational conditions.

$$\sum_{j=1}^{P} x_{ijk}L_k \leq U_{ik} \quad \forall i, k \tag{6.43}$$

The final LP model becomes

$$\text{Minimize } Z = \sum_{i=1}^{M}\sum_{j=1}^{P}\sum_{k=1}^{N} C_{ijk}x_{ijk}$$

Subject to

$$\begin{aligned} &\sum_{i=1}^{M}\sum_{k=1}^{N} x_{ijk}L_k \geq D_j \quad \forall j \\ &\sum_{j=1}^{P}\sum_{k=1}^{N} x_{ijk}L_k \leq S_j \quad \forall i \\ &\sum_{j=1}^{P} x_{ijk}L_k \leq U_{ik} \quad \forall i, k \\ &x_{ijk} \geq 0 \quad \forall i, j, k \end{aligned} \qquad \text{Model (6.15)}$$

6.8.2 Relevant Literature

Ocean transportation is a major logistics activity in the global chemical supply chain industry. Often, the logistics cost can be as high as 20% of the purchasing cost. Efficient routing and scheduling of multi-parcel chemical tankers to reduce logistics expenditure is important for both the chemical and shipping industries. Jetlund and Karimi (2004) considered the maximum-profit scheduling of a fleet of multi-parcel tankers engaged in shipping bulk liquid chemicals. They presented a MILP formulation using variable-length slots. The formulation is generally applicable to all kinds of carriers engaged in the transportation of multiple commodities, and to transportation systems, where frequent schedule updates or a short-term planning horizon is required. They illustrated their approach on a real industrial case study involving 10 tankers, 36 ports, and 79 cargos. Their approach showed an increase of 32.7% in profits compared with the plan actually used by a major chemical shipping company.

Troncoso and Garrido (2005) presented a mathematical model applied to the problem of production and logistics in the forest industry. Specifically, a dynamic model of mixed-integer programming was formulated to solve three common problems in the forest sector: forest production, forest facilities location, and forest freight distribution. The implemented mathematical model allowed the strategic selection of the optimal location and size of a forest facility, in addition to the identification of the production levels and freight flows that will be generated in the considered planning horizon. A practical application of the model was carried out, validating its utility in the location of a sawmill.

A mathematical model for a multi-region distribution problem that addresses the operational benefits of serving a global market using a network of dedicated multimodal cargo facilities (DMCFs) has been introduced by Warsing et al. (2001). The model allows an explicit evaluation of the comparative value of using a dedicated air cargo–based multimodal distribution facility in an established network of supply and demand points as opposed to more traditional methods for interregional shipments. Their model is a large-scale, nonlinear programming model to evaluate the corresponding logistics costs, incorporating the congestion effects of aircraft loading/unloading on dock-to-dock lead times in the network.

The Dead Sea Works Ltd. extracts more than 2 million tonnes of potassium a year from the Dead Sea and ships it by surface and marine transportation to markets around the world. Product mix and market distribution changes have made it necessary to redesign the surface logistics system. This system consists of four sites with a total of more than 300,000 tonnes of roofed storage capacity. The problem was to suggest changes in the storage capacity of each product at each of the sites. Rabinowitz and Mehrez (2001) have designed a model that expresses the expected annual loss for the firm and have solved it under potential market scenarios and management and technological constraints. The annual net benefit from implementing the proposed solution is expected to be $1.7 million.

The Air Force's ability to deploy, employ, and sustain operations in forward operating locations is a key to mission success. An integral part of this new strategy involving forward operating locations is equipment prepositioning, to include vehicles, aircraft support, consumable inventory, and munitions. Proper prepositioning strategies provide a means to deploy forces rapidly without resorting to an increased overseas presence. Johnstone et al. (2004) developed a mathematical model to aid decision makers that determined a strategy for positioning and configuring prepositioned assets. Their research places particular emphasis on the strategic, global prepositioning of the afloat prepositioning fleet (APF), the configuration of these ships with respect to precision guided weaponry, the development of a transportation plan in response to modelled contingencies, and a port selection and distribution strategy once the APF ship is tasked to support a contingency. In addition to the APF assets, the model considers U.S.-based supply points used to augment on-hand and APF-provided munitions assets. The primary objective is to minimize the overall response time involved with offloading these ships and transporting their cargo (the munitions) to the intended point of use.

6.9 Summary

In this chapter, we have introduced the use of the summation sign and subset sign in mathematical modelling. The use of summation and subset signs together with subscripts in variable, already introduced in previous chapter, would make any model mathematically real compact. We have presented a number of well-known problems frequently mentioned in the literature and that have arisen in practice. The types of problems considered were network flow problem, knapsack problem, facility location and layout problem, and production planning problems. The mathematical models for some of these problems were discussed and a brief review on their practical applications was presented.

Exercises

1. Consider the shortest path model (Model 6.6) as shown below:

$$\text{Minimize } Z = \sum_{(i,j)\in A} C_{ij}x_{ij}$$

Subject to

$$\sum_{\{j:(j,i)\in A\}} x_{ji} - \sum_{\{j:(i,j)\in A\}} x_{ij} = \begin{cases} -1 & \text{if } i = s \\ 0 & \text{if } i \neq s \text{ or } d \quad \forall i \in N \\ 1 & \text{if } i = d \end{cases}$$

$$x_{ij} \geq 0 \quad \forall (i,j) \in A$$

If more than one destination node exists, what changes does one need to make to the above model?

2. Consider the maximal flow model (Model 6.7) as shown below:

$$\text{Maximize } Z = f$$

Subject to

$$\sum_{\{j:(j,\,i)\in A\}} x_{ji} - \sum_{\{j:(i,\,j)\in A\}} x_{ij} = \begin{cases} -f & \text{if } i = s \\ 0 & \text{if } i \neq s \text{ or } d \\ f & \text{if } i = d \end{cases} \quad \forall i \in N$$

$$x_{ij} \leq U_{ij} \quad \forall (i, j) \in A$$

$$x_{ij} \geq 0 \quad \forall (i, j) \in A$$

If more than one destination node exists, what changes need to be made to the above model?

3. Consider the multi-commodity flow model (Model 6.8) as shown below:

$$\text{Minimize } Z = \sum_{k} \sum_{(i,\,j)\in A} C_{ijk} x_{ijk}$$

Subject to

$$\sum_{\{j:(j,\,i)\in A\}} x_{jik} - \sum_{\{j:(i,\,j)\in A\}} x_{ijk} = B_{ik} \quad \forall i \in N, \quad k \in K$$

$$\sum_{k} x_{ijk} \leq U_{ij} \quad \forall (i, j) \in A$$

$$x_{ijk} \geq 0 \quad \forall (i, j) \in A, \quad k \in K$$

If certain arcs have unlimited capacity, and if the arc capacity is different for different commodities, what changes need to be made in the above model?

4. Consider the cutting stock problem as presented in Model 6.11. Compare this model with the trim loss problem discussed in the previous chapter.
5. Consider the capacitated facility location model (Model 6.12) as shown below:

$$\text{Minimize } Z = \sum_{j=1}^{M} F_j x_j + \sum_{i=1}^{N} \sum_{j=1}^{M} C_{ij} y_{ij}$$

Subject to

$$\sum_{j=1}^{M} y_{ij} = D_i \quad \forall i$$

$$\sum_{i=1}^{N} y_{ij} - U_j x_j \leq 0 \quad \forall j$$

$$y_{ij} \geq 0 \quad \forall i, j$$

$$x_j \in \{0, 1\} \quad \forall j$$

If we want to limit the number of locations to one or two, what changes would the above model require?

6. Consider the facility layout model (Model 6.13) as shown below:

$$\text{Minimize } Z = \sum_{i=1}^{n} \sum_{j=1}^{m} C_{ij} x_{ij}$$

Subject to

$$\sum_{j=1}^{m} x_{ij} = G_i \quad \forall i$$

$$\sum_{i=1}^{n} x_{ij} = 1 \quad \forall j$$

$$x_{ij} \in \{0, 1\} \quad \forall i, j$$

If the warehouse has more than one dock, what changes would the above model require?

7. Consider the facility layout model (Model 6.14) as shown below:

$$\text{Minimize } Z = \sum_{i=1}^{N} \sum_{t=1}^{T} (C_{it} x_{it} + F_{it} y_{it} + H_{it} I_{it})$$

Subject to

$$\sum_{i=1}^{N} \text{PC}_i x_{it} \leq P_t \quad \forall t$$

$$I_{i,\,t-1} + x_{it} - I_{it} = D_{it} \quad \forall i, t$$

$$x_{it} \leq M_{it} y_{it} \quad \forall i, t$$

$$x_{it}, I_{it} \geq 0 \quad \forall i, t$$

$$y_{it} \in \{0, 1\} \quad \forall i, t$$

How do the constraints of the above model differ from the fixed-charge problem discussed in the previous chapter? Can we combine constraints (1) and (3) in this model? Why or why not?

References

Ahuja, R., Magnanti, T., and Orlin, J., *Network Flows: Theory, Algorithms and Applications*, Prentice Hall, Englewood Cliffs, NJ, 1993.

Balakrishnan, J., Cheng, C.H., Conway, D.G., and Lau, C.M., A hybrid genetic algorithm for the dynamic plant layout problem, *International Journal of Production Economics*, 86, 107, 2003.

Baykasoglu, A. and Gindy, N.N.Z., A simulated annealing algorithm for dynamic layout problem, *Computers and Operations Research,* 28, 1403, 2001.

Bazaraa, M., Jarvis, J., and Sherali, H., *Linear Programming and Network Flows,* John Wiley & Sons, New York, 1990.

Castro, J. and Nabona, N., An implementation of linear and nonlinear multicommodity network flows, *European Journal of Operational Research,* 92, 37, 1996.

Chan, Y., DiSalvo, J.P., and Garrambone, M.W., A goal-seeking approach to capital budgeting, *Socio-Economic Planning Sciences,* 39, 165, 2005.

Chien, C. and Wu, W., A recursive computational procedure for container loading, *Computers and Industrial Engineering,* 35, 319, 1998.

Chien, C. and Wu, W., A framework of modularized heuristics for determining the container loading patterns, *Computers and Industrial Engineering,* 37, 339, 1999.

Correia, M., Oliveira, J., and Ferreira, J., Reel and sheet cutting at a paper mill, *Computers and Operations Research,* 31, 1223, 2004.

Degraeve, Z., Gochet, W., and Jans, R., Alternative formulations for a layout problem in the fashion industry, *European Journal of Operational Research,* 143, 80, 2002.

de Matta, R. and Miller, T., Production and inter-facility transportation scheduling for a process industry, *European Journal of Operational Research,* 158, 72, 2004.

Drezner, T., Location of multiple retail facilities with limited budget constraints—in continuous space, *Journal of Retailing and Consumer Services,* 5, 173, 1998.

Drezner, T., Drezner, Z., and Salhi, S., Solving the multiple competitive facilities location problem, *European Journal of Operational Research,* 142, 38, 2002.

Eben-Chaime, M., Mehrez, A., and Markovich, G., Capacitated location–allocation problems on a line, *Computers and Operations Research,* 29, 459, 2002.

Elhedhli, S., Ranking lower bounds for the bin packing problem, *European Journal of Operational Research,* 160, 34, 2005.

Erkut, E. and Ingolfsson, A., Transport risk models for hazardous materials: Revisited, *Operations Research Letters,* 33, 81, 2005.

Gabrel, V., Knippel, A., and Minoux, M., Exact solution of multicommodity network optimization problems with general step cost functions, *Operations Research Letters,* 25, 15, 1999.

Georgiadis, M.C., Schilling, G., Rotstein, G.E., and Macchietto, S., A general mathematical programming approach for process plant layout, *Computers and Chemical Engineering,* 23, 823, 1999.

Gnoni, M.G., Iavagnilio, R., Mossa, G., Mummolo, G., and Di Leva, A., Production planning of a multi-site manufacturing system by hybrid modelling: A case study from the automotive industry, *International Journal of Production Economics,* 85, 251, 2003.

Göthe-Lundgren, M., Lundgren, J.T., and Persson, J.A., An optimization model for refinery production scheduling, *International Journal of Production Economics,* 78, 255, 2002.

Gue, K.R., A dynamic distribution model for combat logistics, *Computers and Operations Research,* 30, 367, 2003.

Hadjiat, M., Maurras, J., and Vaxes, Y., A primal partitioning approach for single and non-simultaneous multicommodity flow problems, *European Journal of Operational Research,* 123, 382, 2000.

Hall, N., Hershey, J., Dessler, L., and Stotts, R., (1992). A model for making project funding decisions at the national cancer institute, *Operations Research,* 40, 1040, 1992.

Jetlund, A.S. and Karimi, I.A., Improving the logistics of multi-compartment chemical tankers, *Computers and Chemical Engineering*, 28, 1267, 2004.

Johnston, R.E. and Sadinlija, E., A new model for complete solutions to one-dimensional cutting stock problems, *European Journal of Operational Research*, 153, 176, 2004.

Johnstone, D.P., Hill, R.R., and Moore, J.T., Mathematically Modelling munitions prepositioning and movement, *Mathematical and Computer modelling*, 39, 759, 2004.

Karimi, B., Fatemi Ghomi, S.M.T., and Wilson, J.M., The capacitated lot sizing problem: A review of models and algorithms, *Omega*, 31, 365, 2003.

Leighton, T., Makedon, F., Plotkin, S., Stein, C., Tardos, E., and Tragoudas, S., Fast approximation algorithms for multicommodity flow problems, *Journal of Computer and System Sciences*, 50, 228, 1995.

Lodi, A., Martello, S., and Monaci, M., Two dimensional packing problems: A survey, *European Journal of Operational Research*, 141, 241, 2002.

Mak, K.L., Wong, Y.S., and Chan, F.T.S., A genetic algorithm for facility layout problems, *Computer Integrated Manufacturing Systems*, 11, 113, 1998.

Malmborg, C.J. and Krishnakumar, B., Optimal storage assignment policies for multiaddress warehousing systems, *IEEE Transactions on Systems, Man and Cybernetics*, 19, 197, 1989.

Martello, E. and Toth, P., *Knapsack Problems: Algorithms and Implementations*, John Wiley & Sons, New York, 1990.

McGarvey, R.G. and Cavalier, T.M., Constrained location of competitive facilities in the plane, *Computers and Operations Research*, 32, 359, 2005.

Morabito, R. and Garcia, V., The cutting stock problem in a hardboard industry: A case study, *Computers and Operations Research*, 25, 469, 1998.

Ortíz-Gómez, A., Rico-Ramirez, V., and Hernández-Castro, S., Mixed-integer multi-period model for the planning of oilfield production, *Computers and Chemical Engineering*, 26, 703, 2002.

Pattanamekar, P., Park, D., Rilett, L.R., Lee, J., and Lee, C., Dynamic and stochastic shortest path in transportation networks with two components of travel time uncertainty, *Transportation Research Part C: Emerging Technologies*, 11, 331, 2003.

Potts, C.N. and Whitehead, J.D., Workload balancing and loop layout in the design of a flexible manufacturing system, *European Journal of Operational Research*, 129, 326, 2001.

Rabinowitz, G. and Mehrez, A., A multi-echelon multi-commodity, logistic system design at the Dead Sea Works Ltd., *Computers & Industrial Engineering*, 39, 65, 2001.

ReVelle, C.S. and Eiselt, H.A., Location analysis: A synthesis and survey, *European Journal of Operational Research*, 165, 1, 2005.

Rönnqvist, M., A method for the cutting stock problem with different qualities, *European Journal of Operational Research*, 83, 57, 1995.

Schilling, G. and Georgiadis, M.C., An algorithm for the determination of optimal cutting patterns, *Computers and Operations Research*, 29, 1041, 2002.

Sung, K., Bell, M.G.H., Seong, M., and Park, S., Shortest paths in a network with time-dependent flow speeds, *European Journal of Operational Research*, 121, 32, 2000.

Tang, L., Liu, J., Rong, A., and Yang, Z., A multiple traveling salesman problem model for hot rolling scheduling in Shanghai Baoshan Iron & Steel Complex, *European Journal of Operational Research*, 124, 267, 2000.

Troncoso, J.J. and Garrido, R.A., Forestry production and logistics planning: An analysis using mixed-integer programming, *Forest Policy and Economics*, 7, 625, 2005.

Valerio de Carvalho, J.M., LP models for bin packing and cutting stock problems, *European Journal of Operational Research*, 141, 253, 2002.

Warsing, D.P., Souza, G.C., and Greis, N.P., Determining the value of dedicated multimodal cargo facilities in a multi-region distribution network, *European Journal of Operational Research*, 133, 81, 2001.

Weingartner, H.M., *Mathematical Programming and the Analysis of Capital Budgeting Problems*, Prentice-Hall, Englewood Cliffs, NJ, 1963.

Wolsey, L.A., MIP modelling of changeovers in production planning and scheduling problems, *European Journal of Operational Research*, 99, 154, 1997.

Wu, C-C. and Chang, N-B., Corporate optimal production planning with varying environmental costs: A grey compromise programming approach, *European Journal of Operational Research*, 155, 68, 2004.

Yan, H-S., Zhang, X-D., and Jiang, M., Hierarchical production planning with demand constraints, *Computers and Industrial Engineering*, 46, 533, 2004.

7

Modelling Well-Known Problems II

7.1 Introduction

In this chapter, we introduce few more well-known problems frequently mentioned in the literature and that have arisen in practice. The mathematical models of these problems are analyzed and their practical applications are briefly discussed with reference to the literature.

The problems considered in this chapter are the job and machine scheduling, generalized assignment problem (GAP), traveling salesman problem (TSP), vehicle routing problem (VRP), staff rostering, crew scheduling, and timetabling problem.

7.2 Job and Machine Scheduling

Suppose there are n jobs and m machines. Each job must be processed on all machines in a given order. A machine can only process one job at a time, and once a job is started on any machine it must be processed to completion. The objective is to minimize the sum of the completion times of all the jobs.

Parameters:

n = the number of jobs

m = the number of machines

P_{ij} = the processing time of job j on machine i

$j(r)$ = the order of machines/operations for job j (for example, job j must be processed on machine 2 first ($r=1$, $i=2$), and then machine 4 ($r=2$, $i=4$), and so on). For any job j, $r=m$ means the last operation of the job

Variables:

$$t_{ij} = \text{the start time of job } j \text{ on machine } i$$

$$x_{ijk} = \begin{cases} 1 & \text{if job } j \text{ preceeds job } k \text{ on machine } i \\ 0 & \text{otherwise (i.e., if job } k \text{ preceeds job } j \text{ on machine } i) \end{cases}$$

Objective function:
The objective is to minimize the sum of the completion times of all the jobs.

$$\text{Minimize } Z = \sum_{j=1}^{n} t_{j(m),j} \tag{7.1}$$

Note that $j(m)$ is the last operation of job j, which corresponds to a given machine i.

Constraints:
Order of operations: the $(r+1)$th operation on any job j cannot start until the rth operation has been completed.

$$t_{j(r+1),j} \geq t_{j(r),j} + P_{j(r),j} \quad \text{for } r = 1, 2, \ldots, m-1 \text{ and } \forall j \tag{7.2}$$

Note that each $j(r)$ corresponds to a given machine i.

Precedence constraint: on any machine i, either job j precedes job k or job k precedes job j.

$$t_{ik} \geq t_{ij} + P_{ij}, \quad \text{if } x_{ijk} = 1 \tag{7.3}$$

$$t_{ij} \geq t_{ik} + P_{ik}, \quad \text{if } x_{ijk} = 0 \tag{7.4}$$

It is obvious that only one of the above two constraints must hold. Given an upper bound U on $(t_{ij} - t_{ik} + P_{ij})$ for all i, j, and k, the above two constraints can be rewritten as follows:

$$t_{ij} - t_{ik} \leq -P_{ij} + U(1 - x_{ijk}) \quad \forall i, j, k \tag{7.5}$$

$$t_{ik} - t_{ij} \leq -P_{ik} + Ux_{ijk} \quad \forall i, j, k \tag{7.6}$$

The final scheduling model is

$$\text{Minimize } Z = \sum_{j=1}^{n} t_{j(m),j}$$

Subject to

$$\begin{aligned} & t_{j(r+1),j} \geq t_{j(r),j} + P_{j(r),j} \quad \text{for } r = 1, 2, \ldots, m-1 \text{ and } \forall j \\ & t_{ij} - t_{ik} \leq -P_{ij} + U(1 - x_{ijk}) \quad \forall i, j, k \\ & t_{ik} - t_{ij} \leq -P_{ik} + Ux_{ijk} \quad \forall i, j, k \\ & t_{ij} \geq 0 \quad \forall i, j \\ & x_{ijk} \in \{0, 1\} \quad \forall i, j, k \end{aligned} \qquad \text{Model (7.1)}$$

7.2.1 Relevant Literature

The job and machine scheduling problem can be divided into (1) processing identical jobs either on single or parallel machines—single and parallel machine scheduling, (2) processing identical jobs on a number of machines organized as a flow-shop layout, or (3) processing different jobs on a number of machines organized as a job-shop layout. Some models on different job and machine scheduling situations are discussed below.

Pan (2003) has considered a scheduling problem in which n jobs with distinct deadlines are to be scheduled on a single machine. The objective was to find a feasible job sequence that minimized the total weighted completion time. Favorable numerical results were also reported on an extensive set of problem instances of 20–120 jobs. Yang et al. (2003) considered the problem of scheduling n independent jobs on two identical parallel machines, with a limit on the number of jobs that can be assigned to each single machine, so as to minimize the total weighted completion time of the jobs. Zhu and Heady (2000) developed a mixed integer programming formulation for minimizing job earliness and tardiness in a multi-machine scheduling problem. Their formulation allowed setup times to depend on the job-to-job sequence, processing times to depend on the job–machine combination, due dates and cost penalties to vary for each job, and machine characteristics to be nonuniform.

Liao (1995) considered a permutation flow shop where one or more processors were flexible. Two integer programming models were developed: one for the case where the job sequence was given and the other was for the case where the job sequence was to be determined. Bertel and Billaut (2004) developed an integer linear programming (ILP) model for a scheduling problem that came from industry. The workshop was a hybrid flow shop with recirculation and the problem was to perform jobs between a release date and a due date, in order to minimize the weighted number of tardy jobs.

Huq et al. (2004) described the development of a mixed-integer linear programming model for a flow shop with multiprocessor workstations. The primary objective of the model was to minimize the make span through lot-streaming. A secondary objective was to determine workforce size and schedule. A constant daily workload was assumed. The model was simple enough to be understood and be implemented by managers and supervisors using readily available spreadsheet programs. An actual process, at a local insurance company handling a moderate daily level of document and payment processing, was used as a case study. The results of the case study yielded an improvement in the make span of the current process. The model provided to be a useful tool in the document processing industry, and is generic enough to be applied to other multiprocessor flow shops.

Job-shop scheduling problems are acknowledged to be one of the hardest combinatorial optimization problems. Mascis and Pacciarelli (2002) have studied the job-shop scheduling problem with blocking and no-wait constraints. A blocking constraint models the absence of storage capacity between machines, whereas a no-wait constraint occurs when two consecutive operations in a job must be processed without any interruption.

7.3 Assignment and Routing

The GAP and the well-known TSP are discussed in this section. The capacitated routing problem, which is a combination of assignment and TSP, is also described in this section.

7.3.1 Generalized Assignment Problem

The GAP consists of finding a minimal cost (or maximal profit) assignment of n tasks over m capacity-constrained servers, whereby each task has to be processed by only one server.

Parameters:

n = number of tasks

m = number of servers

C_{ij} = cost of assigning task i to server j

b_j = units of resource available to server j

a_{ij} = units of resource required to perform task i by server j

Variables:

$$x_{ij} = \begin{cases} 1 & \text{if task } i \text{ is assigned to server } j \\ 0 & \text{otherwise} \end{cases}$$

The integer linear programming model for the problem is

$$\text{Minimize } Z = \sum_{i=1}^{n} \sum_{j=1}^{m} C_{ij} x_{ij}$$

Subject to

$$\sum_{j=1}^{m} x_{ij} = 1, \quad i = 1, \ldots, n \qquad \text{Model (7.2)}$$

$$\sum_{i=1}^{n} a_{ij} x_{ij} \leq b_j, \quad j = 1, \ldots, m$$

$$x_{ij} \in \{0, 1\}, \quad i = 1, \ldots, n, j = 1, \ldots, m$$

As one can see in the above model, the limits of variable subscripts, the ranges of summation signs, and the binary variables are represented in a slightly different way than has been used so far. This is done so that one is exposed to a slightly different formulation that is also used by some practitioners. One may prefer to express the objective function and constraints as follows.

The objective of the problem is to minimize the total assignment cost. The first constraint ensures that each task will be assigned to one server. The second constraint expresses the fact that the total resources required for all the jobs assigned to a server do not exceed the server's available capacity. This model is different from the assignment model discussed in the previous chapters.

One important application of this model is to assign customers/goods to trucks/ships/trains/cargo aircraft. Amini and Racer (1995) studied the GAP and developed a hybrid heuristic for solving the GAP.

The multilevel generalized assignment problem (MGAP) differs from the classical GAP in that servers can perform tasks at more than one efficiency level. Important manufacturing problems, such as lot sizing, can be formulated as MGAPs; however, the large number of variables in the related 0–1 integer program makes the use of commercial optimization packages impractical. Laguna et al. (1995) presented a heuristic approach to the solution of the MGAP, which consists of a novel application of tabu search (TS).

7.3.2 Traveling Salesperson Problem

The TSP is one of the most well-known optimization problems in the literature. It has attracted the attention of many researchers over the last half a century because of its simple problem description but simultaneously its associated difficulty in obtaining an optimal solution efficiently. The problem can be described as follows: a salesperson, starting from a base, intends to visit each of several cities exactly once and return to the base ensuring a minimum total traveling distance (or cost). There are many ways to formulate the traveling salesperson problem. We present here a simple model based on a network flow problem approach.

Parameters:

n = number of cities/locations/customers to visit. Note that it does not include the base, which is indexed by $i = 0$

C_{ij} = cost/distance of traveling from city i to city j

A = a set of arcs (i, j) that exist. Note that by (i, j) we mean only the existing arcs from node i to node j

Variables:

$$x_{ij} = \begin{cases} 1 & \text{if the salesperson travels from city } i \text{ to city } j \\ 0 & \text{otherwise} \end{cases}$$

y_{ij} = flow from node i to node j (will be discussed later)

Let us assume that the salesperson has n units available at node 0 (source node) and that person must deliver 1 unit to each of the other nodes/cities. Then the model is

Objective function:
The objective is to minimize total travel cost.

$$\text{Minimize } Z = \sum_{(i,j)\in A} C_{ij}x_{ij} \tag{7.7}$$

Constraints:
The salesperson must visit each city i.

$$\sum_{\substack{i=0\\ i\neq j}}^{n} x_{ij} = 1, \quad j = 0, \ldots, n \tag{7.8}$$

After visiting city j, the salesperson must leave for another city i.

$$\sum_{\substack{j=0\\ i\neq j}}^{n} x_{ij} = 1, \quad i = 0, \ldots, n \tag{7.9}$$

These two constraints are not sufficient to define the tours since they can be satisfied by subtours. For example, for the 6 cities problem ($n=5$), $x_{01}=x_{12}=x_{20}=x_{34}=x_{45}=x_{53}=1$ satisfies the above two constraints but does not correspond to a tour. See the details of the constraints below:

For the constraint given by Equation 7.8:

If $j=0$, $x_{10}+x_{20}+x_{30}+x_{40}+x_{50}=1$; satisfies as $x_{20}=1$
If $j=1$, $x_{01}+x_{21}+x_{31}+x_{41}+x_{51}=1$; satisfies as $x_{01}=1$
If $j=2$, $x_{02}+x_{12}+x_{32}+x_{42}+x_{52}=1$; satisfies as $x_{12}=1$
If $j=3$, $x_{03}+x_{13}+x_{23}+x_{43}+x_{53}=1$; satisfies as $x_{53}=1$
If $j=4$, $x_{04}+x_{14}+x_{24}+x_{34}+x_{54}=1$; satisfies as $x_{34}=1$
If $j=5$, $x_{05}+x_{15}+x_{25}+x_{35}+x_{45}=1$; satisfies as $x_{45}=1$

For the constraint given by Equation 7.9:

If $i=0$, $x_{01}+x_{02}+x_{03}+x_{04}+x_{05}=1$; satisfies as $x_{01}=1$
If $i=1$, $x_{10}+x_{12}+x_{13}+x_{14}+x_{15}=1$; satisfies as $x_{12}=1$
If $i=2$, $x_{20}+x_{21}+x_{23}+x_{24}+x_{25}=1$; satisfies as $x_{20}=1$
If $i=3$, $x_{30}+x_{31}+x_{32}+x_{34}+x_{35}=1$; satisfies as $x_{34}=1$
If $i=4$, $x_{40}+x_{41}+x_{42}+x_{43}+x_{45}=1$; satisfies as $x_{45}=1$
If $i=5$, $x_{50}+x_{51}+x_{52}+x_{53}+x_{54}=1$; satisfies as $x_{53}=1$

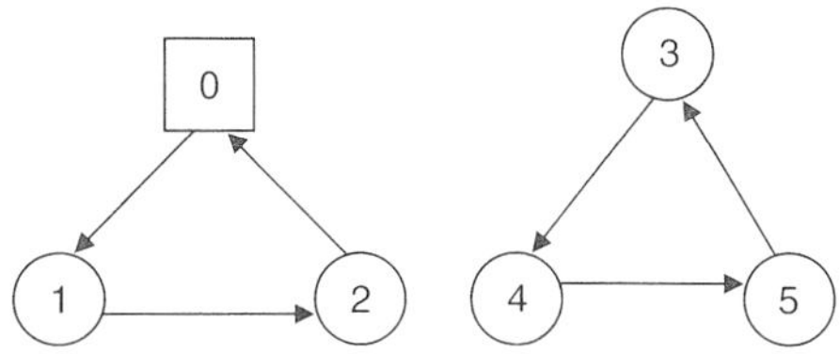

FIGURE 7.1
An example of subtours in TSP.

Figure 7.1 above shows the corresponding solution and the formation of subtours.

So the model needs additional constraints to eliminate the subtours. For the additional constraints, let us introduce a *node–arc incidence matrix* (or simply an incidence matrix). The incidence matrix represents the network as an $n \times m$ matrix N, which contains one row for each node of the network and one column for each arc. The column corresponding to arc (i, j) has only two nonzero elements: it has a $+1$ in the row corresponding to node i and a -1 in the row corresponding to node j. A simple example of an incidence matrix is given below (Figure 7.2).

The node–arc incidence matrix of the above network:

$$\text{Nodes}\ \begin{matrix} \\ 1 \\ 2 \\ 3 \\ 4 \end{matrix} \overset{\text{Arcs}}{\begin{matrix} (1,2) & (2,3) & (2,4) & (3,1) & (4,3) \\ \end{matrix}} \begin{bmatrix} +1 & 0 & 0 & -1 & 0 \\ -1 & +1 & +1 & 0 & 0 \\ 0 & -1 & 0 & +1 & -1 \\ 0 & 0 & -1 & 0 & +1 \end{bmatrix}$$

In this matrix, each column has exactly one $+1$ and one -1. The number of $+1$'s in a row equals the number of outgoing arcs from the corresponding node and the number of -1's in the row equals the number of incoming arcs to that node. Each column of N_{ij} matrix corresponds to the flow variable y_{ij} (representing the flow in arc (i, j)). Then the flow balance constraint for each node is given by

$$\sum_{\substack{j=0 \\ i \neq j}}^{n} N_{ij} y_{ij} = b, \quad i = 0, \ldots, n \tag{7.10}$$

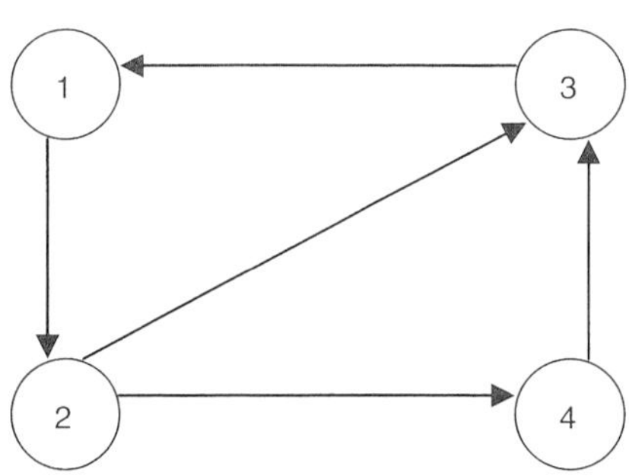

FIGURE 7.2
An example for incidence matrix.

The constraint expresses that the inflow ($-1 \times y_{ij}$) to a node minus the outflow from the node ($+1 \times y_{ij}$) must be equal to the demand of that node (b, which is 1 as per our definition). Alternatively, one can write this constraint as a simple flow balance equation as discussed in multi-commodity flow problem in an earlier chapter. This constraint ensures that all the nodes are connected since we need to send one unit of flow from node 0 to every other node via the existing arcs. However, there must be a link between the binary variables x_{ij} and the flow variables y_{ij} so that if $x_{ij} = 1$ then $y_{ij} > 0$, which is

$$y_{ij} \le n x_{ij} \quad \forall (i, j) \in A \tag{7.11}$$

where n is the maximum unit of supply carried by any arc.

So the final TSP model is

$$\text{Minimize } Z = \sum_{(i,j) \in A} C_{ij} x_{ij}$$

Subject to

$$\begin{aligned}
&\sum_{\substack{i=0 \\ i \ne j}}^{n} x_{ij} = 1, \qquad j = 0, \ldots, n \\
&\sum_{\substack{j=0 \\ i \ne j}}^{n} x_{ij} = 1, \qquad i = 0, \ldots, n \\
&\sum_{\substack{j=0 \\ i \ne j}}^{n} N_{ij} y_{ij} = b, \quad i = 0, \ldots, n \\
&y_{ij} \le n x_{ij} \quad \forall (i, j) \in A \\
&y_{ij} \ge 0 \quad \forall (i, j) \in A \\
&x_{ij} \in \{0, 1\} \quad \forall (i, j) \in A
\end{aligned} \qquad \text{Model (7.3)}$$

Note that the use of incident matrix is not space efficient as it contains too many zero coefficients. A better representation of network data is discussed in a later section.

7.3.3 Relevant Literature on Traveling Salesperson Problem

Order picking in conventional warehouse environments involves determining a sequence in which to visit the unique locations where each part in the order is stored, and therefore can often be modelled as a TSP. With computer tracking of inventories, parts may now be stored in multiple locations, simplifying the replenishment of inventory and eliminating the need to reserve space for each item. In such an environment, order picking requires choosing a subset of the locations that store an item to collect the required quantity. Thus, both the assignment of inventory to an order and the associated sequence in which the selected locations are visited affect the cost of satisfying an order. Daniels et al. (1998) formulated a model for simultaneously determining the assignment and sequencing decisions, and compared it with the previous models for order picking.

Fagerholt and Christiansen (2000) studied a TSP with allocation, time window, and precedence constraints (TSP-ATWPC). The TSP-ATWPC occurs as a subproblem involving optimally sequencing a given set of port visits in a real bulk ship scheduling problem, which is a combined multi-ship pick up and delivery problem with time windows and multi-allocation problem. Each ship in the fleet is equipped with a flexible cargo hold that can be partitioned into several smaller holds in a given number of ways, thus allowing multiple products to be carried simultaneously by the same ship. The allocation constraints of the TSP-ATWPC ensure that the partition of the ship's flexible cargo hold and the allocation of cargoes to the smaller holds are feasible throughout the visiting sequence.

Calvo and Cordone (2003) introduced the overnight security service problem. The model obtained was a single-objective mixed-integer programming problem. It is NP-hard in the strong sense, and exact approaches are not practicable when solving real-life instances. Thus, the model was solved heuristically, through a decomposition into two subproblems. The former was a capacitated clustering problem, the latter a multiple-traveling salesperson problem with time windows.

The time-dependent traveling salesperson problem (TDTSP) is a generalization of the classical TSP, where the cost of any given arc is dependent of its position in the tour. The TDTSP can model several real-world applications (e.g., one-machine sequencing). Gouveia and Voß (1995) presented a classification of formulations for the TDTSP. This framework included both new and old formulations. All previous literature on the TSP assumed that the sites to be visited are stationary. Motivated by practical applications, Helvig et al. (2003) introduced a time-dependent generalization of TSP, which we call moving-target TSP, where a pursuer must intercept in minimum time a set of targets that move with constant velocities.

7.3.4 Vehicle Routing Problem

The VRP is a generic model that practitioners encounter in many planning and decision processes. For example, the delivery of products/goods, garbage collection, the collection of money from vending machines, etc. In reality, the TSP discussed above is a special case of the VRP. A simple VRP can be described as follows:

- A fleet of M capacitated vehicles located in a depot ($i = 1$).
- A set of customer sites (of size $N-1$), each having a demand D_j ($j = 2, \ldots, N$).

- A cost C_{ij} of traveling from location i to location j.
- The problem is to find a set of routes for delivering/picking up goods to/from the customer sites at minimum possible cost.

For simplicity, assume that the vehicle fleet is homogeneous and that each vehicle has a capacity of U units.

Variables:

$$x_{kij} = \begin{cases} 1 & \text{if the vehicle } k \text{ travels on the arc } i \text{ to } j \\ 0 & \text{otherwise} \end{cases}$$

$$y_{ij} = \begin{cases} 1 & \text{if any vehicle travels on the arc } (i, j) \\ 0 & \text{otherwise} \end{cases}$$

Objective function:
The objective is to minimize the total cost of delivery.

$$\text{Minimize } Z = \sum_{k=1}^{K} \sum_{(i,j)\in A} C_{ij} x_{kij} \tag{7.12}$$

Constraints:
Starting from depot (node 1), a vehicle must visit a customer i.

$$\sum_{i=1}^{n} y_{ij} = 1, \quad j = 2, 3, \ldots, n \tag{7.13}$$

After visiting a customer i, the vehicle must leave for another customer j.

$$\sum_{j=1}^{n} y_{ij} = 1, \quad i = 2, 3, \ldots, n \tag{7.14}$$

The number of vehicles going from node 1 to all nodes j must be equal to the total number K.

$$\sum_{j=1}^{n} y_{1j} = K \tag{7.15}$$

The number of vehicles returning to node 1, from all nodes i, must be equal to the total number K.

$$\sum_{i=1}^{n} y_{i1} = K \tag{7.16}$$

Each vehicle must carry less than or equal to its capacity.

$$\sum_{i=2}^{n} \sum_{j=1}^{n} D_j x_{kij} \leq U, \quad k = 1, 2, \ldots, K \tag{7.17}$$

There must be a link between x_{kij} and y_{ij} variables. Each node, except the depot, can only be served once by only one vehicle.

$$\sum_{k=1}^{K} x_{kij} = y_{ij} \quad \forall i, j \tag{7.18}$$

The solution must not contain any cycle using the nodes 2, 3, ..., n (i.e., not contain any subtours on these nodes).

$$\sum_{\substack{(i,j)\in SxS \\ i\neq j}} y_{ij} \leq |S| - 1 \quad \text{for all subsets } S \text{ of } \{2, 3, \ldots, n\} \tag{7.19}$$

where S is any nonempty proper subset of the cities $2, 3, \ldots, n$ and $|S|$ is the size of S. If all the cities are present in a given subset then $|S| = n$ for that subset.

As per the above explanation, it is clear that this constraint eliminates subtours, i.e., tours that do not start and finish at the depot, and it is added for every possible subset of customers, not including the depot.

So the final VRP model is

$$\text{Minimize } Z = \sum_{k=1}^{K} \sum_{(i,j)\in A} C_{ij} x_{kij}$$

Subject to

$$\sum_{i=1}^{n} y_{ij} = 1, \quad j = 2, 3, \ldots, n$$

$$\sum_{j=1}^{n} y_{ij} = 1, \quad i = 2, 3, \ldots, n$$

$$\sum_{j=1}^{n} y_{1j} = K$$

$$\sum_{i=1}^{n} y_{i1} = K \qquad \text{Model (7.4)}$$

$$\sum_{i=1}^{n} \sum_{j=2}^{n} D_j x_{kij} \leq U, \quad k = 1, 2, \ldots, K$$

$$\sum_{k=1}^{K} x_{kij} = y_{ij} \quad \forall i, j$$

$$\sum_{\substack{(i,j)\in SxS \\ i\neq j}} y_{ij} \leq |S| - 1, \quad \text{for all subsets } S \text{ of } \{2, 3, \ldots, n\}$$

$$x_{kij} = 0 \text{ or } 1 \quad \forall (i,j) \in A \text{ and } \forall k$$

$$y_{ij} = 0 \text{ or } 1 \quad \forall (i,j) \in A$$

Constraint (Equation 7.19) can be used in TSP model instead of flow constraint or incidence matrix discussed earlier, but for all subsets S, $|S| = n$.

The VRP may include additional constraints such as the maximum visiting allowance, the total travel time, and the delivery time window.

7.3.5 Relevant Literature on Vehicle Routing Problem

The VRP has been widely studied in the literature, mainly because of the real-world logistics and transportation problems related to it. Toth and Vigo (2002) reviewed the models and the exact algorithms based on the branch-and-bound approach proposed in recent years for the solution of the basic version of the VRP, where only the vehicle capacity constraints are considered. Ruiz et al. (2004) proposed a two-stage exact approach for solving a real problem. In the first stage, all the feasible routes are generated by means of an implicit enumeration algorithm; thereafter, an integer programming model is designed to select in the second stage the optimum routes from the set of feasible routes. The integer model uses a number of 0–1 variables ranging from 2,000 to 15,000 and arrives at optimum solutions in an average time of 60 s (for instances up to 60 clients). The developed system was tested with a set of real instances and, in a worst-case scenario (up to 60 clients), the routes obtained ranged from a 7% to 12% reduction in the distance traveled and from a 9% to 11% reduction in operational costs.

Teixeira et al. (2004) conducted a study of planning vehicle routes for the collection of urban recyclable waste. The aim was to create collection routes for every day of the month, to be repeated every month, minimizing the operation cost. Two important features of the problem were the planning of a relatively long period of time and the separate collection of three types of waste. The collection operation was modelled in accordance to the practice of the company that manages the collection system. Preliminary results suggest that significant economies in collection costs are possible.

The Aeromedical Airlift Wing of the U.S. Air Force is responsible for the transportation of military personnel in need of specialized medical treatment to and from various military hospitals. Over 8 million active and retired personnel, spouses, and dependents benefit from the system. The system operates under a variety of regulations to ensure timely service and safe operation of the aircraft. Ruland (1999) presented a model of the system to assist the route planners in generating solutions minimizing patient inconvenience. This was achieved by assigning patients to sequences of aircraft while minimizing layovers.

Xiong et al. (1998) used the VRP with time windows (VRPTW) to analyze and model the rolling batch planning problem. Kim and Kim (1999) considered a multi-period vehicle scheduling problem (MPVSP) in a transportation system where a fleet of homogeneous vehicles delivers products of a single type from a central depot to multiple (N) retailers. The objective of the MPVSP is to minimize transportation costs for product delivery and inventory holding costs at retailers over the planning horizon.

Wasner and Zäpfel (2004) described why the optimal design of depot and hub transportation networks for parcel service providers makes it necessary to develop a generalized hub location and vehicle routing model (VRM). Analogous problems occur for postal, parcel, and piece goods service providers. A generalized hub location and VRM was developed, which encompassed the determination of the number and locations of hubs and depots and their assigned service areas as well as the routes between the demand points and consolidation points (depots, hubs). The applicability of the model was demonstrated through an Austrian case study. The developed model involved several million binary variables as well as continuous variables and millions of constraints.

7.4 Staff Rostering and Scheduling

Staff scheduling and rostering has become increasingly important tasks as businesses become more service oriented and cost conscious in a global environment. Optimized staff schedules can provide enormous benefits, but require carefully implemented decision support systems if an organization is to meet customer demands in a cost-effective manner while satisfying requirements such as flexible workplace agreements, shift equity, staff preferences, and part-time work. In addition, each industry sector has its own set of issues and must be viewed in its own right.

7.4.1 Staff Scheduling: A Weekly Problem

Many businesses run one shift (9 a.m. to 5 p.m.) a day, 7 days a week. The number of employees required may vary from one day to the next. All the employees work 5 days a week with two consecutive days off. Employees who agree to work on Saturday and/or Sunday, as part of their 5 day-a-week work, will receive a higher wage for those days. The problem is to determine the number of employees required to meet the different daily manpower requirements while minimizing the overall scheduling cost.

For convenient modelling, we define a person–roster type as follows:

Roster Type	Working Days	Days Off
1	Tuesday to Saturday	Sunday and Monday
2	Wednesday to Sunday	Monday and Tuesday
3	Thursday to Monday	Tuesday and Wednesday
4	Friday to Tuesday	Wednesday and Thursday
5	Saturday to Wednesday	Thursday and Friday
6	Sunday to Thursday	Friday and Saturday
7	Monday to Friday	Saturday and Sunday

Parameters:

N = the total number of roster type (in our case seven)
M_j = the set of roster types that will allow working on a day j
R_j = the number of employees required on each day j
C_i = weekly cost per employee assigned to roster type i

Variables:
x_i = the number of employees assigned to roster type i

Objective function:
The objective is to minimize the total assignment cost.

$$\text{Minimize } Z = \sum_{i=1}^{N} C_i x_i \tag{7.20}$$

Constraints:
The number of employees working on any one day must be greater than or equal to the requirements of that day.

$$\sum_{i \in M_j} x_i \geq R_j \quad \forall j \tag{7.21}$$

To understand the above constraint, you need to analyze the set M_j. Suppose $j = 1$ means Monday. Then the roster-type working on Monday can be found from the following table:

Roster Type	**Working Days**	**Working on Monday?**
1	Tuesday to Saturday	No
2	Wednesday to Sunday	No
3	Thursday to Monday	Yes
4	Friday to Tuesday	Yes
5	Saturday to Wednesday	Yes
6	Sunday to Thursday	Yes
7	Monday to Friday	Yes

That indicates the set M_j contains the roster type $i = 3$, 4, 5, 6, and 7. So the corresponding constraint for the day $j = 1$ will be

$$x_3 + x_4 + x_5 + x_6 + x_7 \geq R_1 \tag{7.22}$$

So the final integer programming model becomes

$$\text{Minimize } Z = \sum_{i=1}^{N} C_i x_i$$

Subject to

$$\sum_{i \in M_j} x_i \geq R_j \quad \forall j$$

$$x_i \geq 0 \text{ and integer} \quad \forall i$$

Model (7.5)

7.4.2 Daily Rostering Problem

This problem is an extension of the employee scheduling problem presented in Section 7.4.1. Many businesses run 24 h a day, 7 days a week. The number of employees required may vary from one hour to the next. The employees are assigned to either an 8- or 4-h shift. The possible shift-starting times for 8 h shifts are 6 a.m., 2 p.m., and 10 p.m.; and for 4 h shifts are 6 a.m., 10 a.m., 2 p.m., and 6 p.m. Anyone who works on the night shift will receive a higher pay for that shift. The problem is to determine how many employees to assign to each of the shifts fulfilling the hourly manpower requirements at minimum possible cost.

In this problem, the shift can be defined as follows:

Shift Number	Starting Time	Duration (h)
1	6 a.m.	8
2	2 p.m.	8
3	10 p.m.	8
4	6 a.m.	4
5	10 a.m.	4
6	2 p.m.	4
7	6 p.m.	4

Parameters:

N = the total number of shifts (in our case 7)
M_j = the set of shifts that will work during hour j
R_j = the number of employees required for each hour j
C_i = wage per employee assigned to shift i

Variables:
x_i = the number of employees assigned to shift i

Objective function:
The objective is to minimize the total assignment cost.

$$\text{Minimize } Z = \sum_{i=1}^{N} C_i x_i \tag{7.23}$$

Constraints:
The number of employees working in any hour must be greater than or equal to the manpower requirements for that hour.

$$\sum_{i \in M_j} x_i \geq R_j \quad \forall j \tag{7.24}$$

So the final integer programming model becomes

$$\text{Minimize } Z = \sum_{i=1}^{N} C_i x_i$$

Subject to

$$\sum_{i \in M_j} x_i \geq R_j \quad \forall j$$

$$x_i \geq 0 \text{ and integer} \quad \forall i$$

Model (7.6)

Note that both the models (Models 7.5 and 7.6) are designed to determine the number of employees required to be managed for a given period, and their purpose is not to identify individual employees and their assignment.

7.4.3 Relevant Literature on General Staff Scheduling

There are many computer software packages for solving staff scheduling problems, ranging from spreadsheet implementations of manual processes through to mathematical models using efficient optimal or heuristic algorithms. Ernst et al. (2004) have reviewed rostering problems in specific application areas and the associated models that have been reported in the literature.

Hospitals need to repeatedly produce duty rosters for its nursing staff. Appropriate and considerate scheduling of nurses can have an impact on the quality of health care, the recruitment of nurses, the development of budgets, and other nursing functions. The nurse rostering problem (NRP) has been the subject of many academic studies. Cheang et al. (2003) presented a brief overview, in the form of a bibliographic survey, of the models and methodologies available to solve the NRP.

Bard et al. (2003) presented a full-scale model of the tour scheduling problem as it arises in the U.S. Postal Service, and to examine several scenarios aimed at reducing the size of the workforce. The problem was formulated as a pure integer linear program and was solved using CPLEX.

The objective of project scheduling is to determine start dates and the labor resources to be assigned to each activity in order to complete a project on time. By adjusting start dates within available slack times and altering labor levels, the daily labor-demand profile can be changed. The objective of personnel scheduling is to determine how many of each feasible workday tour are required to satisfy a given labor-demand profile while minimizing labor costs and overheads. Integrating these two problems permits the simultaneous determination of start dates, labor levels, and tours for a minimum-cost and on-time schedule. Bailey et al. (1995) developed single- and multiple-resource optimization models for this integrated problem.

Mattfeld and Kopfer (2003) have described terminal operations for the vehicle transshipment hub in Bremerhaven and have derived an integral decision model for manpower planning and inventory control. One can propose a hierarchical separation of the integral model into submodels and can develop heuristics to solve the arising subproblems.

7.4.4 Crew Planning/Scheduling Problem

Airline schedule planning consists of four major planning subproblems: flight schedule, fleet assignment, aircraft maintenance planning, and crew planning. Crew planning is a common problem in all airline industries. The problem is to assign crews to flights. In this problem, the flight requirements are determined first. For example, flight QA102 from Canberra to Melbourne must be staffed every morning. A second example could be to determine what flights a flight crew can take during a tour. For example, a tour might be flight QA153 from Sydney to Brisbane, flight QA278 from Brisbane to Melbourne, and flight QA945 from Melbourne to Sydney. A tour depends upon the time and the flights that are to be taken, airline regulations, basing, etc. Once a large set of tours has been generated, a mathematical model is used to find a subset of tours, which cover the flight requirements.

Parameters:

N = number of tours generated

$$T_{ij} = \begin{cases} 1 & \text{if tour } j \text{ includes flight } i \\ 0 & \text{otherwise} \end{cases}$$

C_j = cost of tour j

Variables:

$$x_j = \begin{cases} 1 & \text{if tour } j \text{ is used} \\ 0 & \text{otherwise} \end{cases}$$

Objective function:
The objective is to minimize the total assignment cost. However, one may express the objective function as maximization of utility value instead of cost.

$$\text{Minimize } Z = \sum_{j=1}^{N} C_j x_j \tag{7.25}$$

Constraints:
The crews must be assigned to each flight.

$$\sum_{j=1}^{N} T_{ij}x_j = 1 \quad \forall i \tag{7.26}$$

The final crew scheduling model becomes

$$\text{Minimize } Z = \sum_{j=1}^{N} C_j x_j$$

Subject to

$$\sum_{j=1}^{N} T_{ij}x_j = 1 \quad \forall i$$

$$x_j \in \{0, 1\} \quad \forall j$$

Model (7.7)

The aircrew rostering problem entails the assignment of crew members to planned rotations. Airline companies have the monthly task of constructing personalized monthly schedules (rosters). Yan et al. (2002) developed eight scheduling models to minimize crew costs and to plan the proper individual pairings using real constraints for a Taiwanese airline. These models are formulated as integer linear programs, where column-generation-based algorithms were developed to solve them. A case study regarding the operation of a major Taiwanese airline is presented. Dawid et al. (2001) introduced an efficient adaptation of the branch-and-bound technique that solves real-world rostering problems for airline crews. They computed a sample monthly schedule on the basis of a medium-sized European airline's real data.

7.5 Scheduling and Timetabling Problem

Timetabling problems arise in many businesses and organizations such as public transport (bus, tram, train, passenger ship, airplane, and other) scheduling, equipment scheduling, and course and examination timetabling in schools, colleges, universities, and other academic and training institutions. In this section, a simple course timetabling model is presented and the relevant literature on other timetabling problems is discussed.

7.5.1 School Timetabling Problem

In this section, we provide a simple school timetabling model where it is required to allocate classrooms and teaching hours for different courses for each week in a given semester or term. The parameters and decision variables of the model are defined as follows:

Parameters:

N_c = number of courses (index c)
N_r = number of classrooms (index r)
N_h = number of teaching hours available in a week (index h)
N_l = number of courses taught by lecturer l
NH_c = number of teaching hours required for course c in a week
S_c = set of courses to be taught
S_l = set of courses taught by lecturer l
S_y = set of courses belonging to school level/year y

Decision variables:

$$x_{crhd} = \begin{cases} 1 & \text{if course } c \text{ is assigned to room } r \text{ at hour } h \text{ in day } d \\ 0 & \text{otherwise} \end{cases}$$

Constraints:
All lecturers must teach a number of hours equal to all their courses required in a week.

$$\sum_{c \in S_l} \sum_{r} \sum_{h} \sum_{d} x_{crhd} = \sum_{c \in S_l} NH_c N_l \quad \forall l \tag{7.27}$$

Every lecturer teaches at most one course every hour.

$$\sum_{c \in S_l} \sum_{r} x_{crhd} \leq 1 \quad \forall h, l, d \tag{7.28}$$

Every course c is taught exactly NH_c hours in a week.

$$\sum_{r} \sum_{h} \sum_{d} x_{crhd} = NH_c \quad \forall c \tag{7.29}$$

In every room–hour combination, on a given day, at most one course is taught.

$$\sum_{c} x_{crhd} \leq 1 \quad \forall r, h, d \tag{7.30}$$

This will remove the possibility of assigning a course in two different rooms at the same time.

At every hour, at most one course of any school level or year is taught so that any student of a given school year has no clash in the courses.

$$\sum_{c \in S_y} \sum_r x_{crhd} \leq 1 \quad \forall h, d \tag{7.31}$$

The timetabling problem is basically to find a solution that satisfies the constraints (Equations 7.27 through 7.31), if there exists one. So, it is a *constraint programming problem*. If there is no feasible solution, then the problem can be solved as an optimization problem where the objective would be to minimize the constraint violations. An alternative timetabling problem, as optimization model, is also presented below.

7.5.2 University Timetabling

The university timetabling problems deal with the scheduling of regular teaching program. Two different but very much related scheduling problems arise in this context. One is to schedule classes and the other is to schedule examinations in the most efficient way. As of Dimopoulou and Miliotis (2001), another timetabling model is presented below.

Parameters:

- I = set of all subject groups (index i)
- J = set of time groups (index j)
- L = set of classroom groups (index l)
- M = set of subject groups in conflict (index m)
- R_l = subset of subject groups that can be allocated to classroom group l
- T_m = subset of subject groups in conflict; the mth row of the conflict matrix
- A_l = number of classrooms of type l
- S_i = number of courses in subject group i
- C_{ij} = a desirability coefficient of assigning subject groups i to time groups j

Decision variables:

$$x_{ij} = \begin{cases} 1 & \text{if subject group } i \text{ is assigned to time group } j \\ 0 & \text{otherwise} \end{cases}$$

Objective function:
The objective is to maximize the sum of all desirability coefficients of assigning subject groups i to time groups j. If all the coefficients C_{ij} are 0, then a feasible solution is produced—of course if one exists.

$$\text{Maximize } Z = \sum_i \sum_j C_{ij} x_{ij} \tag{7.32}$$

Constraints:
Assign each subject group i with S_i courses to exactly S_i time groups. This takes care of the fact that the subjects in a subject group that are in conflict are assigned to different time groups, and thus the number of conflict constraints is reduced.

$$\sum_{j \in J} x_{ij} = S_i \quad \forall i \in I \tag{7.33}$$

The assignment must be done with the available classrooms.

$$\sum_{i \in R_l} x_{ij} \leq A_l \quad \forall j \in J \quad \forall l \in L \tag{7.34}$$

Assign at most one subject group of the set of subject groups in conflict to a time group (so that for each time group at most one of the subject groups in conflict is assigned).

$$\sum_{i \in T_m} x_{ij} \leq 1 \quad \forall j \in J \quad \forall m \in M \tag{7.35}$$

The final course timetabling model becomes

$$\text{Maximize } Z = \sum_i \sum_j C_{ij} x_{ij}$$

Subject to

$$\begin{aligned}
&\sum_{j \in J} x_{ij} = S_i \quad \forall i \in I \\
&\sum_{i \in R_l} x_{ij} \leq A_l \quad \forall j \in J \quad \forall l \in L \\
&\sum_{i \in T_m} x_{ij} \leq 1 \quad \forall j \in J \quad \forall m \in M \\
&x_{ij} \in \{0, 1\} \quad \forall i \in I \quad j \in J
\end{aligned} \qquad \text{Model (7.8)}$$

For further information on subject groups, time groups, conflict matrix, etc., the readers may consult Dimopoulou and Miliotis (2001).

7.5.3 Relevant Literature

In bus transit operations planning process, the important components are network route design, setting timetables, scheduling vehicles, assignment of drivers, and maintenance scheduling. Haghani and Shafahi (2002) presented a mathematical model to design daily inspection and maintenance schedules for the buses that are due for inspection so as to minimize the interruptions in the daily bus operating schedule, and maximize the utilization of the maintenance facilities. They provided results of an actual 181-bus transit operation.

The setting of timetables and bus routing/scheduling are essential to an intercity bus carrier's profitability, its level of service, and its competitive capability in the market. Yan and Chen (2002) developed a model that will help Taiwanese intercity bus carriers in timetable setting and bus routing/scheduling. The model employs multiple time–space networks that can formulate bus movements and passenger flows and manage the interrelationships between passenger trip demands and bus trip supplies to produce the best timetables and bus routes/schedules. Mathematically, the model is formulated as a mixed integer multiple commodity network flow problem.

Higgins et al. (1996) described the development and use of a model designed to optimize train schedules on single-line rail corridors. The model has been developed with two major applications in mind: as a decision support tool for train dispatchers to schedule trains in real time in an optimal way and as a planning tool to evaluate the impact of timetable changes, as well as railroad infrastructure changes. The model was developed based on a real-life problem and has been described in their paper.

Ghoseiri et al. (2004) developed a multi-objective optimization model for the passenger train–scheduling problem on a railroad network, which includes single and multiple tracks, as well as multiple platforms with different train capacities. In this study, lowering the fuel consumption cost was the measure of satisfaction of the railway company and shortening the total passenger-time was regarded as the passenger satisfaction criterion.

Claessens et al. (1998) considered the problem of cost optimal railway line allocation for passenger trains for the Dutch railway system. A mathematical programming model was developed, which minimized the operating costs subject to service constraints and capacity requirements. The model optimized on lines, line types, routes, frequencies, and train lengths. First, the line allocation model was formulated as an integer nonlinear programming model. The model was then transformed into an integer linear programming model with binary decision variables. The model was solved and applied to a subnetwork of the Dutch railway system for which it showed a substantial cost reduction.

The multi-depot vehicle scheduling problem with time windows (MDVSPTW) consists of scheduling a fleet of vehicles to cover a set of tasks at minimum cost. Each task is restricted to begin within a prescribed time interval and vehicles are supplied by different depots. Desaulniers et al. (1998) formulated the problem as an integer nonlinear multi-commodity network flow model with time variables. This paper considers costs on exact waiting times between two consecutive tasks instead of minimal waiting times. This new and more realistic cost structure gives rise to a nonlinear objective function in the model.

Integer programming has always been used formulating the university timetabling problem. Daskalaki and Birbas (2005) presented a two-stage relaxation procedure that solves efficiently the integer programming formulation of a university timetabling problem. The relaxation is performed in the first stage and addresses the constraints that warrant consecutiveness in multi-period sessions of certain courses. These constraints, which are computationally heavier than the others, are recovered during the second stage and a number of subproblems, one for each day of the week, are solved for local optima.

Bish (2003) addressed a container terminal loading and unloading containers to and from a set of ships, and storing the containers in the terminal yard. Each ship is served by multiple quay cranes, which load and unload containers to and from ships. Containers are moved between the ships and the yard using a fleet of vehicles, each with unit capacity. The problem is (1) to determine a storage location for each unloaded container, (2) to dispatch vehicles to containers, and (3) to schedule the loading and unloading operations on the cranes, so as to minimize the maximum time it takes to serve a given set of ships.

7.6 Summary

In Chapter 6, we introduced a number of well-known problems frequently mentioned in the literature and that have arisen in practice. In this chapter, we have introduced a few more problems of such type. The mathematical models for some of these problems were presented and their practical applications were briefly reviewed. The problems considered in this chapter are also well known for their complexity in solving. These include the traveling salesperson, vehicle routing, rostering, crew scheduling, and timetabling problems.

Exercises

1. Consider the generalized assignment model (Model 7.2) as shown below:

$$\text{Minimize } Z = \sum_{i=1}^{n}\sum_{j=1}^{m} C_{ij}x_{ij}$$

Subject to

$$\sum_{j=1}^{m} x_{ij} = 1, \quad i = 1, \ldots, n$$

$$\sum_{i=1}^{n} a_{ij}x_{ij} \leq b_j, \quad j = 1, \ldots, m$$

$$x_{ij} \in \{0, 1\}, \quad i = 1, \ldots, n, \quad j = 1, \ldots, m$$

How does this model differ from the assignment model?

2. Consider the TSP model (Model 7.3) as shown below:

$$\text{Minimize } Z = \sum_{(i,j)\in A} C_{ij}x_{ij}$$

Subject to

$$\sum_{\substack{i=0 \\ i\neq j}}^{n} x_{ij} = 1, \quad j = 0, \ldots, n$$

$$\sum_{\substack{j=0 \\ i\neq j}}^{n} x_{ij} = 1, \quad i = 0, \ldots, n$$

$$\sum_{\substack{j=0 \\ i\neq j}}^{n} N_{ij}y_{ij} = b, \quad i = 0, \ldots, n$$

$$y_{ij} \leq nx_{ij} \quad \forall (i,j) \in A$$

$$y_{ij} \geq 0 \quad \forall (i,j) \in A$$

$$x_{ij} \in \{0, 1\} \quad \forall (i,j) \in A$$

If we treat the problem as a network flow problem, as discussed earlier, why are the flow balance constraints alone not enough for TSP?

3. Consider the VRP model (Model 7.4). By relaxing the constraints (Equations 7.17 through 7.19), can we use this formulation for an assignment problem and for a knapsack problem but for each vehicle k?
4. Consider the staff scheduling problem (Model 7.5) as shown below:

$$\text{Minimize } Z = \sum_{i=1}^{N} C_i x_i$$

Subject to

$$\sum_{i\in M_j} x_i \geq R_j \quad \forall j$$

$$x_i \geq 0 \text{ and integer} \quad \forall i$$

Can you replace inequality by equality constraints above (Equation 7.21)—why or why not? If there is more than one shift, what changes does the above model require?

5. Consider the staff scheduling problem (Model 7.6) as shown below:

$$\text{Minimize } Z = \sum_{i=1}^{N} C_i x_i$$

Subject to

$$\sum_{i \in M_j} x_i \geq R_j \quad \forall j$$

$$x_i \geq 0 \text{ and integer} \quad \forall i$$

What are the differences between the models for the daily rostering problem and the employee scheduling problem (Section 7.4.1)? How to manage the assignment of individual employees after solving Model 7.5 or Model 7.6?

References

Amini, M.M. and Racer, M., A hybrid heuristic for the generalized assignment problem, *European Journal of Operational Research*, 87, 343–348, 1995.

Bailey, J., Alfares, H., and Lin, W.Y., Optimization and heuristic models to integrate project task and manpower scheduling, *Computers and Industrial Engineering*, 29, 473, 1995.

Bard, J.F., Binici, C., and deSilva, A.H., Staff scheduling at the United States Postal Service, *Computers and Operations Research*, 30, 745, 2003.

Bertel, S. and Billaut, J.-C., A genetic algorithm for an industrial multiprocessor flow shop scheduling problem with recirculation, *European Journal of Operational Research*, 159, 651, 2004.

Bish, E.K., A multiple-crane-constrained scheduling problem in a container terminal, *European Journal of Operational Research*, 144, 83, 2003.

Calvo, R.W. and Cordone, R., A heuristic approach to the overnight security service problem, *Computers and Operations Research*, 30, 1269, 2003.

Cheang, B., Li, H., Lim, A., and Rodrigues, B., Nurse rostering problems—a bibliographic survey, *European Journal of Operational Research*, 151, 447, 2003.

Claessens, M.T., van Dijk, N.M., and Zwaneveld, P.J., Cost optimal allocation of rail passenger lines, *European Journal of Operational Research*, 110, 474, 1998.

Daniels, R.L., Rummel, J.L., and Schantz, R., A model for warehouse order picking, *European Journal of Operational Research*, 105, 1, 1998.

Daskalaki, S. and Birbas, T., Efficient solutions for a university timetabling problem through integer programming, *European Journal of Operational Research*, 160, 106, 2005.

Dawid, H., König, J., and Strauss, C., An enhanced rostering model for airline crews, *Computers and Operations Research*, 28, 671, 2001.

Desaulniers, G., Lavigne, J., and Soumis, F., Multi-depot vehicle scheduling problems with time windows and waiting costs, *European Journal of Operational Research*, 111, 479, 1998.

Dimopoulou, M. and Miliotis, P., Implementation of a university course and examination timetabling system, *European Journal of Operational Research*, 130, 202, 2001.

Ernst, A.T., Jiang, H., Krishnamoorthy, M., and Sier, D., Staff scheduling and rostering: A review of applications, methods and models, *European Journal of Operational Research*, 153, 3, 2004.

Fagerholt, K. and Christiansen, M., A travelling salesman problem with allocation, time window and precedence constraints—an application to ship scheduling, *International Transactions in Operational Research*, 7, 231, 2000.

Ghoseiri, K., Szidarovszky, F., and Asgharpour, M.J., A multi-objective train scheduling model and solution, *Transportation Research Part B: Methodological*, 38, 927, 2004.

Gouveia, L. and Voß, S., A classification of formulations for the (time-dependent) traveling salesman problem, *European Journal of Operational Research*, 83, 69, 1995.

Haghani, A. and Shafahi, Y., Bus maintenance systems and maintenance scheduling: Model formulations and solutions, *Transportation Research Part A: Policy and Practice*, 36, 453, 2002.

Helvig, C.S., Robins, G., and Zelikovsky, A., The moving-target travelling salesman problem, *Journal of Algorithms*, 49, 153, 2003.

Higgins, A., Kozan, E., and Ferreira, L., Optimal scheduling of trains on a single line track, *Transportation Research Part B: Methodological*, 30, 147, 1996.

Huq, F., Cutright, K., and Martin, C., Employee scheduling and makespan minimization in a flow shop with multi-processor work stations: A case study, *Omega*, 32, 121, 2004.

Kim, J.-U. and Kim, Y.-D., A decomposition approach to a multi-period vehicle scheduling problem, *Omega*, 27, 421, 1999.

Laguna, M., Kelly, J.P., González-Velarde, J., and Glover, F., Tabu search for the multilevel generalized assignment problem, *European Journal of Operational Research*, 82, 176, 1995.

Liao, C.-J., Sun, C.-L., and You, W.-C., Flow-shop scheduling with flexible processors, *Computers and Operations Research*, 22, 297, 1995.

Mascis, A. and Pacciarelli, D., Job-shop scheduling with blocking and no-wait constraints, *European Journal of Operational Research*, 143, 498, 2002.

Mattfeld, D.C. and Kopfer, H., Terminal operations management in vehicle transhipment, *Transportation Research Part A: Policy and Practice*, 37, 435, 2003.

Pan, Y., An improved branch and bound algorithm for single machine scheduling with deadlines to minimize total weighted completion time, *Operations Research Letters*, 31, 492, 2003.

Ruiz, R., Maroto, C., and Alcaraz, J., A decision support system for a real vehicle routing problem, *European Journal of Operational Research*, 153, 593, 2004.

Ruland, K.S., A model for aeromedical routing and scheduling, *International Transactions in Operational Research*, 6, 57, 1999.

Teixeira, J., Antunes, A.P., and de Sousa, J.P., Recyclable waste collection planning—a case study, *European Journal of Operational Research*, 158, 543, 2004.

Toth, P. and Vigo, D., Models, relaxations and exact approaches for the capacitated vehicle routing problem, *Discrete Applied Mathematics*, 123, 487, 2002.

Wasner, M. and Zäpfel, G., An integrated multi-depot hub-location vehicle routing model for network planning of parcel service, *International Journal of Production Economics*, 90, 403, 2004.

Xiong, C., Weishui, W., and Xinhe, X., Modelling rolling batch planning as vehicle routing problem with time windows, *Computers and Operations Research*, 25, 1127, 1998.

Yan, S. and Chen, H.-L., A scheduling model and a solution algorithm for inter-city bus carriers, *Transportation Research Part A: Policy and Practice*, 36, 805, 2002.

Yan, S., Tung, T.-T., and Tu, Y.-P., Optimal construction of airline individual crew pairings, *Computers and Operations Research*, 29, 341, 2002.

Yang, H., Ye, Y., and Zhang, J., An approximation algorithm for scheduling two parallel machines with capacity constraints, *Discrete Applied Mathematics*, 130, 449, 2003.

Zhu, Z. and Heady, R.B., Minimizing the sum of earliness/tardiness in multi-machine scheduling: A mixed integer programming approach, *Computers and Industrial Engineering*, 38, 297, 2000.

8

Alternative Modelling

8.1 Introduction

Real-world problems are usually large and complex. It is not always feasible to include all aspects of reality in the model of a problem. In most cases, we deal with a simplified version of the problem that contains only some aspects of reality. Thus, a problem can be modelled in a number of different ways depending on the portion of reality to be included or excluded. In other words, there may be alternative mathematical models for many real-world problems. So the choice of the most appropriate model needs to be analyzed.

The alternative models are developed either as a monolithic model structure under different assumptions and conditions or as a hierarchical model structure. The monolithic approach allows the decision makers to choose a set of assumptions which would make not only the problem simple from a mathematical modelling point of view but also would provide an acceptable approximate solution for a realistic decision-making activity. The hierarchical model, instead of being a full-scale model, provides the advantages of dealing with a number of smaller-sized subproblems and timely information flows. In this chapter, we describe these two alternative modelling approaches with appropriate examples and analysis.

8.2 Modelling under Different Assumptions

To demonstrate the alternative modelling approaches under different assumptions, let us consider two simple examples: (1) a simple coal blending problem and (2) a simple crop planning problem.

8.2.1 A Coal Blending Problem

A simple coal blending problem can be briefly described as follows:

Coals are extracted from mines and upgraded for customers. Each category of raw coal has its own typical quality specifications in terms of percentage of ash, sulfur and moisture, and British Thermal Unit (BTU) content per pound. The BTU content per pound expresses the heating value of coal. The customers specify their required quality parameters for their ordered coal. A coal company usually operates more than one mine. Preparation and blending are the two main coal-upgrading and -processing activities. To keep the problem simple, we consider only a blending process for the upgrade. Production and demand vary from period to period due to technical reasons and the marketing environment. The problem is to determine the coal-mix that maximizes the revenue while satisfying the quality required and the demand on time.

The management has to make one assumption in respect to the raw coal category and also one assumption in respect to the blended coal category when modelling the problem. The possible assumptions that can be made regarding inventories of raw coal and blended coal may state that

1. no inventory of raw coal to be carried over,
2. no inventory of blended coal to be carried over,
3. inventory of raw coal to be carried over from one period to the next,
4. inventory of blended coal to be carried over from one period to the next,
5. inventory of raw coal to be carried over from one period to the next few periods,
6. inventory of blended coal to be carried over from one period to the next few periods,
7. inventory of raw coal to be carried over from one period to any or all future periods within the planning horizon, and
8. inventory of blended coal to be carried over from one period to any or all future periods within the planning horizon.

Based on the above assumptions and conditions, the alternative models that can be developed are shown in Table 8.1.

These assumptions will dictate the nature of the functions, the number of variables, the number of constraints, and the optimal solutions obtained. These models also differ in practical issues such as their capability of handling fluctuating situations, the computational time required for solving the models, and the number of coal storages required. By a fluctuating situation, we mean a variable planning environment.

As the coal extraction, upgrading, and marketing are continuous processes, the multi-period models would be better representations of the problem. Model 1 (Table 8.1) would provide a collection of T single-period models (where T is the number of periods considered under the entire planning horizon). To demonstrate the alternative modelling approach, let us formulate two simple models (Models 2 and 3 as defined

TABLE 8.1

Assumptions and Conditions of Alternative Models

Model	Inventory Assumptions	Model	Inventory Assumptions
1	1 + 2	9	4 + 5
2	2 + 3	10	3 + 8
3	1 + 4	11	4 + 6
4	1 + 6	12	5 + 8
5	2 + 5	13	6 + 7
6	1 + 8	14	5 + 6
7	2 + 7	15	7 + 8
8	3 + 6	—	Any other

in Table 8.1). To keep the model simple, let us assume that only one blended product would be produced in any one period.

8.2.2 First Alternative Blending Model

Let us consider Model 2 first. This model does not consider any blended coal inventory, but the inventory of raw coal can be carried over to the next period only. The inventory of raw coal will be fully consumed as input to the next period.

Parameters:

- T number of periods in the planning horizon (index t)
- I number of sources for input or raw coal (index i)
- J number of quality parameters considered (index j)
- PBC price of blended coal ($ per 1000 BTU content)
- CRC_{it} cost of raw coal from source i in period t ($ per tonne)
- H_i raw coal inventory holding cost per tonne per period ($) for source i
- BTI_{it} BTU content of raw coal from source i in period t
- Q_{ijt} other quality characteristics j (such as ash, sulfur, moisture, etc.) of raw coal from source i in period t
- UQ_j upper bound of required quality characteristics j of blended product
- DUB_t upper limit of demand in terms of BTU content in period t
- DLB_t lower limit of demand in terms of BTU content in period t
- LBTU lower limit of BTU per tonne of blended product produced

Decision variables:

- X_t the amount of blended product (tonnes) produced in period t
- BT_t total BTU supplied to customers in period t

Y_{it} the amount of raw coal (tonnes) from source i, in a period t, used

I_{it} the inventory of raw coal i carried over from period t to $t+1$

The demand and revenue of the blended coal are calculated based on (BT_t), the total BTU supplied.

Objective function:
The objective is to maximize the overall profit which is the total revenue minus the total cost of raw coal and the inventory holding cost.

$$\text{Maximize } Z = \sum_t \text{PBC}(\text{BT}_t) - \sum_t \sum_i \text{CRC}_{it} Y_{it} - \sum_t \sum_i H_i(I_{it}) \quad (8.1)$$

Constraints:
Mass balance: in any period, the total raw coal plus the inventory carried over from the previous period minus the inventory carried over to the next period must be equal to the total blended product produced.

$$\sum_i (Y_{it} + I_{i(t-1)} - I_{it}) = X_t \quad \forall t \quad (8.2)$$

Total BTU produced: the total BTU produced can easily be calculated from the mass balance constraint as follows:

$$\sum_i (\text{BTI}_{it} Y_{it} + \text{BTI}_{i(t-1)} I_{i(t-1)} - \text{BTI}_{it} I_{it}) = \text{BT}_t \quad \forall t \quad (8.3)$$

Lower limit of BTU content per tonne: the BTU content per tonne of blended product must be greater than or equal to the specified lower limit.

$$\sum_i (\text{BTI}_{it} Y_{it} + \text{BTI}_{i(t-1)} I_{i(t-1)} - \text{BTI}_{it} I_{it}) \geq \text{LBTU } (\text{BT}_t) \quad \forall t \quad (8.4)$$

Requirements of quality parameters: all other quality parameters must satisfy the upper limit quality requirements.

$$\sum_i (Q_{ijt} Y_{it} + Q_{ij(t-1)} I_{i(t-1)} - Q_{ijt} I_{it}) \leq \text{UQ}_j X_t \quad \forall j, t \quad (8.5)$$

Demand: the total BTU supplied must be within the upper and lower bounds of total BTU demanded.

$$\text{DLB}_t \leq \text{BT}_t \leq \text{DUB}_t \quad \forall t \quad (8.6)$$

And nonnegativity constraints.

8.2.3 Second Alternative Blending Model

Model 3 does not consider an inventory of raw coal, but the inventory of blended coal can be carried over from one period to the next only. The inventory of blended coal will be an input to the blending process in the next period.

Parameters:

The parameters are the same as for Model 2 except the holding cost of the blended product needs to be defined instead of for raw coal.

H inventory holding cost per 1000 BTU content of blended coal per period ($)

Decision variables:

X_t, Y_{it} are the same as for Model 2. Other variables are defined as follows:

BT_t the BTU content per tonne of blended product in period t

QB_{jt} quality parameter j for blended coal in period t

I_t the inventory of blended product carried over from period t to $t+1$

The demand and revenue of the blended coal are calculated based on (BT_t), the total BTU supplied.

Objective function:
The objective is to maximize the overall profit, which is the revenue minus the total cost of raw coal minus the inventory holding cost.

$$\text{Maximize } Z = \sum_t \mathrm{PBC}(\mathrm{BT}_t X_t) - \sum_t \sum_i \mathrm{CRC}_{it} Y_{it} - \sum_t H(\mathrm{BT}_t I_t) \tag{8.7}$$

Constraints:
Mass balance: in any period, the total raw coal plus the inventory carried over from the previous period minus the inventory carried over to the next period must be equal to the total blended product produced.

$$\sum_i Y_{it} + I_{t-1} - I_t = X_t \quad \forall t \tag{8.8}$$

Total BTU produced: the total BTU produced can easily be calculated from the mass balance constraint as follows:

$$\sum_i \mathrm{BTI}_{it} Y_{it} + \mathrm{BT}_{t-1} I_{t-1} - \mathrm{BT}_t I_t = \mathrm{BT}_t X_t \quad \forall t \tag{8.9}$$

Other quality parameters: other quality parameters can also be calculated from the mass balance constraint as follows:

$$\sum_i Q_{ijt} Y_{it} + \mathrm{QB}_{j(t-1)} I_{t-1} - \mathrm{QB}_{jt} I_t = \mathrm{QB}_{jt} X_t \quad \forall j, t \tag{8.10}$$

BTU content requirements: the BTU content per tonne of blended product must be greater than or equal to the specified lower limit.

$$\sum_i \mathrm{BTI}_{it} Y_{it} + \mathrm{BT}_{t-1} I_{t-1} \geq \mathrm{LBTU}(I_t + X_t) \quad \forall t \tag{8.11}$$

Quality requirements: the blended coal produced is either supplied to the customers or carried over to the next period to use as input to the process. These two coal categories must satisfy the upper limit of quality requirements.

$$\sum_i Q_{ijt} Y_{it} + QB_{j(t-1)} I_{t-1} \leq UQ_j(I_t + X_t) \quad \forall j, t \tag{8.12}$$

Demand: the total BTU supplied must be within the upper and lower bounds of total BTU demanded.

$$\mathrm{DLB}_t \leq \mathrm{BT}_t X_t \leq \mathrm{DUB}_t \quad \forall t \tag{8.13}$$

And nonnegativity constraints.

8.2.4 Comparing the Two Simple Alternative Models

Let us compare the above two models in terms of the number of variables, number of constraints, function properties, and practical issues.

The first model (Model 2) is a linear programming model whereas the second one (Model 3) is a nonlinear model as there are product terms of two variables, which appear in a number of places. The number of variables contained in these models can be calculated as shown in Table 8.2. The number of constraints included in these two models can be calculated as shown in Table 8.3. For a given number of planning periods, inputs, and quality parameters, the actual number of variables and constraints can be calculated as shown in Table 8.4.

TABLE 8.2

Variables in Two Alternative Models

Alternative Model 1		Alternative Model 2	
Variables	**Number of Variables**	**Variables**	**Number of Variables**
X_t	T	X_t	T
BT_t	T	Y_{it}	$I \times T$
Y_{it}	$I \times T$	BT_t	T
I_{it}	$I \times T$	QB_{jt}	$J \times T$
		I_t	T
Total	$2T(1 + I)$	Total	$T(3 + I + J)$

TABLE 8.3

Constraints in Two Alternative Models

Alternative Model 1		Alternative Model 2	
Constraint	Number of Constraints	Constraint	Number of Constraints
8.2	T	8.8	T
8.3	T	8.9	T
8.4	T	8.10	$J \times T$
8.5	$J \times T$	8.11	T
8.6	T	8.12	$J \times T$
		8.13	T
Total	$T(4 + J)$	Total	$T(4 + 2J)$

Model 2 requires a number of storage areas equal to the number of inputs I, whereas the Model 3 requires only one storage area for the blended product. Although, the number of variables and constraints between the two models are not significantly different, the nonlinear model is computationally very expensive. However, the second model requires only one storage space to maintain which is convenient from the practical point of view. Although all the models (Models 1–15) are dealing with the same problem, the problem size will be enormous with Models 14 and 15. The final decision of choosing a model is dependent on the solutions provided (objective function values), computational complexity, practical issues, and management's preference. More details on alternative modelling for coal blending will be covered in a later chapter.

8.2.5 A Crop Planning Problem

A simple crop planning problem can be described as follows.

An agricultural farm has limited areas of land which can be divided into a number of regions. The farm cultivates a number of crops to meet a certain

TABLE 8.4

Parameters in Two Alternative Models

	Alternative Model 1		Alternative Model 2	
Values of T, I, J	Variables $2T(1 + I)$	Constraints $T(4 + J)$	Variables $T(3 + I + J)$	Constraints $T(4 + 2J)$
4, 3, 3	32	28	36	40
4, 5, 3	48	28	44	40
6, 3, 3	48	42	54	60
6, 5, 3	72	42	66	60
12, 3, 3	96	84	108	120
12, 5, 3	144	84	132	120
12, 10, 3	264	84	192	120

portion of the national demand. Different regions require different amounts of input (such as seed, fertilizer, water, insecticide, equipment, man power, etc.) for crop cultivation and they produce at different yield rates. The problem is to determine an appropriate crop-mix for maximizing the profit while not violating any of the constraints such as land limitation, budget, and demand.

The management of the farm may make the following assumptions:

1. All regions are suitable for all crops considered for cultivation.
2. Any number of products can be produced in any region (however, it is convenient if only one crop is cultivated in one region).
3. Profit maximization is the farm's primary goal.
4. Working capital minimization may be considered as a second goal.

8.2.6 Crop Planning Model 1

Let us formulate the model by considering assumptions 1–3.

Parameters:

I	the number of crops to be cultivated (index i)
J	the number of regions of the farm (index j)
K	the number of inputs required for crop cultivation (index k)
A_{ijk}	the amount of input k required per unit land for crop i cultivated in region j
C_{ijk}	cost (\$) of unit input k required for crop i cultivated in region j
Y_{ij}	yield rate of crop i if cultivated in region j
L_j	the area of land in region j
P_i	price of crop i in the market
UD_i	the upper bound of demand for crop i
LD_i	the lower bound of demand for crop i
TB	the total budget available

Variables:
X_{ij} the area of land used in region j for producing crop i

Objective function:
The objective is to maximize the total profit from crop production (total revenue minus total cost).

$$\text{Maximize } Z = \sum_i \sum_j P_i Y_{ij} X_{ij} - \sum_i \sum_j \sum_k C_{ijk} A_{ijk} X_{ij} \tag{8.14}$$

Constraints:
Land limitation: the total land used for different crops in a region must be less than or equal to the available land in that region.

$$\sum_i X_{ij} \leq L_j \quad \forall j \tag{8.15}$$

Working capital: the total investment for all inputs required in all regions must be less than or equal to the working capital available.

$$\sum_i \sum_j \sum_k C_{ijk} A_{ijk} X_{ij} \leq \text{TB} \tag{8.16}$$

Demand: the production of any crop must be greater than (or equal to) the minimum level of demand set and less than (or equal to) the upper bound of demand.

$$LD_i \leq \sum_j Y_{ij} X_{ij} \leq \text{UD}_i \quad \forall i \tag{8.17}$$

And nonnegativity constraints.

Crop planning model 1 is a simple linear programming model.

8.2.7 Crop Planning Model 2

Suppose, we now consider that only one crop will be cultivated in any one region for the convenience of equipment handling and planning of other inputs. We further assume that

- once the crop has been decided for a region, the entire region will be cultivated, and
- number of crops is less than the number of regions (otherwise the problem would be infeasible with a positive lower bound of demand).

Parameters:
The parameters are the same as the previous model.

Variables:
We need binary variables, to identify a crop for a region, as follows:

$$X_{ij} = \begin{cases} 1 & \text{if crop } i \text{ is cultivated in region } j \\ 0 & \text{otherwise} \end{cases}$$

Objective function:
The objective is to maximize the total profit from the crop production (total revenue minus total cost).

$$\text{Maximize } Z = \sum_i \sum_j (P_i Y_{ij} L_j) X_{ij} - \sum_i \sum_j \sum_k (C_{ijk} A_{ijk} L_j) X_{ij} \tag{8.18}$$

Constraints:
Land use: any region must only be planted with one crop. However, one crop can be cultivated in more than one region.

$$\sum_i X_{ij} = 1 \quad \forall j \tag{8.19}$$

Working capital: the total investment for all inputs required in all regions must be less than or equal to the working capital available.

$$\sum_i \sum_j \sum_k (C_{ijk} A_{ijk} L_j) X_{ij} \leq \text{TB} \tag{8.20}$$

Demand: the production of any crop must be greater than (or equal to) the minimum level of demand set and less than (or equal to) the upper bound of demand.

$$LD_i \leq \sum_j (Y_{ij} L_j) X_{ij} \leq \text{UD}_i \quad \forall i \tag{8.21}$$

Nonnegativity constraints: binary

This is a mixed integer linear programming model which is computationally expensive. If one now assumes that a single crop will be cultivated in one region, this does not imply that the entire region may be required. That means a part of the region could be good enough for satisfying the demand. One needs another set of variables to determine the areas of land to be used in each region. The new model is slightly more difficult.

The management may further consider working capital as either a second objective for simultaneous optimization (multi-objective problem) or as a second goal of the problem (goal programming problem). In such cases, more alternative models can be developed. Further crop planning models will be provided in a later chapter.

8.3 Hierarchical Modelling: An Introduction

As reported in the literature, the hierarchical modelling approach is mainly applied to multi-period large-scale production planning and scheduling problems. A hierarchical planning approach (Hax and Meal, 1975, Bitran et al., 1981 and other) partitions a production planning and scheduling problem into a hierarchy of subproblems. In any planning period, the subproblems are solved hierarchically, with the solutions of subproblems from the upper level of the hierarchy imposing constraints on the lower level subproblems. The system only implements the decisions for the immediate period.

Both monolithic and hierarchical production planning (HPP) approaches (Hax and Meal, 1975, Bitran et al., 1981, Hax and Candea, 1984, Saad, 1990) have appeared in the literature. The hierarchical approach has three advantages over the monolithic approach (Graves, 1982). The first advantage is that it is computationally simpler. The second advantage is that the hierarchical approach may require less detailed demand data (i.e., the time-dependent data), in that it needs only aggregate product demand data over the planning horizon, with detailed product demand data over a much shorter scheduling horizon. The monolithic approach usually requires detailed demand data for the entire planning horizon. The third advantage of the hierarchical approach is the extent to which its hierarchical subproblems correspond to the organizational and decision-making echelons in the firm/organization; the consequences of this correspondence are increased interaction between the planning system and the decision makers at each level, and improved coordination of objectives throughout the organization. In addition to these points, Dempster et al. (1981) pointed out that a hierarchical planning approach provides a mechanism for coping with uncertainty in detailed planning. This is due to the fact that much of the data at the detailed level is uncertain at the time aggregate decisions are made. A monolithic (detailed and aggregate combined) model determines the detailed decisions earlier than necessary and hence would be based on less reliable forecasts of the uncertain data. The hierarchical approach postpones the detailed decisions as long as possible so that they can be based on more timely and hence more accurate data.

Hax and Meal's HPP frameworks are very specific for manufacturing industries under various conditions. However, the idea of decomposing the entire planning problem into several smaller planning problems which are to be solved hierarchically will be helpful in developing a decision framework in many productions or service-oriented organizations. In fact, HPP has found its applicability in a wide range of industries like radiator production (Sumichrast and Burch, 1985), fiberglass mat production (Leong et al., 1990), ceramic tile production (Liberatore and Miller, 1985), power generation (Sumichrast and Burch, 1985), iron and steel production (Lin and Moddie, 1989, Tabucanon and Sasiwong, 1991), motor industry (Tsubone and Sugawara, 1987), coal production (Sarker, 2003), and many others.

8.3.1 Hierarchical Modelling in a Manufacturing Context

The basic idea of a hierarchical planning approach is to partition the planning problem into a number of subproblems, and then appropriately linking the resulting subproblems. The number of levels in such process is dependent on the product structure. Hax and Meal (1975) identified three different levels:

1. Individual items—these are the final products to be delivered. A given product may contain a large number of items differing in nonfunctional characteristics such as color, packaging method, labels, etc.
2. Families—a family of items that shares a common, usually major, production setup cost.
3. Types—these are groups of families whose production quantities are to be determined simultaneously.

These three levels are common in many batch-processing manufacturing environments. However, the number of levels depends on the product structure and problem complexity.

The first step in this approach is to allocate available production capacity among different product types by using an aggregate production planning model. The planning horizon is usually 1 year in order to take into consideration the demand fluctuation. The second step is to allocate the production quantities for each type, among families belonging to that type, by disaggregating the results of the aggregate model for only the first period. Finally, the family production allocation is distributed among the individual items belonging to each family.

The mathematical models for a hierarchical decision process in the context of manufacturing are presented below.

8.3.2 Aggregate Model

Parameters:

I	the total number of product types (index i)
T	the length of planning horizon (index t)
L	the length of production lead time
PC_{it}	the unit production cost (excluding labor)
HC_{it}	the inventory carrying cost per unit per period
CR_t	the cost per man-hour
RH_t	the total availability of hours in period t
$\mathrm{PD}_{i,t+L}$	the effective demand for type i during period $t+L$
M_i	the inverse of the productivity rate for type i in hours/unit

Decision variables:

PX_{it}	the number of units to be produced of type i during the period t
$\mathrm{PI}_{i,t+L}$	the number of units of inventory of type i left over at the end of period $t+L$
RT_t	man-hours used during period t

Objective function:
The objective is to minimize the sum of the production, inventory holding, and the labor costs associated.

$$\text{Minimize } Z = \sum_{i}\sum_{t}(\text{PC}_{it}\text{PX}_{it} + \text{HC}_{i,t+L}\text{PI}_{i,t+L}) + \sum_{t}(\text{CR}_t\text{RT}_t) \tag{8.22}$$

Constraints:
The total production for i in period t plus the inventory carried over from the previous period minus the inventory carried over for the next period must be equal to the demand of i in the same period.

$$\text{PX}_{it} - \text{PI}_{i,\,t+L} + \text{PI}_{i,\,t-L} = \text{PD}_{i,\,t+L} \quad \forall i, t \tag{8.23}$$

In any period, the total production hours must be less than or equal to the total man-hours used.

$$\sum_{i} \text{M}_i\text{PX}_{it} \le \text{RT}_t \quad \forall t \tag{8.24}$$

The total man-hours used for production, in any period, must be less than or equal to the total hours available.

$$\text{RT}_t \le \text{RH}_t \quad \forall t \tag{8.25}$$

The final aggregate production planning model becomes

$$\text{Minimize } Z = \sum_{i}\sum_{t}(\text{PC}_{it}\text{PX}_{it} + \text{HC}_{i,\,t+L}\text{PI}_{i,\,t+L}) + \sum_{t}(\text{CR}_t\text{RT}_t)$$

Subject to

$$\begin{aligned}
&\text{PX}_{it} - \text{PI}_{i,\,t+L} + \text{PI}_{i,\,t-L} = \text{PD}_{i,\,t+L} \quad \forall i, t \\
&\sum_{i} \text{M}_i\text{PX}_{it} \le \text{RT}_t \quad \forall t \\
&\text{RT}_t \le \text{RH}_t \quad \forall t \\
&\text{PX}_{it}, \text{PI}_{i,t+L} \ge 0 \quad \forall i, t \\
&\text{RT}_t \ge 0 \quad \forall t
\end{aligned} \qquad \text{Model (8.1)}$$

8.3.3 Family Scheduling Model

The prime condition to be fulfilled at the family scheduling level for a logical disaggregation is the equality between the sum of the different productions under the families in a product type and the amount obtained from the higher level for the product type. This equality condition will guarantee consistent planning between the aggregate production plan and the family disaggregation plan. This is usually achieved by determining the production quantities for each family that minimize the total setup cost among different families.

We now present a family disaggregation model similar to one proposed by Bitran and Hax (1981).

Parameters:

- J^0 the families which trigger during the current period (that implies that the current inventory is not enough to meet the expected demand)
- SC_j the setup cost for family j
- FD_j the forecast demand for family j
- XA_i the total amount to be allocated among all the families belonging to type i. XA_i has been determined by the aggregate planning model and corresponds to the optimum value of the variable PX_{i1} since only the first period result of the aggregate model is to be implemented

Decision variables:
FY_j the number of units of family j to be produced in each run

Objective function:
The objective is to minimize the sum of all setup costs.

$$\text{Minimize } Z = \sum_{j \in J^0} (\mathrm{SC}_j \mathrm{FD}_j)/\mathrm{FY}_j \tag{8.26}$$

Constraint:
The sum of productions of the families in a product type must be equal to its amount dictated by the aggregate model.

$$\sum_{j \in J^0} \mathrm{FY}_j = \mathrm{XA}_i \tag{8.27}$$

In addition, the production quantity for each family must be in between its assigned lower and upper bounds.

8.3.4 Individual Item Scheduling Model

The item disaggregation model, presented below, is similar to Hax and Candea (1984). They proposed a strictly convex knapsack problem for each family j.

Parameters:

- AI_k the available inventory of item k
- ISS_k the safety stock of item k

FD_{kt} the forecast demand for item k in period t

K^0 the set of indices of all items belonging to family j

YD_j the total amount to be allocated for all items belonging to family j as determined by family disaggregation model

Decision variables:
NI_k the number of units to be produced of item k

Objective function:

$$\text{Minimize } Z = \sum_{k \in K^0} \left[\frac{\text{YD}_j + \sum_{k \in K^0} (\text{AI}_k - \text{ISS}_k)}{\sum_{k \in K^0} \sum_{t=1}^{L+1} \text{FD}_{kt}} - \frac{\text{NI}_k + \text{AI}_k - \text{ISS}_k}{\sum_{t=1}^{L+1} \text{FD}_{kt}} \right]^p \tag{8.28}$$

The first term inside the square brackets represents the production run out time for a family and the second represents the run out time for an item belonging to that family. The value of p greater than or equal to 2 will force those quantities to be as close as possible.

Constraint:
The sum of all items produced in a family j must be equal to the number dictated by the family disaggregation model.

$$\sum_{k \in K^0} \text{NI}_k = \text{YD}_j \tag{8.29}$$

The upper and lower bound of an item produced must also be set.

Interested readers may find further information regarding hierarchical planning model for manufacturing context in Hax and Candea (1984). The disaggregation model can be based on time only rather than product structure and time (Sarker, 2003).

8.4 Summary

This chapter has addressed the realism issue since real-world problems are usually large and complex and it is not always possible to include all aspects of reality in the model describing a problem. It has been shown that a problem can be modelled in a number of different ways depending on the portion of reality to be included or excluded. Having addressed the issue of how to formulate a mathematical model, we will, in the next chapter, briefly analyze possible solution approaches for different mathematical models that have so far discussed in this book.

References

Bitran, G.R., Haas, E.A., and Hax, A.C., Hierarchical production planning: A single stage system, *Operations Research*, 29, 717, 1981.

Dempster, M.A., Fisher, M.L., Jansen, L., Lageweg, B.J., Lenstra, J.K., and Rinnooy Kan, A.H.G., Analytical evaluation of hierarchical planning systems, *Operations Research*, 29, 707, 1981.

Graves, S.C., Using Lagrangian techniques to solve hierarchical production planning problems, *Management Science*, 28, 260, 1982.

Hax, A.C. and Candea, D., *Production and Inventory Management*, Prentice-Hall, Englewood Cliffs, NJ, 1984.

Hax, A.C. and Meal, H.C., Hierarchical integration of production planning and scheduling. In Geisler M.A. (Ed.), *Studies in Management Science*, Vol. I, Logistics, North Holland-American Elsevier, New York, 1975, p. 53.

Leong, G.K., Oliff, M.D., and Markland, R.E., Improved hierarchical production planning, *Journal of Operations Management*, 8, 90, 1989.

Liberatore, M.J. and Miller, T., A hierarchical production planning system, *Interfaces*, 15, 1, 1985.

Lin, C.W. and Moddie, C.L., Hierarchical production planning for a modern steel manufacturing system, *International Journal of Production Research*, 27, 613, 1989.

Saad, G.H., Hierarchical production planning system: Extensions and modifications, *Journal of Operational Research Society*, 41, 609, 1990.

Sarker, R.A. (2003) *Operations Research Applications in a Mining Company*, dissertation. de Verlag in Internet, Berlin, Germany.

Sumichrast, R.T. and Rurch, E.E., Understanding hierarchical production planning and scheduling models, *Industrial Management*, 27, 6, 1985.

Tabucanon, M.T. and Sasiwong, B., Production planning and scheduling for an iron and steel production system, *Applied Mathematical Modelling*, 15, 164, 1991.

Tsubone, H. and Sugawara, M., A hierarchical production planning system in the motor industry, *OMEGA: International Journal of Management Science*, 15, 113, 1987.

Section III

Model Solving

9

Solution Approaches: An Overview

9.1 Introduction

In this chapter, the solution approaches for the different mathematical models outlined in this book will be briefly discussed and their relative advantages and disadvantages analyzed in order that users get an exposure to these approaches without having to go into the details of all the algorithms. As has been discussed earlier, this book does not intend to cover the detailed algorithms and their computational aspects.

The solution approaches for solving optimization problems can be divided into two major groups: (1) the classical optimization techniques and (2) the modern heuristic techniques. Although the popular techniques from both groups are discussed, the main emphasis is on the classical techniques and some of the associated available software. In addition to the algorithms, we briefly discuss the algorithmic complexity and complexity classes in this chapter.

9.2 Complexity and Complexity Classes

The term "algorithm" is used interchangeably in many circumstances with the term "solution approach" in optimization. An algorithm is a step-by-step procedure for solving a problem. In some instances, the efficiency of an algorithm may be judged on its complexity, in other circumstances, the problems may be classified in regard to their complexity of solution, known as complexity classes. We discuss these two interesting computational aspects in this section.

9.2.1 Complexity of Algorithms

In any algorithm, the operations that are performed include assigning values to variables, undertaking arithmetic operations (such as addition, subtraction, multiplication, and division), and carrying out comparisons.

The total number of operations may vary with problem size, and even for similar-sized problems; the number can differ from one instance to another. One way of evaluating the complexity of an algorithm is to count, in the worst case, the total number of operations performed. The operation count may be denoted by the big O notation. There are other notations used such as the big Ω (big omega) to specify the lower bound on computational time, and the big θ (big theta) to provide both a lower and an upper bound on an algorithm's performance.

In the big O notation, the constant terms in the complexity analysis are ignored and the most dominant terms in computational effort represent the complexity. For example, if A is an $n \times n$ matrix, Gaussian elimination applied to the system $Ax = b$ requires $n^3/3$ addition and multiplication operations and $n^2/2$ division operations. So the total number of operations required is $(n^3/3) + (n^2/2)$. For a sufficiently large n, the growth of the second term is insignificant in comparison to the growth of the first term. Therefore, the first term dominates the second. Hence, the complexity of Gaussian elimination is $O(n^3)$ for the matrix A. The referred time complexity function measures the upper bound of the rate of growth in solution time as the problem size increases.

An algorithm is recognized as "good"(/of acceptable performance) if its worst-case complexity is bounded by a polynomial function of the problem's parameters. Any such algorithm is said to be a polynomial-time algorithm. Examples of polynomial-time bounds are $O(n^2)$, $O(nm)$, and $O(\log n)$. An algorithm is said to be an exponential-time algorithm if its worst-case computational time grows as a function that cannot be polynomially bounded by the input length. Some examples are $O(2^n)$, $O(n!)$, and $O(n^{\log n})$. The polynomial-time algorithms are always preferred as it is asymptotically superior to any exponential-time algorithm, even in extreme cases. For example, n^{4000} is smaller than $n^{0.1 \log n}$ if n is sufficiently large (i.e., $n \geq 2^{100,000}$). For more on time complexity, see Chapter 3 of Ahuja et al. (1993) and Appendix-B of Martin (1999).

9.2.2 Complexity Classes

The complexity of algorithms for solving optimization problems has been briefly discussed above; however, decision problems may also be classified into varying sets of comparable complexity, called complexity classes. The complexity class P is the set of decision problems that can be solved using a deterministic approach in polynomial time. This class corresponds to a group of problems that can be effectively solved, even in the worst of cases, using an intuitive approach or guess. A problem is identified as NP (nondeterministic polynomial) if its solution (if one exists) can be guessed and verified in polynomial time; nondeterministic means that no particular rule is followed to make the guess.

The NP-complete problems are the toughest problems to be encountered of the NP type, in the sense that they are the ones most likely not to belong to the P class. This is because any problem of type NP can be transformed, in polynomial time, into an instance of a specific NP-complete problem. For instance, the decision problem version of the traveling salesman problem is NP-complete. So any instance of any problem in NP can be transformed mechanically into an instance of the traveling salesman problem, in polynomial time. So, if the traveling salesman problem turned out to be in P, then P = NP! The traveling salesman problem is one of many such NP-complete problems. If any NP-complete problem is in P, then it would follow that P = NP. Unfortunately, many important problems have been shown to be NP-complete and no single expedient algorithmic solution method for any of them has been derived.

For further reading on complexity classes, consult the interesting materials available in Chapter 8 of Papadimitriou and Steiglitz (1998).

9.3 Classical Optimization Techniques

In the classical optimization domain, the solution approach for a given type of model is determined by the problem classification discussed previously. The solution approaches for nonlinear programming models are different from linear programming (LP) models. In this section, the solution approaches to linear, integer, goal, nonlinear, and multi-objective models from the classical optimization point of view are discussed.

9.3.1 Linear Programming

There are basically three popular approaches to solve LP models. They are (1) the graphical method, (2) the simplex method and its variants, and (3) the interior point method. All three methods determine the feasible solution space of the problem, which can be defined as the space bounded by the constraints and variable bounds. The optimal point (the best candidate solution) is then identified from the solution space.

The main purpose of graphical method is to illustrate the concepts of acceptable solutions and search boundary. The method has a practical value when solving small problems with two decision variables and only a few constraints (Turban and Meredith, 1994). The graphical method is demonstrated in Chapter 12. The simplex method is an algebraic procedure. Its underlying concepts are geometric that provide a strong intuitive feeling for how it operates and what makes it an efficient method. The details of the simplex method can be found in Hillier and Lieberman (2005). The simplex method will also be demonstrated in Chapter 12 with an example. The search process in the simplex method evaluates solutions obtained from corner points (intersections of constraint equations) to determine the optimal valued corner point via the boundary/surface of the feasible space (the candidate solution moves between adjacent corner points). In contrast, the candidate solution moves through the interior of the feasible space in the interior point method. The details of the interior point method can be found in Sierksma (1996). The current capability of computing power facilitates the simplex and interior point methods in solving very large LP models without significant difficulty. These two methods are briefly compared below. For further details on the comparison between the two methods, see the article published by Illes and Terlaky (2002).

The simplex method proceeds from one feasible solution (corner point) to another until it arrives at an optimal solution. The number of solutions might be an exponential function of the problem dimension and in the worst case, the simplex method might visit all those solutions. The total number of arithmetic operations for each pivot (iteration) is $O(mn)$ and the simplex algorithm may need to visit up to $\binom{n}{m} = \frac{n!}{m!(n-m)!} > \left(\frac{n}{m}\right)^m$ vertices, which represents at least 2^m whenever $n \geq 2m$ (for details see Papadimitriou and Steiglitz, 1998). Here, n is the number of variables and m is the number of constraints. Although the simplex method is not a polynomial-time algorithm, the most practical problems require solution efforts approximately equivalent to problems of polynomial time complexity. The interior point method enjoys a polynomial time worst-case complexity of $O(m^2n)$. So far, the best known iteration complexity bound is $O(\sqrt{n}L)$ where L is the bit-length of data. The details of L can be found in Martin (1999, pp. 231–232). The interior point method found to be efficient in practice has a weaker, worst-case complexity bound. Thus, there is a gap between theory and practice in the estimate of complexity.

A basic solution is called *primal degenerate* if it contains a zero coordinate. A severe consequence of degeneracy is that the objective value remains the same in subsequent iterations. This property opens up the possibility for cycling, i.e., starting from a certain basic solution, the same set of basic solutions are revisited again and again. Various tools are available to avoid cycling, such as the lexicographic simplex method or the utmost simplex least-index rule. Another implication of degeneracy is that multiple optimal solutions occur. However, all the multiple optimal solutions may not be identified and there is limited or no control at all over which optimal solution will be provided by the simplex method. Degeneracy is not an issue in the interior point method.

The simplex method is very flexible; it allows the implementation of various heuristics to enhance its performance. Although no polynomial time version of the simplex method is known and exponential examples are known for most variants, the method allows the implementation of various heuristics to enhance practical performance. The interior point method is specifically efficient for solving very-large-scale problems. When solving huge problems, possibly involving millions of variables, and solving highly degenerate problems the interior point method outperforms simplex-method-based codes.

Restarting a simplex algorithm (for a modified problem with slightly changed data) from a previous optimal solution allows quick solution of the modified problem. This ability might favor the simplex method when performing well-structured sensitivity and parametric analysis. However, sensitivity analysis in the presence of degeneracy is not easy. The interior point method does not exhibit such efficiency in resolving slightly modified problems as the simplex method does.

For small LP models, the interior point algorithm requires relatively extensive calculations and then, after many iterations, it may only obtain an approximation of the optimal solution. By contrast, the simplex method requires only a few quick iterations to find the optimal solution. For large-scale LP models, the interior point method is much more efficient but provides only an approximate solution.

9.3.2 Integer Programming: The Curse of Dimensionality

It is well known that integer and mixed-integer linear models are difficult to solve. This is due to the fact that the number of alternative solutions increases much faster (usually exponentially) than the size of the problem. That makes the large-scale integer program extremely difficult to solve using the existing algorithms. Let us consider a few examples to show this curse of dimensionality.

Consider an assignment problem of optimally matching 10 candidates to 10 jobs (one-to-one basis). The number of feasible combinations of matching can be calculated as follows:

- Theoretically, there are $n! = 3{,}628{,}800$ different combinations (where $n = 10$).
- If $n = 11$ (a 10% increase in the number of candidates and jobs), the number of alternative matchings increases by 1,100%, i.e., to 39,916,800.
- If $n = 12$ (a 20% increase), the number of alternative matchings increases by 13,200%, i.e., to 480,000,000.

Now, consider the well-known traveling salesman problem (TSP) where a salesman wants to find the least costly (or shortest) route when visiting clients in n different cities, visiting each city exactly once before returning home. Although this problem is very simple to state, it becomes extremely difficult to solve as the number of cities increase.

- In general, for an n city TSP, there are $(n - 1)!$ possible routes the salesman can take.
- The Table 9.1 shows the value of $(n - 1)!$ for several n.

Take an example of a multidimensional 0–1 knapsack problem (Sakawa and Kato, 2003). As can be seen in Table 9.2, the computational time increases exponentially as the problem size increases.

TABLE 9.1

Alternative Solutions in TSP

n	$(n - 1)!$
3	2
5	24
9	40,320
13	479,001,600
17	20,922,789,888,000
20	121,645,100,408,832,000

9.3.3 Integer Linear Program: Solution Approaches

It is known that integer programming (IP) is an extension of the general LP problem. IP borrows many concepts and techniques from LP when developing solution approaches. The common solution approaches for solving IP models are complete enumeration, graphical method, rounding the non-integer solution, branch-and-bound, cutting plane, and branch-and-cut method.

The complete enumeration approach is inefficient and it would be impossible to apply such a method for large-scale problems. Like LP, the graphical approach is for only problems consisting of two variables and few constraints. The IP problem may be solved using the simplex method by relaxing the integrality conditions. Then the non-integer solutions can be rounded off to meet the integer requirements. However, the rounding-off procedure does not guarantee optimality and it could provide infeasible solutions.

The cutting plane and the branch-and-bound techniques both rely on the simplex method by repeatedly solving a series of linear programs. In the branch-and-bound method, the problem is first solved using the simplex method by relaxing the integer requirements. If the integer solutions are found, the process does not need to proceed any further. If not, based on a current non-integer variable, the original problem is branched or

TABLE 9.2

Solutions of Multidimensional 0–1 Knapsack Problem

Problem Instance	Size ($V \times C$)	Computational Time (Seconds)
1	30×10	0.090
2	50×20	0.891
3	100×30	113.00
4	150×40	4440.00
5	200×50	11000.00

Note: V, number of variables; C, number of constraints.

partitioned into mutually exclusive subproblems. Note that in this branching process, part of the continuous solution space is deleted by introducing an integer condition on the non-integer variable under consideration. One then proceeds by solving each of the branched subproblems as LPs using the simplex method. Once the optimum solution of the feasible solution subspace satisfies the integrality constraints then it stops. Otherwise the subproblem must be partitioned further into two subproblems. The branching of subproblems is continued until all subproblems have been explored. The branch-and-bound method is demonstrated in Chapter 12.

In the cutting plane method, the problem is first solved using the simplex method by relaxing the integer requirements. If the integer solutions are found, the process does not need to proceed any further. If not, then one must add an additional cut constraint to the problem and solve it using LP again. The new cut constraint will reduce the feasible space such that regions not containing any integers are removed. This allows the found solution to converge toward an integer solution. Additional cut constraints are generated and the LPs with the added cut constraints are solved until the optimum integer solution is found. The details of branch-and-bound and cutting plane algorithms can be found in Nemhauser and Wolsey (1999). The branch-and-cut approach uses a combination of three techniques: automatic problem preprocessing, the generation of cutting planes, and clever branch-and-bound techniques. The details of this technique can be found in Hillier and Lieberman (2005).

From the above discussion, it is clear that both branch-and-bound and cutting plane methods require applying the simplex method many times to get the optimal integer solution. As indicated earlier, the simplex method is not a polynomial-time algorithm. That means, solving a large-scale IP would be a difficult job using these two methods. However, the good news is that the efficient warm-start capability of simplex method helps to solve the subsequent subproblems very quickly. As the interior point method (IPM) is a polynomial-time algorithm for large-scale LP, you may think to implement the IPM instead of the simplex method in branch-and-bound and cutting plane methods for solving large-scale integer programs. However, the lack of efficient warm-start procedures makes IPM less efficient when solving IP problems.

The mixed-integer cutting plane algorithm and its finite convergence require the assumption that the objective function variable must be an integer. This is an unreasonable assumption, which is unsatisfactory for computational purposes (for more see Nemhauser and Wolsey, 1999). Without this assumption, no finite cutting plane algorithm for mixed-integer LP problems is known. As a consequence, the branch-and-bound method is generally regarded as being the more effective approach than the cutting plane method when solving mixed-integer programming problems of a practical size, and for this reason most of the currently available commercial IP packages are based on the branch-and-bound method.

It should be noted, however, that not all IP problems can be solved by using the branch-and-bound method due to its computational complexity and the current capability of computer resources.

9.3.4 Special Linear Programming Models

Most network problems fall under the category of specially structured linear problems, for example, transportation, transshipment, and assignment type problems. Although almost all specially structured network problems require integer solutions, the corresponding models can be solved using the simplex method (well known for continuous solutions). In these problems, the entries in the constraint matrix are either zero or one. If the right-hand sides (RHSs) of the constraints are integers, then the simplex method would provide integer solutions. This is due to the fact that the fractional solutions are introduced by the matrix operation of the fractional entries in the constraint matrix of the subsequent iterations. In these problems, there are no fractions in the constraint matrix. As a result, the time-consuming branch-and-bound and cutting plane methods are not required for these types of problems.

The specially structured LP models usually have their own specialized solution approaches, which are much more efficient than the simplex method, for example, the transportation simplex for transportation problems, the Hungarian method for assignment problems, and the shortest path algorithm for shortest path problems. The details of these algorithms can be found in Turban and Meredith (1994) and Hillier and Lieberman (2005). Any variation of the special structure, such as adding side-constraints, does not permit the use of specialized solution approaches. In such cases, the general methods such as simplex or branch-and-bound should be used.

Many network problems can be represented as special cases of the generalized network flow problem known as the minimum cost flow problem. The transportation, assignment, transshipment, maximal flow, shortest path, and other problems can be derived as special cases of the minimum cost flow problem. The powerful network simplex method developed for the minimum cost flow problem can then be used as a single method for all those special cases. The network simplex method and further details on the minimum cost flow problem are discussed in Bazaraa et al. (1990).

9.3.5 Goal Programming

The linear goal programming models can be divided into nonpreemptive and preemptive. In the nonpreemptive goal programming problems, all the goals are of roughly comparable importance. In the preemptive goal programming, there is a hierarchy of priority levels for the goals, so that the goals of primary importance receive first-priority attention, those of secondary importance receive second-priority attention, and so on. The nonpreemptive goal programming can be solved using the simplex method since the coefficients of the objective function are known numerical values.

The preemptive goal programming models are solved either by using a sequential or a streamlined method.

In the sequential method, the goal programming problem is solved by solving a sequence of LP models. In the first stage of the procedure, the only goals included in the LP are the first-priority goals, and the simplex method is applied in the usual way. If the resulting optimal solution is unique, the procedure stops immediately without considering any additional goals. If $Z^* = 0$, the deviation variables must be equal to zero (fully achieved). These deviation variables are completely deleted when the second-stage problem is considered. If $Z^* > 0$, the second-stage model simply adds the second-priority goals to the first-stage model (as if these additional goals actually were first-priority goals), and it also adds the constraint that the first-stage objective function equals Z^* (which allows one to again delete the terms involving first-priority goals from the second-stage objective function). In the streamlined procedure, the goal programming problem is solved by solving just one LP model. The penalty weights are set as *big M* for first-priority level and 1 for second-priority level. The details of goal programming approaches can be found in Hillier and Lieberman (2005).

9.3.6 Nonlinear Programming

Nonlinear models are classified as unconstrained or constrained models. The constrained models are further divided into models classified as linearly constrained, quadratic, convex, non-convex, separable, geometric, and fractional programming. Each of the nonlinear models has its own well-defined set of characteristics and requires very specific solution algorithms. That means no one algorithm would solve all classes of nonlinear models. However, many approaches are based on either the concept or the direct extension of the simplex method.

The unconstrained problem approaches are divided into single variable and multivariable problems with and without using derivatives. The one-dimensional search is the backbone of many algorithms for solving a nonlinear programming problem. There are a number of line search procedures described in the literature, for solving unconstrained problems of one variable, with or without using derivatives. A line search method using derivatives assumes differentiability. All the procedures find a sequence of trial solutions that leads toward an optimal solution. At each iteration, the current trial solution is used to conduct a systematic search and culminates by identifying a new improved trial solution. Multidimensional search can also be performed with or without using derivatives. Examples of the multidimensional search without using derivatives are the cyclic coordinate method, the method of Hooke and Jeeves, and Rosenbrock's method. Examples of multidimensional search using derivatives include the steepest decent method and the method of Newton. In addition, the methods using conjugate directions are also used in practice. For more details, see Bertsekas (1995).

The penalty function method is a well-known general approach for solving constrained nonlinear optimization problems. The approach converts the constrained problem into an equivalent unconstrained problem and then solves the problem using a suitable search algorithm. The constraints are placed into the objective function via a penalty parameter in such a way that the parameter penalizes any violation of the constraints. The solution to the penalty problem can be made arbitrarily close to the optimal solution of the original problem by choosing a sufficiently large penalty parameter (μ). However, if we choose a very large μ and attempt to solve the penalty problem, we may get into some computational difficulties of ill-conditioning (Bazaraa et al., 1990). With a large μ, more emphasis is placed on feasibility, and most procedures for unconstrained optimization will move quickly toward a feasible point. Even though this point may be far from optimal, premature termination could occur. As a result of the above difficulties associated with large penalty parameters, most algorithms use penalty functions that employ a sequence of increasing penalty-parameter values. With each new value of the penalty parameter, an optimization (search) technique is employed, starting with the optimal solution corresponding to the previously chosen parameter value. There are many other exiting algorithms for solving nonlinear models such as barrier function, gradient projection, reduced gradient, method of Zoutendijk, and the convex–simplex method. For more details see Bertsekas (1995).

9.3.7 Multi-Objective Models

Many real-world problems involve simultaneous optimization of several incommensurable and often competing objectives. In such problems, there are no single optimal solutions, but rather a set of alternative solutions. These solutions are optimal in the wider sense that no other solutions in the search space are superior to them when all the objectives are considered simultaneously. They are known as *Pareto-optimal* solutions. In multi-objective optimization, the *Pareto* type solutions are expected to provide flexibility for the human decision-maker (Coello, 1999; Coello et al., 2002).

Traditionally, there are several methods described in the OR literature that can be used when solving multi-objective optimization problems as mathematical programming models. Most methods formulate a composite objective function and then repeatedly solve the model to generate a number of alternate solutions. Such methods include weighting, ε-constraint, hybrid, goal attainment, value function, lexicographic ordering, interactive surrogate worth trade-off method, Geoffrin–Dyer–Feinberg method, sequential proxy optimization techniques, Tchebycheff method, step method, reference point method, satisfying trade-off method, light beam search, and the reference direction approach.

The weighting method is very simple and a simple linear combination of the objectives, by varying the weights, generates the trade-off surface. In the ε-constraint method, one of the objective functions is selected to be optimized and all other objective functions are converted into constraints by setting an upper bound to each of them. The hybrid method combines the weighting method and the ε-constraint method. Details of other methods can be found in Miettinen (1999). From our experience, these methods solve the class of problems that generate continuous and monotonically increasing and decreasing Pareto frontiers. However, if the Pareto frontier is noncontinuous (e.g., a set of discretely spaced continuous subregions) or nonuniform (higher density of solutions in one region than in another), it is impossible to find the appropriate Pareto frontier. In addition, none of the above methods treat all the objectives simultaneously, which is a basic requirement in most multi-objective problems. However, there exists a simplex based method for solving linear multi-objective models. For details on the method see Eiselt et al. (1987, Part III, Chapter 2). The appropriate methodologies for simultaneous optimization are discussed in a later section.

9.4 Heuristic Techniques

The word *heuristic*, which originated from Greek, means to discover. In optimization problem-solving, a heuristic is a rule-of-thumb approach that may not guarantee convergence and optimality. However, in most cases, they work well and produce solutions of acceptable quality. The use of a heuristic approach in optimization is not new. However, in the past, heuristics were developed based on the concept of either conventional optimization techniques or traditional artificial intelligence techniques. Nowadays, heuristics are also inspired by biology, physics, neuroscience, and other disciplines. The field of heuristics is growing very rapidly. Some of the widely used heuristics are discussed briefly in this section.

9.4.1 Hill Climbing

Hill climbing is the greediest heuristic yet encountered. The idea is simply not to accept a new solution unless it is better than the best solution found so far. This represents a pure search intensification without any chance for search exploration; therefore, the algorithm is more likely to end up with a local optimum and, therefore, can be very sensitive in regard to the starting point.

9.4.2 Simulated Annealing

Simulated annealing (SA) is a stochastic search algorithm based on the concept called "annealing." Annealing is a heat treatment process. The process involves heating and cooling. Heating a solid up to a point where its atoms can move freely and then cooling it down allows the atoms to rearrange themselves. The cooling rate is vital in the process. Too fast or too slow cooling may not facilitate expected crystal formations.

In complex optimization, many conventional techniques usually get trapped in a local optimum. As a mechanism of escaping a local optimum, SA accepts a lower-quality solution in an iteration with some probability depending on a parameter called "temperature." The algorithm behaves like a random search at high temperature (using a higher probability) and like a greedy hill-climbing at low temperature (with a probability close to zero). In the algorithm, a cooling schedule with an initial temperature must be defined by the user, which is not an easy task. The SA algorithm grows exponentially with respect to the size of the problem. The details of the algorithm can be found in Van Laarhoven and Aarts (1987).

9.4.3 Tabu Search

Tabu search (TS) is an iterative process and it has the capability of escaping local optima like SA. Similar to SA, the neighborhood length is a critical parameter in TS. Unlike the conventional hill-climbing approach, TS may allow lower-quality solutions in any intermediate iteration. TS also forbids reverse moves to avoid cycling. The forbidden movements are recorded in a data structure called a tabu list, which is updated in every iteration. Because of the tabu list, the search is able to escape local optima. However, the performance of TS is sensitive to the size of the tabu list in many practical applications. The detailed algorithm can be found in Glover and Laguna (1997).

TS has been widely and successfully used in combinatorial optimization. However, its use in continuous search space is not common due to the difficulties of performing neighborhood movements in continuous search spaces.

9.4.4 Genetic Algorithms

Genetic algorithms (GAs) are a stochastic search procedure for solving optimization problems that uses ideas from natural evolution and adaptation. The heuristics discussed so far start from a single initial point or solution and move to another single solution. However, GAs start with a randomly generated population (a set of solutions) and then move from one population to another. This process continues until the stopping criteria are met. At each iteration, the new population is generated applying various search operators.

GAs and their general field of evolutionary algorithms (EAs) can be regarded as a population-based version of generate-and-test search. They use search operators like crossover and mutation to generate new solutions, and use evaluation to test which solutions are better. The better solutions are usually selected (using a selection process) to form the next generation of solutions. It can be noted here that there is no strict rule to use crossover and mutation to generate new solutions in EAs. In principle, one may use any search procedure to generate new solutions that will increase the probability of finding a global optimum. This is also true for selection. The details of GAs and their different parameters and operators can be found in Mitchell (1998) and Sarker et al. (2003).

EAs have some advantages over the conventional mathematical programming techniques. Some of the advantages are highlighted below, which justify the use of EAs in solving optimization problems.

- Consideration of convexity/concavity and continuity of functions are not necessary in EAs; however, these attributes of functions are a real concern in most mathematical programming techniques.
- It is not difficult to implement EAs because they do not require any rich domain knowledge. However, domain knowledge can be incorporated into evolutionary computation (EC) techniques.
- EAs are a class of general purpose (domain independent) search methods that strike a remarkable balance between exploration and exploitation of the search space. This property helps to improve the solution by skipping from the local optima, and is particularly useful when solving multi-modal problems.
- The most favorable point of using EAs is that they provide quick approximate solutions. In most cases, EAs make a significant improvement within the first few generations.
- EAs are more suitable for multi-objective optimization, than conventional optimization techniques, because of their capability of simultaneous optimization of conflicting objective functions and generation of a number of alternative solutions in a single run.
- Optimization under changing environments (dynamic or nonstationary or real time) can be handled nicely by EC techniques. EAs are also good for dealing with noisy objective problems.

9.4.5 Ant Colony Optimization

One approach, using natural system concepts for optimization, is known as *ant colony optimization* and is a branch of *swarm intelligence*. Swarm intelligence is a field that studies the emergent collective intelligence of groups of simple agents. In groups of insects, such as ants and bees, that live in colonies, an individual can only do simple tasks on its own whereas the colony's cooperative work is the main reason in determining the intelligent behavior the colony shows.

When ants travel seek out certain targets, such as food sources, they deposit a chemical substance on the ground (path) called a *pheromone*. Over time, the shortest paths tend to show a higher pheromone growth rate. All ants prefer to take those paths where there is a larger amount of pheromone. This intelligent concept can be used for solving optimization problems by applying the algorithm known as ant colony optimization. The details of the algorithm can be found in Dorigo and Sttzle (2004).

9.4.6 Memetic Algorithms

A memetic algorithm uses local search heuristics with a population-based strategy. The algorithm can be seen as a hybrid population-based algorithm. Basically, the local search heuristics are combined with crossover operators. When a local search is combined with GAs (usually known as genetic local search), the algorithm provides a much better performance than GAs alone can do. Any constructive heuristics or exact method may be combined with a population-based algorithm (not necessarily GAs) to develop a memetic algorithm. These algorithms tend to be very suitable for parallel computers and distributed computing systems (including heterogeneous systems).

The method is gaining wide acceptance, in particular, in well-known combinatorial optimization problems where large instances have been solved to optimality and where other metaheuristics have failed. For further reading on the topic, see the paper by Moscato and Norman (1992).

9.4.7 Other Heuristics

There are a number of other heuristics such as immune system, particle swarm optimization, cultural algorithms, and cooperative search. A brief description of these algorithms can be found in Coello et al. (2002).

9.5 Optimization Software

There are many commercial software packages available in the market for solving mathematical models. All software packages have the basic components as shown in Figure 9.1.

The software package is basically the solver engine, which contains one or more algorithms for solving a certain class or a number of different classes of mathematical models, such as the simplex and the interior point algorithms for solving LP models. The following are some of the widely used software packages:

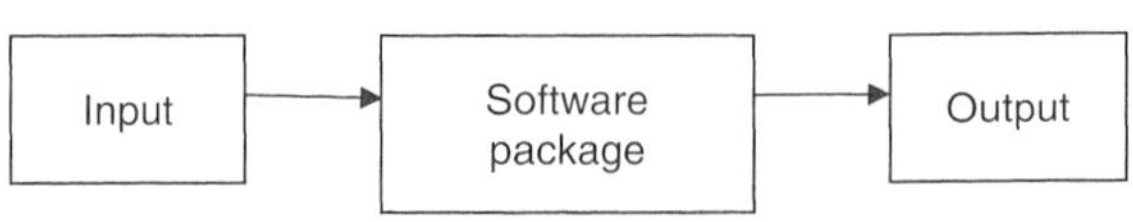

FIGURE 9.1
Model solving using a software package.

- LINGO/LINDO
- GAMS
- OptiMax 2000
- CPLEX and XPRESS
- MINOS
- Solver and Premium Solver
- Others

These software packages are briefly described below.

9.5.1 LINGO/LINDO

LINGO (Linear, Integer, Nonlinear, and Global Optimization) is a comprehensive tool designed to make building and solving linear, nonlinear, and IP models faster, easier, and more efficient. LINGO provides a completely integrated package that includes a powerful language for expressing mathematical models, a full featured environment for building and editing problems, and a set of fast built-in solvers.

LINDO provides powerful solvers for linear, nonlinear (convex and nonconvex), quadratic, quadratically constrained, and integer optimization problems. The use of LINGO and LINDO in solving mathematical problems is demonstrated in Appendix-9A. The free demo and trial versions for LINGO and LINDO can be downloaded from the Web site of LINDO Systems Inc. (http://www.lindo.com/). The demo/trial versions allow the solution of models with 300 variables (including up to 30 integers) and 150 constraints. However, LINGO and LINDO-API allow up to 30 nonlinear variables. The extended commercial version apparently can handle any number of variables (of any type) and constraints.

9.5.2 MPL with OptiMax 2000, CPLEX, and XPRESS

OptiMax 2000, CPLEX, and XPRESS are the solver engines. These packages require additional interfaces called mathematical programming language (MPL) packages for inputting the model, whereas, the user interface for inputting the models is an integral part of LINDO/LINGO and GAMS.

Maximal Software (http://www.maximalsoftware.com/) allows one to download the free student/trial versions of MPL for Windows and OptiMax with CPLEX solver. The student/trial versions are limited in size (300 constraints) but are otherwise fully functional versions of the software. The use of MPL with OptiMax and CPLEX in solving mathematical model is demonstrated in Appendix-9B.

9.5.3 GAMS

The general algebraic modelling system (GAMS) is a high-level modelling system for mathematical programming problems. It consists of a language compiler and a stable of integrated high-performance solvers. GAMS is tailored for complex, large-scale modelling applications, and allows to build large maintainable models that can be adapted quickly to new situations. The package GAMS is specifically designed for modelling linear, nonlinear, and mixed-integer optimization problems. Without a valid GAMS (http://www.gams.com/) license, the system will operate as a free demonstration system with the limitations of 300 variables (including up to 50 discrete variables) and 300 constraints. GAMS is further discussed in Appendix-9C.

9.5.4 Solver and Premium Solver

Solver is an integral part of Microsoft Office Excel. Frontline Systems (http://www.solver.com/) developed the Solvers/Optimizers not only in Microsoft Excel but also in Lotus 1-2-3 and Quattro Pro. The standard Solver can solve linear, integer, and nonlinear optimization models of up to 200 variables and 200 constraints. The Premium Solver products (additional Add-Ins) work with Excel Solver models, offer new features such as Evolutionary Solver, and handle problems of virtually unlimited size. The use of Excel Solver in solving mathematical model is demonstrated in Appendix-9D.

9.5.5 Win QSB

Win QSB (quantitative systems for business) is a windows-based decision-making tool. Win QSB is an educational software tool that allows solving small-scale decision problems. The size of the optimization problems one can handle using Win QSB is approximately similar to LINGO, Solver, or any other student/trial version. The software has neither a commercial version nor a Web site for free download. The software comes with a book called Win QSB by Chang (2003). Further details on Win QSB can be found in Appendix-9E.

9.5.6 MINOS

MINOS (modular incore nonlinear optimization system) is a software package for solving large-scale optimization problems (linear and nonlinear programs). With sufficient memory, MINOS can process large LP models similar to those solved by commercial systems such as CPLEX and OSL. MINOS can also process models with large numbers of smooth nonlinear constraints. The functions need not be convex. MINOS is highly effective for problems with a nonlinear objective function and large numbers of sparse linear constraints (as well as bounds on the variables). It has an optional data input capability from MPS files. It can be linked with GAMS, AMPL, Matlab, and CUTE interfaces. For more details on MINOS, visit http://www.ici.ro/camo/nonlin/minos.htm.

Almost all of these commercial packages (excluding Solver and Win QSB) have been developed using different computer languages and are available on a number of different computer platforms. The input requirements and formats are usually different for different packages, which lead to inconvenience when using different packages for solving a given mathematical model. This aspect is discussed in a later chapter. The outputs are also different from the different packages.

9.6 Summary

We have briefly presented the algorithmic complexity and complexity classes. A number of well-known classical optimization and modern heuristic techniques are discussed. A number of widely used classical optimization-based software packages are described. For ease of implementation, the use of some of these packages is demonstrated in the appendices of the chapter.

References

Ahuja, R., Magnanti, T., and Orlin, J., *Network Flows: Theory, Algorithms and Applications*, Prentice Hall, Englewood Cliffs, NJ, 1993.

Bazaraa, M., Jarvis, J., and Sherali, H., *Linear Programming and Network Flows*, John Wiley & Sons, New York, 1990.

Bertsekas, D., *Nonlinear Programming*, Athens Scientific, Belmont, 1995.

Chang, Y.-L., *Win QSB: Decision Support Software for MS/OM*, version 2.0, John Wiley & Sons, New York, 2003.

Coello, C.A.C., A comprehensive survey of evolutionary-based multiobjective optimization techniques, *Knowledge and Information Systems: An International Journal*, 1, 269–308, 1999.

Coello, C.A.C., Van Veldhuizen, D.A., and Lamont, G.B., *Evolutionary Algorithms for Solving Multi-Objective Problems*, Kluwer Academic Publishers, New York, 2002.

Dorigo, M. and Sttzle, T. *Ant Colony Optimization*, The MIT Press, Cambridge, MA, 2004.

Eiselt, H.A., Pederzoli, G., and Sandblom, C.-L., *Continuous Optimization Models*, Walter deGruyter, Berlin, 1987.

Glover, F. and Laguna, M., *Tabu Search*, Kluwer Academic Publishers, London, 1997.

Hillier, F. and Lieberman, G., *Introduction to Operations Research*, 8th ed., McGraw-Hill, Boston, MA, 2005.

Illes, T. and Terlaky, T. Pivot versus interior point methods: Pros and cons, *European Journal of Operational Research*, 140, 170–190, 2002.

Martin, R.K., *Large Scale Linear and Integer Optimization*, Kluwer, Boston, MA, 1999.

Miettinen, K., *Nonlinear Multiobjective Optimization*, Kluwer's International Series in OR/MS, Kluwer Academic Publishers, Boston, 1999.

Mitchell, M., *An Introduction to Genetic Algorithms (Complex Adaptive Systems)*, MIT Press, Cambridge, MA, 1998.

Moscato, P. and Norman, M.G., A 'memetic' approach for the traveling salesman problem, implementation of a computational ecology for combinatorial optimization on message-passing systems, *Parallel Computing and Transputer Applications*, M. Valero, E. Onate, M. Jane, J.L. Larriba, and B. Suarez, Eds., IOS Press, Amsterdam, 1992, pp. 187–194.

Nemhauser, G. and Wolsey, L., *Integer and Combinatorial Optimization*, John Wiley & Sons, New York, 1999.

Papadimitriou, C. and Steiglitz, K., *Combinatorial Optimization: Algorithms and Complexity*, Dover Publications, New York, 1998.

Sakawa, M. and Kato, K., Genetic algorithms with double strings for 0–1 programming problems, *European Journal of Operational Research*, 144, 581–597, 2003.

Sarker, R., Kamruzzaman, J., and Newton, C., Evolutionary optimization (EvOpt): A brief review and analysis, *International Journal of Computational Intelligence and Applications*, 3, 311–330, 2003.

Sierksma, G., *Linear and Integer Programming: Theory and Practice*, Marcel Dekker, Inc., New York, 1996.

Turban, E. and Meredith, J., *Fundamentals of Management Science*, 6th ed., Irwin McGraw-Hill, Boston, MA, 1994.

Van Laarhoven, P.J.M. and Aarts, E.H.L., *Simulated Annealing: Theory and Applications*, Kluwer Academic Publishers, Dordrecht, 1987.

Appendix-9A LINGO: An Introduction

9A.1 Introduction

LINGO (Linear, Integer, Nonlinear, and Global Optimization) is a mathematical modelling language designed particularly for formulating and solving a wide variety of optimization problems, including linear programming, integer programming, and nonlinear programming. In this section, we introduce LINGO modelling language for solving simple mathematical programming models.

The demo versions of LINGO can be downloaded from LINDO/LINGO Web site. The demo versions are limited in size as indicated in Section 9.5.1 and usually valid for 40 days. However, as per the current policy, the demo license can be extended (at no cost) repeatedly after expiration date.

Let us start with a simple problem such as in Example 3.1. For ease of explanation, in this section, we will let the variables TABLE and CHAIR denote the number of tables and chairs to produce, respectively. The model with the revised variable definitions is as follows:

Maximize $Z = 100\ \text{TABLE} + 60\ \text{CHAIR}$

Subject to

$5\ \text{TABLE} + 2\ \text{CHAIR} \leq 270$	Machining	Model (9A.1)
$4\ \text{TABLE} + 3\ \text{CHAIR} \leq 250$	Sanding	
$3\ \text{TABLE} + 4\ \text{CHAIR} \leq 200$	Assembly	
$\text{TABLE} \geq 0$ and $\text{CHAIR} \geq 0$	Nonnegativity	

To solve the above model using LINGO, open LINGO and choose New from the File menu. Now you are ready to enter the above model in LINGO model window.

9A.2 Inputting Model in LINGO

We tell LINGO we want to maximize an objective function by preceding it with MAX =. Therefore, our objective function is written on the first line of our model window as

```
MAX=100 * TABLE + 60 * CHAIR;
```

Each line in LINGO is terminated with a semicolon. The semicolon is necessary without which LINGO will not solve the model.

Enter the constraints just below the objective function as follows:

```
5 * TABLE + 2 * CHAIR <= 270;
4 * TABLE + 3 * CHAIR <= 250;
3 * TABLE + 4 * CHAIR <= 200;
```

Since most computers do not have less than or equal to keys (≤), LINGO has adopted the convention of using the two character symbol <= to denote ≤. As an alternative, you may simply enter < to mean less than or equal to. In a similar manner, >= or > are used to mean greater than or equal to (≥).

After entering the above expressions and entering comments to improve readability of the model, your model window should look like Figure 9A.1 (also see Figure 9A.4).

An expression may be broken up into as many lines as you want, however, it must be terminated with a semicolon. As an example, we may use two lines rather than just one to represent the objective function as follows:

```
MAX = 100 * TABLE
      + 60 * CHAIR;
```

In the above model, we have entered some comments to improve its readability. Comments begin with an exclamation point (!) and end with a semicolon (;). All text between an exclamation point and terminating semicolon are assumed as nonessential for mathematical model hence ignored by LINGO. Comments can occupy more than one line and can share lines with other LINGO expressions. For example,

```
X1 = 25*X2 + X3/3*X4; !A comment at the end of constraint;
X1 = 25* !A comment in the middle of a constraint; X2 +
    X3/3*X4;
```

So far, we have used all uppercase letters for our variable names. However, LINGO does not distinguish between uppercase and lowercase in

```
! The total profit objective function;
MAX = 100 * TABLE + 60 * CHAIR;

! Machining Constraint;
5 * TABLE + 2 * CHAIR <=  270;

! Sanding Constraint;
4 * TABLE + 3 * CHAIR <=  250;

! Assembling Constraint;
3 * TABLE + 4 * CHAIR <=  200;
```

FIGURE 9A.1
Input model.

variable names. For example, the following variable names would all be considered equivalent:

```
TABLE
Table
table
tABLE
```

LINGO requires variable names begin with an alphabetic character (A–Z) and subsequent characters may be either alphabetic, numeric (0–9), or the underscore (_). Names can be given up to 32 characters in length.

9A.3 Solving the Model

The simple model has already been entered and it is now ready to be solved. To begin solving the model, select the Solve command from the LINGO menu, or press the Solve button on the toolbar at the top of the main frame window. LINGO will begin compiling the model that determines the model's conformity to all syntax requirements. If the syntax conformity test fails, LINGO will provide an error message by listing the line of the model and pointing to the place in the line where it occurred.

9A.3.1 Solver Status Window

If there are no syntax errors during the compilation phase, LINGO will call the appropriate solver module to begin searching for the optimal solution. When the solver module starts, it displays a solver status window on your computer screen similar to the one in Figure 9A.2.

The solver status window is useful for monitoring the progress of the solver module and the size of the model. The Variables box shows the total number of variables in the model. The Variables box also displays the number of integer and nonlinear variables involved in the model.

In variable counts, LINGO does not include any variables for which determines are fixed in value. For instance, consider the following constraints:

```
X1 = 5;
X1 + X2 = 9;
```

LINGO determines from the first constraint that *X*1 is fixed at the value of 5. Using this information in constraint 2, it determines *X*2 is fixed at a value of 4. *X*1 and *X*2 will then be substituted out of the model and they will not contribute to the total variable count.

The Constraints box reports the total constraints and the number of these constraints that are nonlinear. LINGO searches your model for fixed

FIGURE 9A.2
LINGO solver status.

constraints. A constraint is considered fixed if all the variables in the constraint are fixed. Fixed constraints are substituted out of the model and do not add to the total constraint count.

The Nonzeros box reports the total nonzero coefficients in the model and the number of these that appear on nonlinear variables. In a given constraint, there usually appears only a small subset of the total variables. So, one can view the total nonzero coefficient count as a tally of the total number of times variables appear in all the constraints.

The Generator Memory Used box lists the amount of memory LINGO's model generator is currently using from its memory allotment. One may change the size of the generator's memory allotment using the LINGO|Options command. The Elapsed Runtime box shows the total time used so far to generate and solve the model.

When LINGO finishes solving the model, there will be a new window created on your screen titled Solution Report containing the details of the solution. The solution report should appear as Figure 9A.3 (also see Figure 9A.4).

9A.3.2 LINGO Special Features

LINGO's modelling language lets you express the problem in a natural manner that is very much similar to standard mathematical notation. Rather than entering each variable and constraint explicitly, you can express similar

```
Global optimal solution found.
  Objective value:                        5671.429
  Total solver iterations:                3

        Variable            Value          Reduced Cost
          TABLE          48.57143            0.000000
          CHAIR          13.57143            0.000000

             Row     Slack or Surplus        Dual Price
               1         5671.429            1.000000
               2         0.000000            15.71429
               3         15.00000            0.000000
               4         0.000000            7.142857
```

FIGURE 9A.3
Solution report.

variables and constraints in a compact manner. This leads to models that are much easier to maintain and scale up. Another convenient feature of LINGO's modelling language is the data section. The data section allows you to isolate required data from the formulation. In fact, LINGO can even read data from a separate file such as spreadsheet, database, or text file. With data independent of the model, it is much easier to make changes, and there is less chance of error when you develop larger models. To demonstrate the compactness and convenience of LINGO modelling language, we provide an example of transportation problem below.

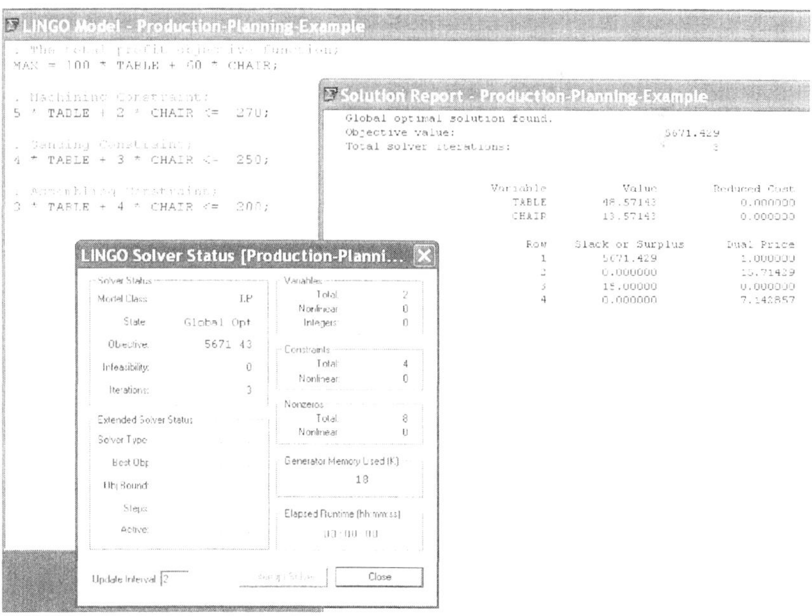

FIGURE 9A.4
Model, solution report, and status windows.

9A.4 Another Example

Suppose the transportation example provided earlier (in Example 4.13), Austral Steel Company now has five plants supplying to seven distribution centers. Since each plant can ship to each distribution center, there are a total of 35 possible shipping paths, or arcs. We will need a variable for each of these arcs to represent the amount shipped on the arc. For this problem, the following data are available:

Plant Capacity Data

Plant	Capacity
1	80
2	75
3	45
4	65
5	40

Distribution Center Demand

Center	Demand
1	40
2	45
3	35
4	50
5	30
6	40
7	25

Shipping Cost per Tonne ($)

	DC1	DC2	DC3	DC4	DC5	DC6	DC7
P1	6	2	6	7	4	2	5
P2	4	9	5	3	8	5	8
P3	5	2	1	9	7	4	3
P4	7	6	7	3	9	2	7
P5	2	3	9	5	7	2	6

9A.4.1 Objective Function

Our objective is to minimize the total shipping costs. We will let the variable *VOLUME_I_J* denote the number of tonnes shipped from warehouse *I* to vendor *J*. Then, if we were to explicitly write out our objective function using scalar variables we would have

```
MIN= 6 * VOLUME_1_1 + 2 * VOLUME_1_2 +
6 * VOLUME_1_3 + 7 * VOLUME_1_4 +
4 * VOLUME_1_5 + . . . + 7 * VOLUME_5_5
+ 2 * VOLUME_5_6 + 6 * VOLUME_5_7;
```

We have shown here only 8 out of the 35 terms in the objective. As one can see, entering such a lengthy formula would be tedious and prone to errors. You can imagine the length of the objective function where the number of plants and distribution centers is in the scale of thousands.

As discussed earlier, we can express this long equation in a much more compact manner as follows:

$$\text{Minimize} \sum_{ij} COST_{ij} \cdot VOLUME_{ij}$$

In a similar way, LINGO's modelling language allows to express the objective function in a form that is short, easy to type, and easy to understand. The equivalent LINGO statement is

```
MIN= @SUM( LINKS(I,J): COST(I,J) * VOLUME(I,J));
```

In words, this says to minimize the sum of the shipping COST per tonne times the VOLUME of rod (in tonne) shipped for all LINKS between the plants and distribution centers. The following table compares the mathematical notation to the LINGO syntax for the above objective function.

Mathematical Notation	LINGO Syntax
Minimize	MIN =
$\Sigma_i\Sigma_j$	@SUM(LINKS(I,J):
$COST_{ij}$	COST(I,J))
•	*
$VOLUME_{ij}$	VOLUME(I,J))
End of equation	;

9A.4.2 Constraints

There are two sets of constraints in this model: the demand constraints and the capacity constraints. In the first demand constraint for distribution center 1, we need to sum up the shipments from all the plants to distribution center 1 and set them equal to distribution center 1's demand of 40 widgets. Thus, if we were using scalar based notation, we would need to construct the following:

```
VOLUME_1_1 + VOLUME_2_1 + VOLUME_3_1 +
VOLUME_4_1 + VOLUME_5_1 = 40;
```

You would then need to type six additional demand constraints, in a similar form, to cover all seven distribution centers. Again, as you can see, this would be a tedious and error-prone process. However, LINGO modelling language can simplify your task. Using mathematical notation, all seven demand constraints can be expressed in the single statement:

Σ_i $\mathrm{VOLUME}_{ij} = \mathrm{DEMAND}_j$, for all j

The corresponding LINGO modelling statement appears as follows:

```
@FOR (DISTRIBUTION_CENTER(J):
@SUM (PLANT(I): VOLUME(I, J)) = DEMAND(J));
```

This LINGO statement replaces all seven demand constraints. In simple words, this says for all DISTRIBUTION CENTERS, the sum of the VOLUME shipped from each of the PLANTS to a given distribution center must equal the corresponding DEMAND of the distribution center. As you can see, this statement closely resembles the mathematical notation above as shown in the following table:

Mathematical Notation	LINGO Syntax
for all j	*@FOR(DISTRIBUTION_CENTER(J):*
Σ_i	*@SUM(PLANT(I):*
VOLUME_{ij}	*VOLUME(I, J))*
=	=
DEMAND_j	*DEMAND(J))*
End of equation	;

In standard mathematical notation, the five capacity constraints can be expressed as follows:

Σ_j $\mathrm{VOLUME}_{ij} <= \mathrm{CAP}_i$, for all i

The equivalent LINGO statement for all capacity constraints would be

```
@FOR( PLANT( I):
  @SUM(DISTRIBUTION_CENTER ( J):
    VOLUME( I, J))<= CAPACITY( I));
```

In simple words, this says, for each member of the set PLANTS, the sum of the VOLUME shipped to each of the DISTRIBUTION CENTERS from that plant must be less than or equal to the CAPACITY of the plant.

9A.4.3 Complete LINGO Model

Putting together everything we have done so far yields the following LINGO model:

```
MODEL:
  MIN= @SUM( LINKS( I, J):
    COST( I, J) * VOLUME( I, J));
  @FOR(DISTRIBUTION_CENTER ( J):
    @SUM(PLANT ( I): VOLUME( I, J))
      = DEMAND( J));
  @FOR(PLANT ( I):
    @SUM(DISTRIBUTION_CENTER ( J): VOLUME( I, J))
      <= CAPACITY( I));
END
```

However, we still need to define the data of the model, which are inputted in the sets section and the data section.

9A.4.4 Defining the Sets

In modelling some real-life situations, there are typically one or more sets of related objects. Examples of such things are plants, warehouses, customers, vehicles, and employees. Usually, if a constraint applies to one member of a set, then it will apply equally to each other member of the set. This simple idea is at the core of the LINGO modelling language. LINGO allows defining the sets of related objects in the sets section. The sets section begins with the keyword SETS: on a line by itself and ends with ENDSETS on a line by itself. Once the set members are defined, LINGO has a group of set looping functions (e.g., @FOR), which apply operations to all members of a set using a single statement.

In the case of our transportation example, we have constructed the following three sets:

- Plants
- Distribution centers
- Shipping arcs from each plant to distribution center

The three sets are defined in the model's sets section as follows:

```
SETS:
  PLANT/P1 P2 P3 P4 P5/: CAPACITY;
  DISTRIBUTION_CENTER/DC1 DC2 DC3 DC4 DC5 DC6 DC7/:DEMAND;
  LINKS(PLANT, DISTRIBUTION_CENTER): COST, VOLUME;
ENDSETS
```

The second line indicates that the set *PLANT* has members *P1, P2, P3, P4*, and *P5*, each with an attribute called *CAPACITY*. The seven distribution centers are defined similarly in the third line, each with an attribute called *DEMAND*. The final set, titled *LINKS*, represents the 35 links in the transportation network. Each link has a *COST* and a *VOLUME* attribute associated with it. The syntax used to define this set differs from the previous two. By specifying

```
LINKS(PLANT, DISTRIBUTION_CENTER)
```

we are expressing to LINGO the *LINKS* set is derived from the *PLANT* and *DISTRIBUTION_CENTER* sets. In this case, LINGO generates each ordered (plant, distribution center) pair. Each of these 35 ordered pairs becomes a member in the *LINKS* set. To help clarify this, we list selected members from the LINKS set below.

Member Index	Shipping Arc
1	P1 -> DC1
2	P1 -> DC2
3	P1 -> DC3
...	...
34	P5 -> DC6
35	P5 -> DC7

One can explicitly enter each member arc in the sets section, but LINGO saved us considerable work by generating the set members for us.

9A.4.5 Inputting the Data

LINGO allows the user to isolate data within the data section of the model. In our transportation example, we have the following data section:

```
DATA:
  CAPACITY = 80 75 45 65 40;
  DEMAND = 40 45 35 50 30 40 25;
  COST = 6  2  6  7  4  2  5
         4  9  5  3  8  5  8
         5  2  1  9  7  4  3
         7  6  7  3  9  2  7
         2  3  9  5  7  2  6;
ENDDATA
```

The data section begins with the keyword *DATA:* on a line by itself and ends with *ENDDATA* on a line by itself. Both the *CAPACITY* attribute of the set *PLANT* and *DEMAND* attribute of the set *DISTRIBUTION_CENTER* are initialized in a straightforward manner. The *COST* attribute of the two-dimensional set *LINKS* is a bit tricky. When LINGO initializes a multidimensional array in a data section, it increments the outer index first. Thus, in our example, *COST(P1, DC1)* is initialized first, followed by *COST(P1, DC2)*, through to *COST(P1, DC7)*, and then *COST(P2, DC1)*, and so on.

Putting together the data section, the sets section, the objective, and the constraints, the complete LINGO model is as follows:

```
MODEL:
  !A 5 Plants 7 Distribution centers Transportation Problem;
  !Here, the total demand is less than the total capacity;

SETS:
  PLANT/P1 P2 P3 P4 P5/: CAPACITY;
  DISTRIBUTION_CENTER/DC1 DC2 DC3 DC4 DC5 DC6 DC7/:DEMAND;
  LINKS( PLANT, DISTRIBUTION_CENTER): COST, VOLUME;
ENDSETS

!The objective;
MIN= @SUM( LINKS( I, J):
  COST( I, J) * VOLUME( I, J));
!The demand constraints;
@FOR ( DISTRIBUTION_CENTER(J):
  @SUM ( PLANT(I):
    VOLUME(I, J)) = DEMAND(J));
!The capacity constraints;
@FOR(PLANT ( I):
  @SUM(DISTRIBUTION_CENTER ( J):
    VOLUME( I, J)) <= CAPACITY(I));
!Here is the data;
DATA:
  CAPACITY= 80 75 45 65 40;
  DEMAND= 40 45 35 50 30 40 25;
  COST= 6  2  6  7  4  2  5
        4  9  5  3  8  5  8
        5  2  1  9  7  4  3
        7  6  7  3  9  2  7
        2  3  9  5  7  2  6;
ENDDATA
END
```

In this example, we have placed the data directly within the model's data section. LINGO also has the ability to import data from external sources. More specifically, a LINGO model can import data from external text and spreadsheet files. This can be a very useful feature when solving a model in which data changes frequently.

9A.5 LINGO Syntax

In the above sections, we have provided two simple examples of solving mathematical programming models using LINGO. More examples and LINGO syntax can be found in later chapters. For advanced LINGO syntax and details on the use of external data sources, one should consult LINGO user manual (available online). Some of the frequently used LINGO syntax are given below:

Expression for	LINGO Syntax
Binary variable (X is binary)	@BIN(X)
Integer variable (X is integer)	@GIN(X)
Real value, positive, or negative (X is unrestricted)	@FREE(X)
Logical operators	#NOT#
	#NE#...
Mathematical functions	@ABS(X)
	@EXP(X)
	@LOG(X)
	@SQRT(X)
	@SIN(X)
Probability function	@PBN(P, N, X)
Interface function	@FILE('filename')
Miscellaneous function	@IF(logical_condition, true result, false result)

Appendix-9B MPL: An Introduction

9B.1 Introduction

MPL (mathematical programming language), a product of Maximal Software, Inc., for Windows is a modelling system that allows the model developer to formulate optimization models in an efficient way. MPL is capable of handling problems with millions of variables and constraints. MPL works with optimization engines, such as CPLEX and XPRESS, and many other industrial strength solvers.

The student/trial versions of MPL for Windows and CPLEX can be downloaded from Maximal Software site. The student/trial versions can be used for up to 300 constraints and for limited time only:

9B.2 Use of MPL

To demonstrate the use of MPL, let us consider the following example:

Maximize $Z = 100\ \text{TABLE} + 60\ \text{CHAIR}$

Subject to

Constraint		
$5\ \text{TABLE} + 2\ \text{CHAIR} \leq 270$	Machining	Model (9B.1)
$4\ \text{TABLE} + 3\ \text{CHAIR} \leq 250$	Sanding	
$3\ \text{TABLE} + 4\ \text{CHAIR} \leq 200$	Assembly	
$\text{TABLE} \geq 0$ and $\text{CHAIR} \geq 0$	Nonnegativity	

To model the above problem using MPL, you need to go through the following steps:

- Start the MPL application.
- Choose New from the File menu to create a new empty model file.
- Choose Save As from the File menu and save the file as *ProdPlan.mpl*.
- Enter the model in MPL language. The model editor in MPL is a standard text editor, which allows you to enter the model and

perform various editing operations on the model text. In the model editor, enter the following model formulation:

```
TITLE ProductionPlanning;
MAX
  Profit = 100 Table1 + 60 Chair1;
SUBJECT TO
  5 Table1 + 2 Chair1 <= 270;
  4 Table1 + 3 Chair1 <= 250;
  3 Table1 + 4 Chair1 <= 200;
END
```

It is mandatory to use semicolon, ;, after the objective function and after each constraint. This allows MPL to separate the constraints. The spacing used between entries and lines in MPL is not rigid. It is recommended to use spaces and extra lines to make the model formulation easier to read and understand.

- When you have finished entering the model, choose Save from the File menu to save the model.

After you have entered the formulation in the model editor, you can check the model for syntax errors. If MPL finds a mistake it will report in the Error Message window showing the erroneous line in the model, along with a short explanation of the problem. The cursor is automatically positioned at the error in the model file, with the offending word highlighted.

- To check the syntax, choose Check Syntax from the Run menu. If there are no errors found, MPL will respond with a message stating that the syntax of the model is correct. If there is an error in the model, MPL will display the Error Message window.

Suppose, in the model editor, the semicolon at the end of the first constraint is missing as follows:

```
SUBJECT TO
  5 Table1 + 2 Chair1 <= 270   !note the missing semicolon
  4 Table1 + 3 Chair1 <= 250;
  3 Table1 + 4 Chair1 <= 200;
```

If you choose Check Syntax from the Run menu now, MPL will go through the model and find the missing semicolon when it is parsing the second constraint. When you press the OK button you are returned to the model editor. The cursor will automatically be positioned at the location where MPL found the error, which in our case is at the "<=" in the second constraint. Now you can reenter the semicolon for the first constraint and if you check the syntax again, MPL will report back with message that the syntax is correct.

The next step is to solve the model *ProdPlan.mpl*. To solve the model follow the steps below:

- Choose Solve CPLEX from the Run menu or press the Run Solve button in the Toolbar.
- When solving the model, the status window appears; providing you with information about the solution progress.

If everything goes well, MPL will display the message Optimal Solution Found. After solving the model, MPL automatically creates a standard solution file containing various elements of the solution to the model. The solution file is created with the same name as the model file but with the extension .sol. In our case, the solution file will be named *ProdPlan.sol*. After you have solved the model, you can display the solution file in a view window by pressing the View button at the bottom of the status window.

The first part of the solution file contains various statistics for the model, such as the filename, date and time the model was solved, the solver used, the value of the objective function, and the size of the model. The next part of the solution file contains the solution results.

9B.3 Using Vectors and Indexes in MPL

Indexes define the domains of the model, encapsulate the problem dimensions, and make it easy to quickly adjust the problem size. *Vectors* are basically aggregations of elements in the model that share common characteristics and purpose. Once you have defined the indexes in a model, you can use them to define vectors that contain the data, variables, and constraints for the model.

Data Vectors are used when the coefficients or statistics from the problem come in lists or tables of numerical data. When an index is defined, there is one value for each value of the index and the data vectors allow you to group collections of data together in the model. *Variable Vectors* can be defined in a similar way as data vectors, to form a collection of variables defined over a certain index. *Constraint Vectors* are defined over indexes, which MPL expands to a collection of simple constraints when generating the model. A vector constraint can be defined in this way, over a number of indexes, such as periods and products. *Data Constants* are used in the model to aid readability, and make the model easier to maintain. They are assigned a specific value, but not defined over a specific index.

One of the operations usually done on vectors is to sum or add all the values for each element of the vector. This is done in MPL by using the keyword SUM surrounding the vector expression to be added together.

The expression is prefixed by a list of indexes, over which the sum hinges. The sum expression contains a single variable vector per term, possibly multiplied by one or more data vectors. Examples are

$$SUM(product: Price * Sales);$$
$$SUM(product, month: ProdCost * Produce);$$

9B.4 A Product-Mix Model with Three Variables

Consider a production planning problem with three products, which will be called P1, P2, and P3. The problem is to distribute production capacity between products, and to determine the production level, given the demand. We have a total of 22 production days available. Other data for these products are given in the table below:

Parameters	Product P1	Product P2	Product P3
Production rate (unit/day)	600	400	500
Demand (unit)	4800	4000	5200
Production cost ($/unit)	70	50	60
Selling price ($/unit)	150	120	130

For these three products we are going to create an index, and then create a variable vector that represents how much of each of these products need to be produced.

Let us create MPL formulation for this problem. You should now have an empty editor window where you can enter your MPL formulation. To enter the title for the ProdPlanX model, type in the following text in the model editor:

```
TITLE
  Production_PlanningX;
```

The first section in an MPL model is usually the INDEX section where you define the indexes for the model. In the model editor, directly below the title, add an INDEX section with a definition for the product index as follows:

```
INDEX
  product:= (P1, P2, P3);
```

The next section is usually the DATA section where you define the Data Vectors and Data Constants for the model. Directly below the index definition, add a DATA section with a definition for the data vector S_Price followed by the index [product] inside brackets:

```
DATA
  S_Price[product] := (150.00, 120.00, 130.00);
```

Following the declaration, you enter an assignment symbol ":=" and then a list of numbers containing the prices for each product. Surround the list with parentheses and separate each number by a space, comma, or both. There should be a semicolon after each data vector definition to separate it from the other definitions in the model.

The problem description also listed data for the demand, production cost, and production rate. To enter this data, add the following definitions to the DATA section directly below the S_Price data vector:

```
Demand[product]    := (4800, 4000, 5200);
Prod_Cost[product] := (70.00, 50.00, 60.00);
Prod_Rate[product] := (600, 400, 500);
```

Add the following data constant definition for the production days available directly below the Prod_Rate data vector:

```
Prod_Days_Avail := 22;
```

Usually, the next section will be the VARIABLES section where you define the variables for the model. We define a vector variable named Produce over the index product. In the model editor, directly below the data definitions add the VARIABLES section with a definition for the Produce vector variable as follows:

```
VARIABLES
  Produce[product] -> Prod;
```

The name that appears after the "->" (read becomes) sign is an optional abbreviation of the vector name used to offset the name size limitations of most LP solvers. This allows you to use long and descriptive names for variables in your model.

In deriving the objective function, the calculations for total revenue and total cost can be found separately as macros. In the model editor, directly below the variable definition, enter the following macrodefinitions in the MACROS section.

```
MACROS
  Total_Revenue := SUM(product: S_Price * Produce);
  Total_Cost    := SUM(product: Prod_Cost * Produce);
```

The model part in MPL is where you define the actual objective function and the constraints for the model. You will be using the macros, defined above, to create the objective function by referring to the macronames where you need to use the summations. In the model editor, enter the keyword MODEL to note the start of the model part, followed by the definition for the objective function:

```
MODEL
  MAX Profit = Total_Revenue - Total_Cost;
```

The formula for the objective function is quite simple, as we are using macros to contain to actual summations.

Following the objective function you need to define the constraints for the model in the SUBJECT TO section. In the model editor, add the SUBJECT TO heading, followed by the production capacity constraint definition:

```
SUBJECT TO
  Prod_Capacity -> PCap:
    SUM(product: Produce/Prod_Rate)
      <= Prod_Days_Avail;
```

The BOUNDS section is used to define the upper and lower bounds on the variables in the model. Enter the following upper bound on the Produce variable in the BOUNDS section.

```
BOUNDS
  Produce <= Demand;
END
```

Please note that in most LP models all variables have an implied lower bound of zero. These lower bounds are handled automatically by MPL and do not have to be specified unless they are nonzero.

At the end of the model enter the keyword END to note the end of the model. After you have finished entering the model, you should save it by choosing Save from the File menu.

A full listing of the model formulation is shown below:

```
TITLE
  Production_PlanningX;
INDEX
  product := (P1, P2, P3);
DATA
  S_Price[product]    := (150.00, 120.00, 130.00);
  Demand[product]     := (4800, 4000, 5200);
  Prod_Cost[product]  := (70.00, 50.00, 60.00);
  Prod_Rate[product]  := (600, 400, 500);
  Prod_Days_Avail     := 22;
```

```
VARIABLES
  Produce[product] -> Prod;
MACROS
  Total_Revenue := SUM(product: S_Price * Produce);
  Total_Cost    := SUM(product: Prod_Cost * Produce);
MODEL
  MAX Profit=Total_Revenue - Total_Cost;
SUBJECT TO
  Prod_Capacity -> PCap:
    SUM(product: Produce/Prod_Rate) <= Prod_Days_Avail;
BOUNDS
  Produce <= Demand;
END
```

For further syntax and examples visit the Web site of Maximal Software, Inc.

Appendix-9C GAMS: An Introduction

9C.1 Introduction

The general algebraic modelling system (GAMS) is a modelling approach for mathematical programming and optimization. It consists of a language compiler and an integrated high-performance solver. GAMS (http://www.gams.com/) is available for use on personal computers, workstations, mainframes, and supercomputers.

9C.2 An Example

Consider the transportation problem of Appendix-9A. The mathematical model of the problem can be formulated as follows:

Indices:

$i =$ plants
$j =$ distribution centers

Given data:

$S_i =$ supply at plant i
$D_j =$ demand at distribution center j
$C_{ij} =$ shipping cost per unit from plant i to distribution center j

Decision variables:

$x_{ij} =$ amount to be shipped from plant i to distribution center j

Objective function:

$$\text{Minimize} \sum_i \sum_j C_{ij} x_{ij}$$

Constraints:

$$\sum_j x_{ij} \leq S_i \quad \forall i \text{ supply limitation at plant } i$$

$$\sum_i x_{ij} \geq D_j \quad \forall j \text{ demand at distribution center } j$$

$$x_{ij} \geq 0 \quad \forall i, j$$

The GAMS model for this problem is shown below:

```
SETS
  I  plants/P1, P2, P3, P4, P5 /
  J  distributioncenters/DC1, DC2, DC3, DC4, DC5, DC6, DC7 /;
PARAMETERS
  S(I)   capacity of plant i
     /   P1  80
         P2  75
         P3  45
         P4  65
         P5  40 /
  D(J)   demand at distribution center j
     /   DC1  40
         DC2  45
         DC3  35
         DC4  50
         DC5  30
         DC6  40
         DC7  25 /;
TABLE C(I,J)   transport cost per unit
       DC1  DC2  DC3  DC4  DC5  DC6  DC7
  P1   6     2    6    7    4    2    5
  P2   4     9    5    3    8    5    8
  P3   5     2    1    9    7    4    3
  P4   7     6    7    3    9    2    7
  P5   2     3    9    5    7    2    6;
VARIABLES
  X(I,J)   shipment quantities
  Z        total transportation costs;
POSITIVE VARIABLE X;
EQUATIONS
  COST          define objective function
  SUPPLY(I)     observe supply limit at plant i
  DEMAND(J)     satisfy demand at distribution center j;
COST ..         Z  =E=  SUM( (I,J), C(I,J)*X(I,J));
SUPPLY(I) ..    SUM(J, X(I,J))  =L= S(I);
DEMAND(J) ..    SUM(I, X(I,J))  =G= D(J);
MODEL TRANSPORTATION /ALL/;
SOLVE TRANSPORT USING LP MINIMIZING Z;
```

Let us now explain the sections of the above GAMS model with the syntax used. In the SET section, GAMS specifies indices in a straightforward way: declare and name the set (here, *I* and *J*), and enumerate their elements.

In the PARAMETER section, data are entered as indexed parameters *S(I)* and *D(J)*, and values simply are listed. GAMS allows explanatory text (shown in lower case) throughout your model, as you develop it. Your comments are automatically incorporated into the output report, at the appropriate places.

Data can be entered in convenient table form. In the TABLE section, GAMS lets you input data in their basic form—transformations are specified algebraically.

A constant simply can be declared as a SCALAR, and its value specified. We do not have any scalar in this model. When data values are to be calculated, you first declare the parameter (i.e., give it a symbol and, optionally, index it), then give its algebraic formulation. GAMS will automatically make the calculations. For example,

```
C(I,J)=F * T(I,J)/1000;
```

Where
F is a scalar
$T(I, J)$ is parameter.

In the VARIABLES section, decision variables are expressed algebraically, with their indices specified. From this general form, GAMS generates each instance of the variable in the domain. Variables are specified as to type: FREE, POSITIVE, NEGATIVE, BINARY, or INTEGER. The default is FREE. The objective variable (z, here) is simply declared without an index.

In the EQUATIONS section, objective function and constraint equations are first declared by giving names to these. Then their general algebraic formulae are described. GAMS now has enough information (from data entered and from the algebraic relationships specified in the equations) to generate each individual constraint statement. An extensive set of tools enables you to model any expression that can be stated algebraically: arithmetic, indexing, functions, and exception-handling log (e.g., if-then-else and such-that constructs). The type of constraint is expressed as follows:

$=E=$ indicates equal to

$=L=$ indicates less than or equal to

$=G=$ indicates greater than or equal to

The model is given a unique name (here, TRANSPORTION), and the modeller specifies the equations that should be included in this particular formulation. In this case, we specified ALL, which indicates that all equations are part of the model. This would be equivalent to MODEL TRANSPORTION/COST, SUPPLY, DEMAND/. This equation selection enables you to formulate different models within a single GAMS input file, based on the same or different given data.

The solve statement tells GAMS which model to solve, selects the solver to use (here an LP solver), indicates the direction of the optimization, either MINIMIZING or MAXIMIZING, and specifies the objective variable.

The full GAMS output report contains many aids for interpreting and diagnosing your model. You also can modify the output format to suit your particular purposes. GAMS is capable of generating management-style reports. In GAMS, data manipulation, model specification, and report writing are done in one single environment.

Appendix-9D Excel Solver: An Introduction

9D.1 Introduction

We can solve many mathematical programming models using Excel Solver. The solver module is an integral part of standard Microsoft Office Excel product. However, you may need to add in Solver (if it does not exist in the drop-down menu under Tool) using the function Add-Ins under Tool menu.

Solver is readily available, since MS Office has become an integral part of our life, and it is very easy to use. However, it has a number of limitations, which we will discuss later.

9D.2 Solving Linear Programs with Solver

Enter the input data and construct relationships among data elements in a readable and easy to understand way. When building these information and relationships, think ahead about the optimization model you will be developing. Make sure there is a cell in the spreadsheet for each of the following:

- The quantity you wish to optimize (maximize or minimize)
- Every decision variable
- Every quantity that you might want to constrain (include both sides of the constraints)

If you do not have any particular initial values you want to enter for your decision variables, you can start by just entering a value of 0 (zero) in each decision variable cell.

Consider the following LP example:

Maximize $Z = 100\text{ TABLE} + 60\text{ CHAIR}$

Subject to

$5\text{ TABLE} + 2\text{ CHAIR} \leq 270$	Machining	Model (9D.1)
$4\text{ TABLE} + 3\text{ CHAIR} \leq 250$	Sanding	
$3\text{ TABLE} + 4\text{ CHAIR} \leq 200$	Assembly	
$\text{TABLE} \geq 0$ and $\text{CHAIR} \geq 0$	Nonnegativity	

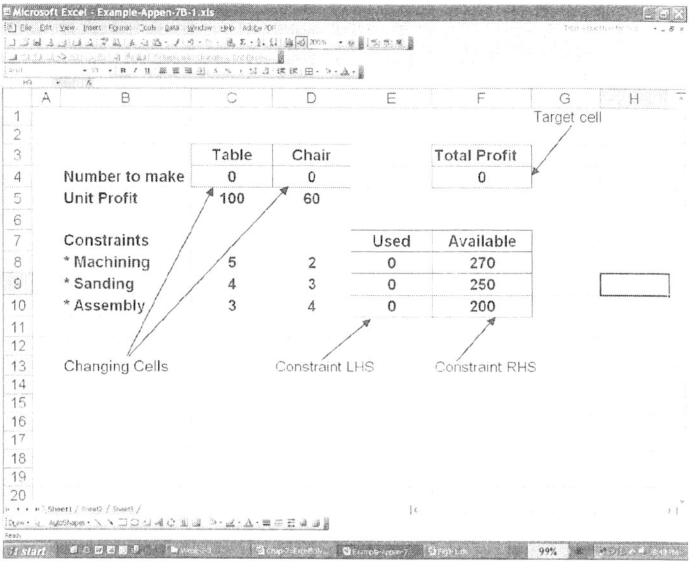

FIGURE 9D.1
Input data.

Enter the required information in a plain Excel sheet as Figure 9D.1. The required functions/equations to represent the objective function and constraints' left-hand side (LHS) in terms of Excel expressions are shown in Figure 9D.2. You may

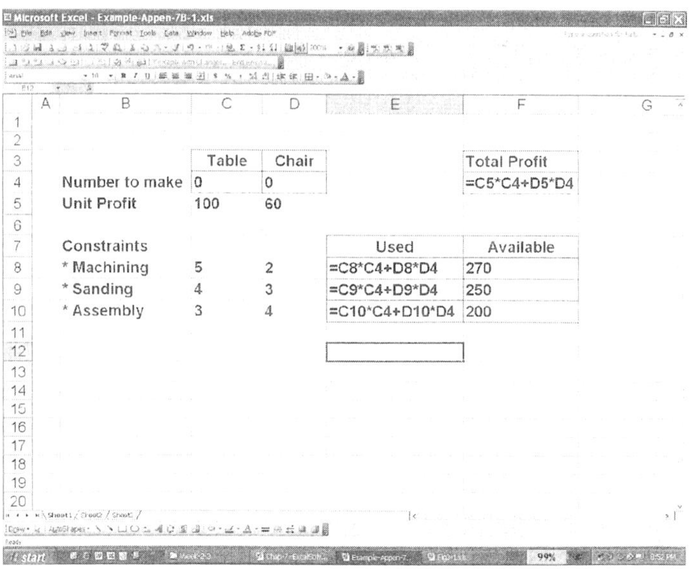

FIGURE 9D.2
Input data with the equations/functions.

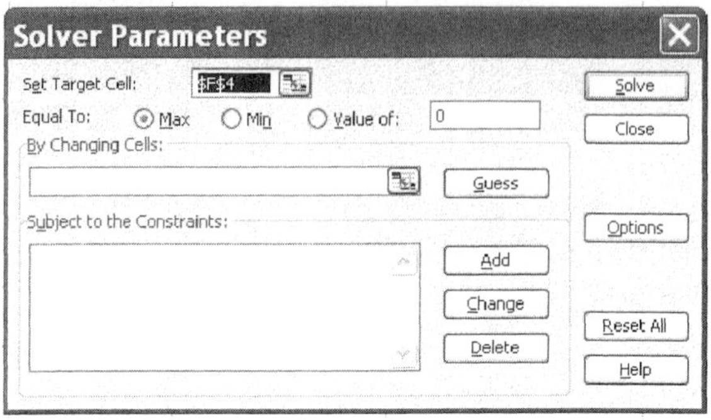

FIGURE 9D.3
Setting target cell and function type in Solver dialogue box.

organize the layout differently and write the text differently. However, the functions/equations should be represented correctly in the Excel sheet. For LP, you should try to always use the SUMPRODUCT function (or SUM) for the objective function and constraints, instead of simple product and sum function we used in the example, as this guarantees that the equations will be linear.

Now select Solver from the drop-down menu under Tool. It will open up a Solver dialogue box.

9D.2.1 Defining the Target Cell (Objective Function)

To select the cell you wish to optimize, select the Set Target Cell window within the Solver dialogue box, and then either type the address of the cell or click on the cell you wish to optimize (see Figure 9D.3). Choose either Max or Min depending on whether the objective is to maximize or minimize the target cell.

Note:

- The target cell should contain an equation that defines the objective and depends on the decision variables.
- The target cell must be a single cell (there can only be one objective).

9D.2.2 Identifying the Changing Cells (Decision Variables)

You next tell Excel which cells are decision variables. That means the cells that Excel is allowed to change when optimizing. Move the cursor to the By Changing Cells window, and either

- drag the cursor across all cells you wish to treat as decision variables (see Figure 9D.4), or
- type the addresses of every cell you wish to treat as a decision variable, separating them by commas.

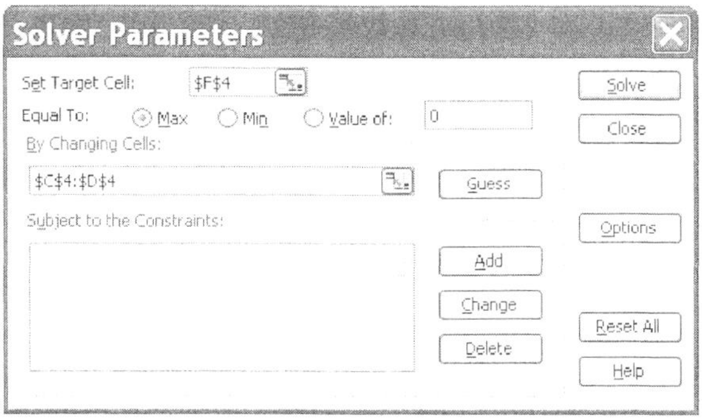

FIGURE 9D.4
Setting changing cells.

If you wish to use the dragging method, but the decision variables do not all lie in a connected rectangle in the spreadsheet, you can drag them in one group at a time by

- dragging the cursor across one group of decision variables,
- putting a comma after that group in the By Changing Cells window,
- dragging the cursor across the next group of decision variables, and so on.

9D.2.3 Adding Constraints

To begin entering constraints, click on the Add button to the right of the constraints window. A new dialogue box will appear. The cursor will be in the Cell Reference window within this dialogue box. Then

- click on the cell that contains the quantity you want to constraint (constraint LHS), or
- type the cell address that contains the quantity you want to constrain as shown below (Figure 9D.5).

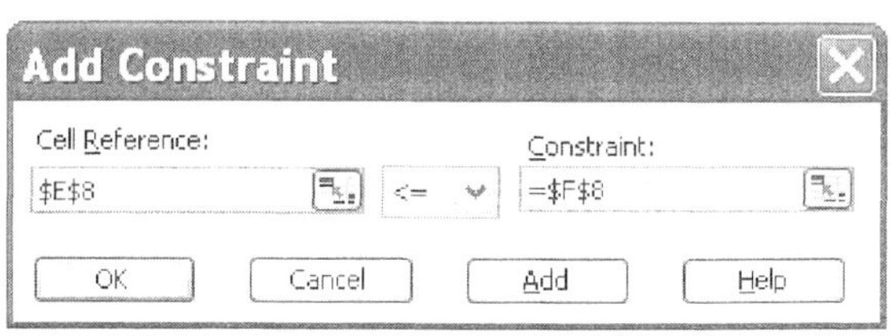

FIGURE 9D.5
Adding constraint.

The default inequality that first appears for a constraint is <=. To change this,

- click on the arrow beside the "<=" sign and
- select the inequality (or equality) you wish from the list provided.

Notice that you may also force a decision variable to be an integer or binary (that is either 0 or 1) using this window. We will use this feature later in the book.

After setting the inequality, move the cursor to the Constraint window and

- click on the cell you want to use as the constraining value for that constraint (constraint RHS),
- type the number or the cell reference you want to use as the constraining value for that constraint, or
- type a number that you want to use as the constraining value.

After you are satisfied with the constraints,

- click the Add button if you want to add another constraint, or
- click the OK button if you want to go back to the original dialogue box.

After adding constraints, our example problem will be as follows (Figure 9D.6):

You may define a set of similar constraints (e.g., all <= constraints, or all >= constraints) in one step if they are in adjacent rows. Simply select the range of cells for the set of constraints in both the Cell Reference and Constraint window. An example of such entry is given below (Figure 9D.7).

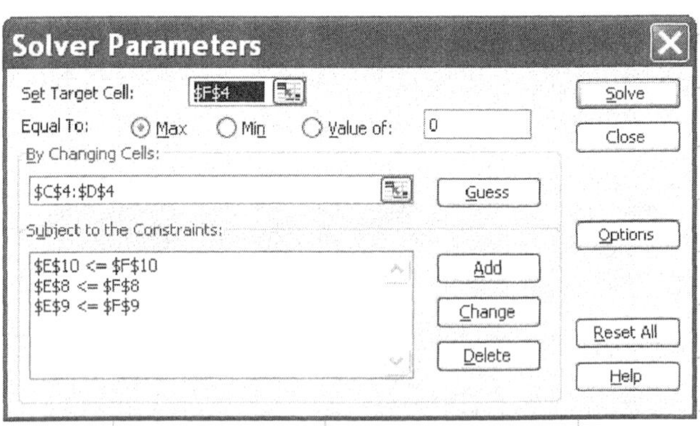

FIGURE 9D.6
After setting constraints.

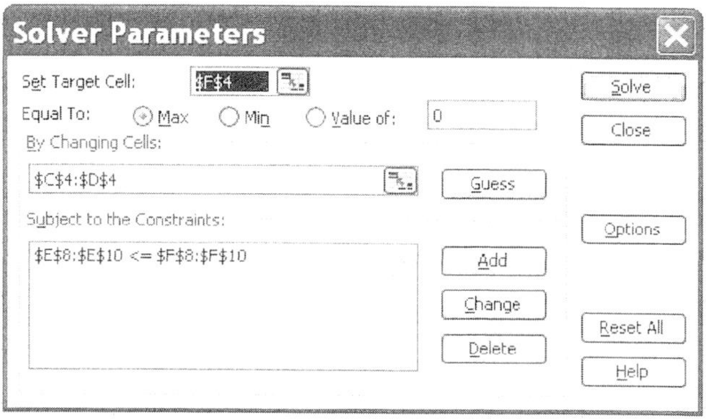

FIGURE 9D.7
After setting all constraints together.

9D.2.4 Some Important Options

The Solver dialogue box now contains the optimization model, including the target cell (objective function), changing cells (decision variables), and constraints.

Once you are satisfied with the optimization model you have set, there is one more very important step. Click on the Options button in the Solver dialogue box, and click in both the Assume Linear Model and the Assume Nonnegative box (see Figure 9D.8).

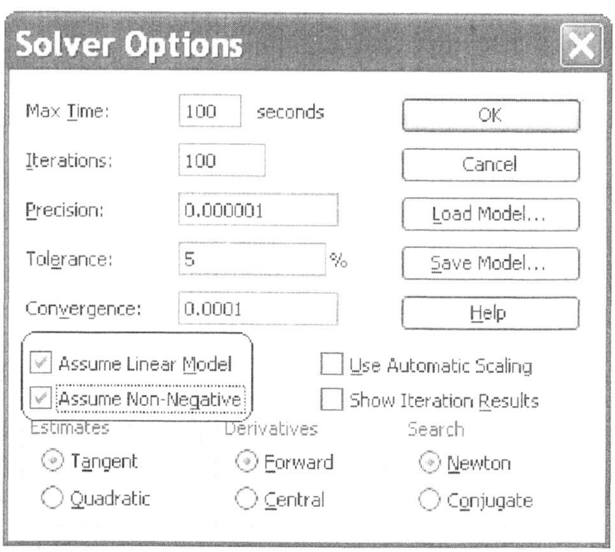

FIGURE 9D.8
Setting linear model and nonnegativity.

The Assume Linear Model option tells the Excel Solver that it is a linear program that is being solved. This speeds the solution process, makes it more accurate, and enables the more informative sensitivity report. The Assume Nonnegative box adds nonnegativity constraints to all of the decision variables.

9D.2.5 The Solution

After setting up the model, and selecting the appropriate options, it is time to click Solve. When it is done, you will receive one of four messages:

- "Solver found a solution. All constraints and optimality conditions are satisfied." This means that Solver has found the optimal solution.
- "Cell values did not converge." This means that the objective function can be improved to infinity. You may have forgotten a constraint (perhaps the nonnegativity constraints) or made a mistake in a formula.
- "Solver could not find a feasible solution." This means that Solver could not find a feasible solution to the constraints you entered. You may have made a mistake in typing the constraints or in entering a formula in your spreadsheet.
- "Conditions for Assume Linear Model not satisfied." You may have included a formula in your model that is nonlinear. There is also a slim chance that Solver has made an error. (This bug shows up occasionally.)

If Solver finds an optimal solution, you have some options (see Figure 9D.9). First, you must choose whether you want Solver to keep the optimal values in the spreadsheet (you usually want this one) or go back to the original numbers you typed in. Click the appropriate box to make your selection. You also need to choose what kind of reports you want. For your analysis, you will often want to select Sensitivity Report. Once you have made your selections, click on OK. To view the sensitivity report, click on the Sensitivity Report tab in the lower-left-hand corner of the window.

As you can see, Solver produces three reports: Answer, Sensitivity, and Limits. These reports for the example problem are shown in Figures 9D.10 through 9D.12.

As shown in Figure 9D.10, the value of the objective function (target cell) is \$5671.428571. The value of the decision variable (adjustable or changing cells) in the optimal solution is 48.57142857 Tables and 13.57142875 Chairs. To get integer numbers, you need to declare each of changing cell integer using constraint setting step discussed earlier. In this report, Solver also provides constraint LHS values (cell value), constraints' status, and slack.

The sensitivity report (Figure 9D.11) is divided into two parts. The first part is for adjustable (changing) cells, which reports final value of variable,

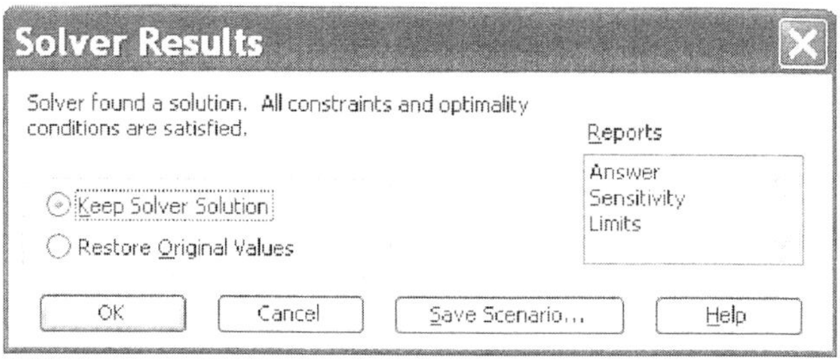

FIGURE 9D.9
Solver results dialogue box.

Microsoft Excel 11.0 Answer Report
Worksheet: [Example-Appen-7B-1.xls]Sheet1
Report Created: 29/08/2006 3:49:13 PM

Target Cell (Max)

Cell	Name	Original Value	Final Value
$F$4	Number to make Total Profit	0	5671.428571

Adjustable Cells

Cell	Name	Original Value	Final Value
$C$4	Number to make Table	0	48.57142857
$D$4	Number to make Chair	0	13.57142857

Constraints

Cell	Name	Cell Value	Formula	Status	Slack
$E$8	* Machining Used	270	$E$8<=$F$8	Binding	0
$E$9	* Sanding Used	235	$E$9<=$F$9	Not Binding	15
$E$10	* Assembly Used	200	$E$10<=$F$10	Binding	0

FIGURE 9D.10
Answer report.

Microsoft Excel 11.0 Sensitivity Report
Worksheet: [Example-Appen-7B-1.xls]Sheet1
Report Created: 29/08/2006 3:49:13 PM

Adjustable Cells

Cell	Name	Final Value	Reduced Cost	Objective Coefficient	Allowable Increase	Allowable Decrease
$C$4	Number to make Table	48.57142857	0	100	50	55
$D$4	Number to make Chair	13.57142857	0	60	73.33333333	20

Constraints

Cell	Name	Final Value	Shadow Price	Constraint R.H. Side	Allowable Increase	Allowable Decrease
$E$8	* Machining Used	270	15.71428571	270	30	170
$E$9	* Sanding Used	235	0	250	1E+30	15
$E$10	* Assembly Used	200	7.142857143	200	30	38

FIGURE 9D.11
Sensitivity report.

Microsoft Excel 11.0 Limits Report
Worksheet: [Example-Appen-7B-1.xls]Limits Report 1
Report Created: 29/08/2006 3:49:13 PM

Cell	Target Name	Value
$F$4	Number to make Total Profit	5671.428571

Cell	Adjustable Name	Value	Lower Limit	Target Result	Upper Limit	Target Result
$C$4	Number to make Table	48.57142857	0	814.2857143	48.57142857	5671.428571
$D$4	Number to make Chair	13.57142857	0	4857.142857	13.57142857	5671.428571

FIGURE 9D.12
Limits report.

reduced cost, objective coefficient, and allowable increase and allowable decrease for objective coefficient. The reduced cost is the increase in the objective function value per unit increase in the value of a zero-valued variable. The reduced cost is zero here as both the variables are at the positive level in the optimal solution. The allowable increase and decrease specify the range of the cost coefficients in the objective function for which the current solution (value of the variables in the optimal solution) will not change.

The second part of the sensitivity report is for constraints, which reports the constraint LHS value, shadow price, RHS value (as data entered), and allowable increase and decrease. The final value column reports the usage of the resource (constraint LHS) in the optimal solution. The shadow price is the change in the value of the objective function per unit increase in the RHS of the constraint—that is $\Delta Z = (\text{shadow price}) \times (\Delta \text{RHS})$. The allowable increase and decrease define the range of values of the RHS for which the shadow price is valid and hence for which the new objective function value can be calculated (Not the range for which the current solution will not change). The Constraint RHS column indicates the current value of the RHS of the constraint (the amount of the resource available).

The limits report (Figure 9D.12) is self-explanatory.

For further details on solver use help menu in MS Office Excel and visit Frontline Systems (http://www.solver.com/).

Appendix-9E Win QSB: An Introduction

9E.1 Introduction

Win QSB (quantitative systems for business) is a windows-based decision-making tool. It contains a number of modules covering almost all the basic techniques of operations research and management science. Although Win QSB is an educational software tool, it can conveniently be used for solving small-scale decision problems. The size of the optimization problems one can handle using Win QSB is approximately similar to LINGO, Solver, or any other student/trial version. The software has neither a commercial version nor a Web site for free download. The software comes with a book called Win QSB by Chang (2003). It is very simple, interactive, and easy to use. In this section, we briefly demonstrate the use of linear and integer programming module.

9E.2 Problem Solving with Win QSB

Let us consider a simple problem such as in Example 3.1, which was demonstrated in Appendix 9A as below:

Maximize $Z = 100\ \text{TABLE} + 60\ \text{CHAIR}$

Subject to

$$
\begin{array}{lll}
5\ \text{TABLE} + 2\ \text{CHAIR} \le 270 & \text{Machining} & \\
4\ \text{TABLE} + 3\ \text{CHAIR} \le 250 & \text{Sanding} & \text{Model (9A.1)} \\
3\ \text{TABLE} + 4\ \text{CHAIR} \le 200 & \text{Assembly} & \\
\text{TABLE} \ge 0 \text{ and } \text{CHAIR} \ge 0 & \text{Nonnegativity} &
\end{array}
$$

To solve the above model using Win QSB, first choose Linear and Integer Programming from Win QSB and then choose New Problem from the FILE menu. It will open an LP–ILP Problem Specification window as shown in Figure 9E.1.

Now you are ready to enter the model in Win QSB. After entering the problem title, number of variables, and number of constraints, if you click OK, it will bring a matrix style model entry form as given in Figure 9E.2.

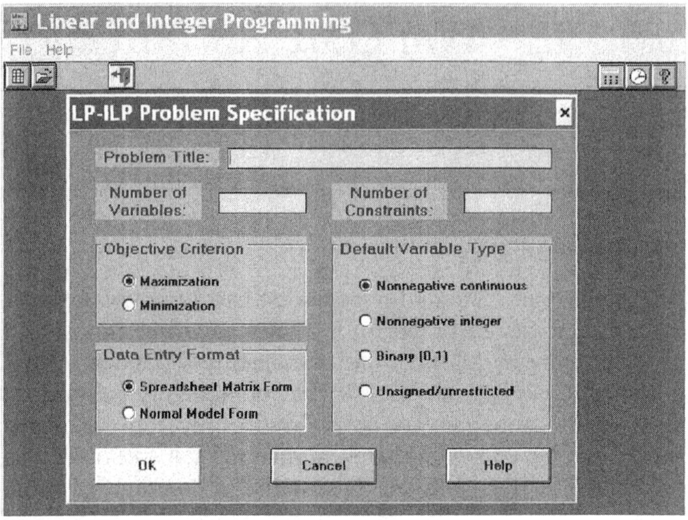

FIGURE 9E.1
LP–ILP problem specification window.

Variable -->	X1	X2	Direction	R. H. S.
Maximize				
C1			<=	
C2			<=	
C3			<=	
LowerBound	0	0		
UpperBound	M	M		
VariableType	Continuous	Continuous		

FIGURE 9E.2
LP–ILP model entry matrix.

Variable -->	TABLE	CHAIR	Direction	R. H. S.
Maximize	100	60		
Machining	5	2	<=	270
Sanding	4	3	<=	250
Assembly	3	4	<=	200
LowerBound	0	0		
UpperBound	M	M		
VariableType	Continuous	Continuous		

FIGURE 9E.3
Entering model.

16:12:04		Saturday	September	02	2006		
Decision Variable	Solution Value	Unit Cost or Profit c(j)	Total Contribution	Reduced Cost	Basis Status	Allowable Min. c(j)	Allowable Max. c(j)
1 TABLE	48.5714	100.0000	4,857.1430	0	basic	45.0000	150.0000
2 CHAIR	13.5714	60.0000	814.2857	0	basic	40.0000	133.3333
Objective	Function	(Max.) =	5,671.4290				
Constraint	Left Hand Side	Direction	Right Hand Side	Slack or Surplus	Shadow Price	Allowable Min. RHS	Allowable Max. RHS
1 Machining	270.0000	<=	270.0000	0	15.7143	100.0000	300.0000
2 Sanding	235.0000	<=	250.0000	15.0000	0	235.0000	M
3 Assembly	200.0000	<=	200.0000	0	7.1429	162.0000	230.0000

FIGURE 9E.4
Win QSB solution report.

After editing the names of variables and constraints and entering model data, we get the input window similar to Figure 9E.3.

Now you choose Solve the Problem from Solve and Analyze menu. It provides a message that the problem has been solved and optimal solution is achieved. The solution report is shown in Figure 9E.4.

Further details on the use of Win QSB can be found in Chang (2003).

9E.3 Reference

Chang, Y.-L., *Win QSB: Decision Support Software for MS/OM*, version 2.0, John Wiley & Sons, New York, 2003.

10

Input Preparation and Model Solving

10.1 Introduction

Once a mathematical model has been formulated for a given problem, the next stage of the decision-making process (as of Figure 2.1, Chapter 2) is to solve the model. In solving a model, one has to either select an appropriate solution approach from existing methodologies or develop a new approach. In some situations, the available solution approach or software package influences the development of a particular class of model. Otherwise, the model cannot be solved. A mathematical model usually dictates what data are needed for solving that model and its implementation. However, the data may not be available in the right form as required by the model. It is usually necessary to either prepare the data as required by the model or modify (reformulate) the model because of unavailability of data. Sometimes, the preprocessing of some data may simplify the mathematical model to be developed.

Although data preparation is a crucial factor in solving any optimization model, this topic has not been addressed very well in the literature. In fact, very little has been reported in the literature about data preparation and analysis. In this chapter, the basic concept of data preparation is discussed and some examples of problem solving are demonstrated.

10.2 Data and Data Collection

Data express or represent certain things about events, activities, and objects. Data can be numeric, alphanumeric, figures, signals, sounds, or images. Data are recorded and stored for use in the decision-making process. However, an individual piece of data may not convey any specific meaning.

The model developer must know what data are available or could be made available that would be used in the developed model for a given decision-making problem or process. Some of these data are standard and readily available in most organizations such as historical sales data, production capacity, and manpower availability. Some data can be made available if required such as future demand, price, future availability of resources, and manpower through estimation or forecasting.

Raw data can be collected manually or by instruments and sensors. Data are composed using the following methods:

- Time series studies: Time series data usually vary with time such as product demand and price. Depending on the requirements, these data are collected at discrete intervals over a long period or over a known period.
- Company registers: Traditionally, all companies keep records of certain items on a daily or transactional basis such as inventory level and material/product purchased. These records are usually preserved for future use.
- Surveys: Data can be collected from a group of people using questionnaires, surveys, and interviews, such as from customer surveys of product quality in a shopping center or from telephone surveys investigating respondents' behaviors, habits, or preferences.
- Observations: Collecting data using video cameras. For example, the photographing of every vehicle passing through a red light in a busy downtown intersection.
- Soliciting information from experts: Collecting data using interviews with domain or subject matter experts, for example, estimating the project duration for a new project.
- Sensors and scanners: Data can also be collected using barcodes or radio-frequency identification codes, for example, the use of price tags and security tags.

The data must be collected, preserved, and analyzed to make them useful to the model and any other decision-making process. The quality of data is extremely important as the quality determines the usefulness of data as well as the quality of model solution based on these data. However, the data quality issue is often overlooked. As a result, there are frequent complaints from data users, such as the data are not correct, or timely, or that the required data do not exist. To minimize these complaints the following steps may be taken:

- During data collection, one must ensure that the data are of the right type, of the right format, and within the specified limits. Specially developed software packages may be used to check these problems while conducting data capture.
- Predict the data that may be needed in the future and collect them in advance.
- Generate new data or estimate them using appropriate techniques. Data can also be generated by using sampling and representative statistical distributions, or through simulation.

10.3 Data Type

The data required by a mathematical model could be one or more of the following types:

Stationary data: Technical data such as existing machine capacity, storage capacity, transportation costs and quantities, machine reliability, etc., which are fixed or constant for a number of future periods, can be termed as stationary data. Examples of stationary data are given in Table 10.1.

These data could be time varying in many situations. Accounting data such as cost, revenue, overheads, taxes, insurance, and planning environment data such as the length of the planning horizon, number of periods, and length of each period are usually considered as stationary data.

Time-varying data: Data such as future period demand, budget, and price, which may vary from one period to another can be recognized as time-varying data. These data could be known at the time of modelling (in advance) or may be required to be generated or estimated. Examples of time-varying data are provided in Table 10.2.

Forecasted data: Data such as future market demand, price, and production volume that must be forecasted for the model are known as forecasted data. Depending on the data precision requirements, these data may be generated using models varying from very simple to highly sophisticated forecasting/mathematical techniques. Suppose the demand of a product follows a simple linear relationship, which can be found from time series data. An example of forecasting in such a situation is shown below. Suppose a relationship can be expressed as the following equation:

$$Y_t = a + bt \tag{10.1}$$

where

Y_t = demand forecast for period t
a, b = model parameters
t = period

Most forecasting techniques require historical data as input to generate future forecasts. Historical data are discussed in the following subsection.

TABLE 10.1

Stationary Demand and Capacity Data

Data	Period								
	1	2	3	4	5	6	7	...	30
Demand (tonnes)	40	40	40	40	40	40	40	...	40
Production capacity (tonnes/day)	50	50	50	50	50	50	50	...	50

TABLE 10.2

Time-Varying Demand and Capacity Data

Data	Period								
	1	2	3	4	5	6	7	...	30
Demand (tonnes)	30	20	40	50	15	60	20	...	65
Production capacity (tonnes/day)	40	30	50	30	50	40	35	...	50

Due to the shortage of historical data and other necessary information, the model data may have to be generated from subject matter experts' opinions or from trends or expected patterns identified by domain experts.

Historical data: Time series data such as historical sales figures, manpower usage, and maintenance history are known as historical data. The details of time series data may be found in Turban and Meredith (1994). These data must be systematically collected for a reasonable period and then prepared as per the requirements of the models or forecasting techniques to be used.

Stochastic data: In some cases, data/situations are expressed as either simple probabilities or as probability distributions. Although probability distributions are not normally used as input to deterministic modelling, they are an important part of stochastic programming. Probability distributions may be applied in deterministic modelling when using a number of discrete points in conjunction with binary variables.

Data from surveys and interviews: Where data are not available, for example, the demand for a new product or the estimated duration of a new project, they may be generated from results of conducted surveys or interviews using domain experts. However, the preparation of questionnaire instruments for surveys and interviews and the abstraction of data from those surveys and interviews are not always easy, in particular, if the questions are open ended. Details of how to prepare questionnaires and compile such data can be found in Neuman (2003).

Text: In the modelling of optimization problems, sometimes we need information expressed in plain text, for example, answering questions or providing opinions such as yes/no responses or as positive/negative answers. It was seen in Chapter 4 that one can model such situations using binary variables. Text information involving multiple options and opinions can also be analyzed using binary and discrete variables.

10.4 Data Preparation

To make better use of available information, it must be collected systematically, preserved in an appropriate form and medium, and prepared in an appropriate form as per the model or system requirements. The amount of information to be collected for a given project and the data-collection format may vary from organization to organization. Although the data-preparation process may seem very simple, it can be a tedious task as it may involve huge volumes of information. Before preserving or using data, it is expected that all data will have been verified with great care.

Data cleaning is an important task when we use collected data for solving any model or addressing any decision-making process. It is basically a process for checking and treating for errors and unusual figures. The errors must be corrected and the unusual figures need to be considered as to being either excluded from the data set because of reasons of being irrelevant to the process or being considered as extreme data and may require some auxiliary treatment. Note that inappropriate data may lead to misleading solutions. Data cleaning and preparation is a complex and time-consuming task.

The quality of solutions of any model is highly correlated to the appropriate identification of data requirements and the data preparation. To understand the data-identification and data-preparation processes in solving optimization problems, consider the following example.

Example 10.1: A simple logistics problem

The South-West Pacific has been hit by an enormous tsunami, which has devastated several South-West Pacific nations. Australia has been asked to provide emergency food, clothing, and relief support. The nations seeking help are Kiribati, Tuvalu, Vanuatu, and Tonga. The Australian Government is willing to respond to those nations' requests and provide relief assistance. How, as a newly recruited coordinator for the relief program would you start the planning process?

One of the first phases would be to identify the possible sources of supplies and the carriers to transport the supplies. In many cases, this information can be found in the respective government departments or charitable organizations, through their contact officers, Web sites, publications, and in their libraries or archives. From such previous projects, you are aware that supplies can be obtained from all capital cities (such as Perth, Adelaide, Melbourne, Sydney, and Brisbane) situated on the mainland. Assume in this case that the Australian Government tasks the Royal Australian Air Force (RAAF) to move the supplies to the devastated nations. The RAAF after considering what supplies to move, decide on using Hercules C130 aircraft.

What data do you think you require in order to make an appropriate delivery plan and how would you obtain that data?

From the initial problem-description, it is clear that there are a number of sources for supplying the required goods and a number of destinations for receiving the deliveries of goods. These goods will be transported from the sources to the destinations based on the availability of goods at each source and the requirements at each destination. The transport carrier has been identified and can be considered as ready for transportation. It looks like a logistics/transportation type problem. To solve the problem using the mathematical modelling approach, different types of data would be required. These are briefly discussed below.

10.4.1 Data Requirements

The requirement of each item, for each of the tsunami-affected nations, is estimated based on the level of damage and affected population size. The estimations are performed by the government of the affected nation or government of the donor country or relief agency. To be on the safe side, it is usually the trend to overestimate the relief requirements. Suppose, the number of people affected in one of the trouble nations is n thousands. From the demographic data, the average number of people in the different age-groups can easily be determined for those people. Based on the age-groups, they will require clothing of different sizes and food of different types and quantities. If we estimate the number of pieces of clothing per person and the number of days people require food supplies, it is not difficult to calculate the overall requirements of clothing and food. A simple example for calculating food required for a given nation is shown below.

A_j = fraction of population in each age-group j (where $\Sigma_j A_j = 1$)

F_{ij} = daily average amount of food item i required per person of age-group j

N = affected population size

D = number of days supplies required

$$\text{The daily food requirements for item } i = N\sum_j A_j F_{ij} \tag{10.2}$$

$$\text{The daily food requirements for all items} = N\sum_i \sum_j A_j F_{ij} \tag{10.3}$$

$$\text{Total food requirements for } D \text{ days} = D \cdot N\sum_i \sum_j A_j F_{ij} \tag{10.4}$$

That means the final requirement is the sum of all individual requirements. Initially, the amount of food, clothing, and relief support required, for each nation, is expressed in weights and volumes. Later, they may be converted into equivalent Hercules C130 loads. An example of requirements is Kiribati requires eight Hercules C130 loads of food, five Hercules C130 loads of clothing, and two Hercules C130 loads of other goods.

Availability: There must be some supplies (food, clothing, and relief support) available at each source. If there are no supplies at a source, then that source should not be considered further in the modelling. That information is usually available and supplied by the government or relief agency. The agencies may collect goods from individual donors and buy from open markets to meet their target supplies. The total availability of supplies (for each item) is usually higher than the total requirement. However, that constraint is not necessary as one would supply relief up to the level of his or her ability. An example of availability is Perth can offer 15 Hercules C130 loads of food, 10 Hercules C130 loads of clothing, and 6 Hercules C130 loads of other goods. However, these Hercules loads must have detailed age-wise food and clothing information. Otherwise, an incorrect mix of relief supplies may be delivered to a given nation. Alternatively, each item may be packaged in boxes and organized as either pallet loads or requirements of each nation.

Transportation: The movement will be carried out using Hercules C130 aircraft. The number of Hercules C130 aircraft available for this purpose and their existing locations are important for the proposed transportation plan. If there are sufficient numbers of Hercules C130s available for this operation, then one trip by a number of Hercules C130s may be enough to transport all goods. However, if the Hercules C130 numbers are restricted due to unavailability of appropriate personnel involved in Hercules C130 operations then there will be a constraint relating to the number of available aircraft. Another factor is related to the issue of sufficient aircraft fuel being available for the operation. The speed of Hercules C130 aircraft is also a factor for transportation time calculations.

Distance: The flight distances from each source to each destination are required. This information can be gained from the RAAF or there are a number of sites on the Internet, which provide distances based on longitude and latitude information. The Hercules C130 aircraft may need to refuel at a different location before flying on to their South-West Pacific destinations. This information is very important for refueling facility planning and also for mathematical modelling.

So what impression does one gain from the above example about data identification and preparation?

Although all the data are not discussed in detail here, the task is not a simple one even for a small optimization problem as presented above. It is clear that the required data are collected from different sources and then compiled as required by the problem/model. However, the details and amount of data required depend on the objectives and constraints to be formulated for the problem. For example, it can be assumed that there is required aircrew for Hercules C130 aircraft, and loading and unloading equipment that may be needed for the above project. Details of the requirements and availability must be sorted out and then set out as constraints in the model.

10.4.2 Data Aggregation

Data aggregation here refers to any data roll-up process, using basic descriptive statistics tools such as averaging and summation, in which information can be expressed in a summary form. The individual data items or data sets are aggregated under the assumption that they are homogeneous across all the elements/data items (Garrett, 2003; Fadlalla, 2005). However, this is not true in all cases. That means we may solve a model with inappropriate data sets. For example, in Example 10.1, the food requirements may be estimated based on the average food requirements per person. However, what is the average food requirement per person here? The population is a mix of different age-groups of unequal numbers. This mix could be different in different regions. The food items required by different age-groups could be different. However, data aggregation is still useful and widely used in many practical situations. This is because of the following facts:

- For decision making at the tactical level, the decision is to be made based on aggregate data as the manager/decision maker is not interested in the detailed data. For example, RAAF Hercules C130 people are interested in the number of Hercules loads to be carried and not of the content of items in each crate.
- Aggregate data provide better decisions where detailed data are uncertain.
- The use of aggregate data reduces the size of the corresponding mathematical model, which indeed reduces the computational effort.

A few examples of data aggregation and problem situations are discussed below.

Example 10.2: A crop planning problem
Consider the crop planning problem described in Chapter 8. The corresponding mathematical model requires data A_{ijk}, which are the amount of input k required per unit land for crop i in region j. How does one get access to these data?

There are several inputs for crop cultivation. The requirement of input k is never uniform for a crop in the whole region. To take this uniformity into consideration, one can divide a region into a number of subregions of equal or different sizes. Although this approach will increase the model size and complexity, the resulting model would produce better quality solutions. However, the practice is to find the average data as the following equation:

$$A_{ijk} = \frac{\text{Total input of } k \text{ required in the whole region}}{\text{Total area of the region}} \tag{10.5}$$

The total or subregion-wise detailed input data are available from agricultural departments and the bureau of statistical publications.

Example 10.3: A multi-period production planning problem

Consider a simple production planning problem where the production capacity is fixed in each period; however, the demand varies from one period to the next (Table 10.3). Suppose the planning horizon is 12 periods long. Can we use aggregated demand and production data in the planning model?

The answer is "yes" or "no" as it depends on the level of aggregation. That is whether it is for the entire planning horizon or for individual periods. This issue is briefly analyzed below.

Suppose, the total production capacity in 12 periods is 600 units and the total demand is 540 units. Technically, the problem is feasible as the total capacity is greater than the total demand. To meet the demand, the average production rate required is 45 units per period. Let us fix this average rate (45 units per period) as our target production rate for every period. Based on the average production rate, the production–inventory balance equation will be as follows:

(Average production required in period t)
+ (Inventory carried from period $t - 1$)
− (Inventory carried to period $t + 1$) = (Demand in period t)

If the demand is to be met in each period separately, the use of the above equation would result in a shortage in period 2. Work out why this is so? To remove shortages, the equation must be modified as follows:

(Production in period t) + (Inventory carried from period $t - 1$)
− (Inventory carried to period $t + 1$) = (Demand in period t)

That shows that aggregated information may not be useful in all cases. However, the demand in a period here is the sum of all daily demands within that period. The same is true for total production calculation in a period. Note that "the production in period t" in the above equation is now a decision variable, which is less than or equal to the production capacity of that period.

In the above problem, if required, the length of a period can be reduced to being a day long. However, this would lead to a substantial number

TABLE 10.3

Time-Varying Demand and Fixed-Capacity Data

	Period								
Data	1	2	3	4	5	6	7	...	12
Demand (units)	30	70	20	60	30	70	20	...	60
Production capacity (units/day)	50	50	50	50	50	50	50	...	50

of periods that would increase the model size due to additional period-dependent variables and constraints. That is why almost all models allow a certain level of aggregation. The level of aggregation is dependent on many factors such as the data availability, data quality, imposed constraints, available solver's capability, level of decision-making process, developer's choice, and the users' preference.

Example 10.4: A power generation planning problem

A strategic level power generation planning process requires a decision regarding the installation of new generating plants and the operational level of existing plants to fulfill the electricity demand on a real-time basis for a number of years (usually 10 to 20 years planning horizon) by minimizing the sum of capital and operational costs. The depreciation of plants' value, and maintenance and operational costs vary year to year. In addition, the inflation and time value of money make the case difficult to solve using simple sum or average figures for overall cost parameters in the corresponding mathematical model. How should we prepare the cost data for such a situation?

The simple way to handle this data problem is to convert all cost data using appropriate discount factors with respect to a given reference point in time. For example, converting all future values to a present value at the beginning of the planning period. The following formulae are used for such conversions.

To convert a future amount \$$F$, which is n years ahead, to a present value (at time $t = 0$) \$$P$ with an interest rate T%, we use the relationships given by the following equation:

$$P = \frac{F}{\left(1 + \frac{T}{100}\right)^n} \quad \text{or} \quad F = P\left(1 + \frac{T}{100}\right)^n \tag{10.6}$$

The present value of a series of amounts $r_1, r_2, \ldots, r_n$ over the next n years at an interest rate T% can be represented as a present value \$$P$ as the following equations:

$$P = \frac{r_1}{\left(1 + \frac{T}{100}\right)} + \frac{r_2}{\left(1 + \frac{T}{100}\right)^2} + \cdots + \frac{r_n}{\left(1 + \frac{T}{100}\right)^n} \tag{10.7}$$

$$\text{Alternatively, } P = \sum_{i=1}^{n} \frac{r_i}{\left(1 + \frac{T}{100}\right)^i} \tag{10.8}$$

$T/100$ is known as the discount rate and $1/(1 + T/100)$ as the discount factor.

Example 10.5: A hierarchical production planning model
Consider the hierarchical production planning model discussed in Chapter 8. For convenience of modelling, the items are grouped into families and families are further grouped into product type. The aggregate model deals with product type at the upper level, the family disaggregation model deals with product families at the mid level, and the item disaggregation model with individual items at the lower level. So, it is obvious that the family disaggregation model must use data that are consolidated from individual items and the aggregate model must use data consolidated from individual items via families of items. This sort of consolidation of data is necessary as the model structure is rigid and designed for using such data.

10.5 Data Preprocessing

Preprocessing refers to the elementary operations that can be performed to improve or simplify the formulation by fixing values, adding additional information, analyzing existing information, tightening bounds on variables, and removing redundant constraints. So the preprocessing can be performed during the model formulation, as an additional task in data preparation or data refinement, or after formulation but before the solution approach is applied.

To demonstrate the preprocessing operations during formulation (as an additional task in data preparation) let us consider a rostering problem.

Example 10.6: A rostering problem
A Canberra-based small business employs six staff to share the workload fairly. The workplace has a morning and an evening shift of 8 h each weekday, and a single 10 h shift on Saturdays and Sundays. There are six staff, who are paid to work 38 h per week. One staff member works during weekdays only, and only on the morning shift. The remaining staff members work on weekends and during the week. However, no one is allowed to work in two consecutive shifts in a day. At least two staff members are required on each shift during weekdays and only two over the weekend. On weekdays, more staff are preferred in the afternoon shift in case of shortage. Staff must have an equal number of weekend shifts. Prepare a staff roster satisfying all the conditions and restrictions. The roster can cover as many weeks as are needed to give a fair distribution of load.

The problem can be modelled as an integer program by defining the variables as follows:

$$X_{ijkl} = \begin{cases} 1 & \text{if staff } i \text{ works in shift } j \text{ on day } k \text{ in week } l \\ 0 & \text{otherwise} \end{cases}$$

Here $i=1$ represents the staff member 1, who works only on weekday morning shifts (i.e., $k=1$, 5, and $j=1$). The value $k=6$ means Saturday and $k=7$ Sunday. When $k=6$ or 7, only $j=1$ allowed.

The constraints of the problem are as follows:

For each shift there must be at least two staff members available during weekdays.

$$\sum_{i=1}^{6} X_{ijkl} \geq 2 \quad \text{for all } j \text{ and } l\text{, and } k = 1, 5 \tag{10.9}$$

There must be two staff members available over the weekend.

$$\sum_{i=2}^{6} X_{ijkl} = 2 \quad \text{for } k = 6 \text{ and } 7, j = 1 \text{ (only), and all } l \tag{10.10}$$

The total load of staff member ($i=1$) must be equal to an average of 38 h per week.

$$\sum_{l=1}^{L} \sum_{k=1}^{5} 8\, X_{i=1,\, j=1,\, k,\, l} = 38L \tag{10.11}$$

Here, L is the number of weeks considered for the planning period.

The total load of the other five staff members must also be equal to an average of 38 h per week.

$$\sum_{l=1}^{L} \left(\sum_{k=1}^{5} \sum_{j=1}^{2} 8\, X_{ijkl} + \sum_{k=5}^{6} 10\, X_{ijkl} \right) = 38L \quad \text{for } i = 2, 6 \tag{10.12}$$

No one is allowed to work on both shifts on a given weekday.

$$\sum_{j=1}^{2} X_{ijkl} \leq 1 \quad \text{for all } l,\, i = 2, 5, \text{ and } k = 1, 5 \tag{10.13}$$

There must be an equal number of weekend shifts for each of the five staff members.

$$\sum_{l=1}^{L} \sum_{k=5}^{6} X_{ijkl} = M \quad \text{for all } i = 2, 6 \text{ and } j = 1 \tag{10.14}$$

Here, M is the number of weekend shifts attended by each staff member.

The number of staff members on the second shift $j=2$ must be greater than or equal to the number of staff members on the first shift $j=1$ for weekdays.

$$\sum_{i=1}^{6} X_{ij_1kl} \le \sum_{i=1}^{6} X_{ij_2kl} \quad \text{for all } l,\ k = 1, 5, \text{ and } j_1 = 1 \text{ and } j_2 = 2 \tag{10.15}$$

Both L and M are given values.

From the information available from the problem, we can analyze the following:

- For 5 weekdays, two shifts per day, 8 h per shift, and at least two staff members per shift, we require at least $5 \times 2 \times 8 \times 2 = 160$ staff-hours.
- Over a weekend, we need $2 \times 2 \times 10 = 40$ staff-hours.
- In a week, a minimum of $160 + 40 = 200$ staff-hours.
- The minimum possible number of working hours per staff member $= 200/6 = 33.33$. That means it is feasible to run the business with six staff members.
- On average six staff members would work $38 \times 6 = 228$ h per week.
- To maintain equal weekend shifts for five staff members, we need 5 week periods so that each works for four weekend shifts.
- Since the shifts are 8 and 10 h long, a 20 weeks planning period is required in order to have each member working 38 h work per week on average. So the minimum value of L in the above model would be 20.
- In a 20 week period, each staff member has 16 weekend shifts. That means $M = 16$, if we consider $L = 20$.
- Staff member 1 knows that he or she has to work 19 days (morning shift only) in 4 weeks or 95 days in 20 weeks.
- Staff members 2–6 know that each of them must work 16 weekend shifts and 75 weekday shifts in a 20 week periods. Each staff member would work for only $75 + 16 = 91$ days out of 140. They are entitled to have 49 days off in a 20 week period.

As the holidays of staff members 2–6 are not evenly spread, management thinks it is better to offer them flexibility in planning their own holidays. If they are asked to submit their holiday plans for half (or a certain number) of the days they are entitled to, then an availability matrix can easily be generated. In such a case, the rest of the holidays will be determined by the solution of the model. Otherwise, the problem could be infeasible by having too many staff members away for certain shifts. Management can organize a meeting to fix minor conflicts and avoid infeasibility. In developing the availability matrix, the number of staff members available in any shift should not be less than the number required. Thus, it is to ensure feasibility at the data preparation level. The staff availability matrix can be denoted by A_{ijkl}. Now, let us revise the model with the new information/preprocessed data.

For each shift there must be at least two staff members available during weekdays.

$$\sum_{i=1}^{6} A_{ijkl}X_{ijkl} \geq 2 \quad \text{for all } j \text{ and } l,\ k = 1, 5 \tag{10.16}$$

There must be exactly two staff members available over the weekend.

$$\sum_{i=2}^{6} A_{ijkl}X_{ijkl} = 2 \quad \text{for } k = 6 \text{ and } 7,\ j = 1, \text{ and all } l \tag{10.17}$$

The total load of staff member ($i = 1$) must be equal to an average of 38 h per week.

$$\sum_{l=1}^{L}\sum_{k=1}^{5} 8\ A_{i=1,\ j=1,\ k,\ l}X_{i=1,\ j=1,\ k,\ l} = 38L \tag{10.18}$$

The total load of the other five staff members must also be equal to an average of 38 h per week.

$$\sum_{l=1}^{L}\left(\sum_{k=1}^{5}\sum_{j=1}^{2} 8\ A_{ijkl}X_{ijkl} + \sum_{k=5}^{6} 10\ A_{ijkl}X_{ijkl}\right) = 38L \quad \text{for } i = 2, 6 \tag{10.19}$$

No one is allowed to work on both shifts on a given day. This constraint is no longer required as the staff availability matrix will ensure this condition.

There must be an equal number of weekend shifts for each of the five staff members.

$$\sum_{l=1}^{L}\sum_{k=5}^{6} A_{ijkl}X_{ijkl} = M \quad \text{for all } i = 2, 6 \text{ and } j = 1 \tag{10.20}$$

The number of staff members on shift $j = 2$ must be greater than or equal to the number of staff members on shift $j = 1$ for weekdays.

$$\sum_{i=1}^{6} A_{ijkl}X_{ij_1kl} \leq \sum_{i=1}^{6} A_{ijkl}X_{ij_2kl} \quad \text{for all } l,\ k = 1, 5, \text{ and } j_1 = 1 \text{ and } j_2 = 2 \tag{10.21}$$

TABLE 10.4

Sample Staff Availability Matrix Generated

			Staff					
Week	**Day**	**Day and Shift**	**1**	**2**	**3**	**4**	**5**	**6**
1	1	Monday—shift 1	1	0	1	0	0	0
		Monday—shift 2	0	0	0	1	1	1
	2	Tuesday—shift 1	1	0	1	0	0	0
		Tuesday—shift 2	0	0	0	1	1	1
	3	Wednesday—shift 1	1	0	0	1	0	0
		Wednesday—shift 2	0	0	0	0	1	1
	4	Thursday—shift 1	1	0	0	0	1	0
		Thursday—shift 2	0	1	0	0	0	1
	5	Friday—shift 1	1	0	0	0	0	1
		Friday—shift 2	0	1	1	0	0	0
	6	Saturday	0	1	1	1	0	0
	7	Sunday	0	1	1	1	0	0
2	8	Monday—shift 1	1	1	1	0	0	0
		Monday—shift 2	0	0	0	1	0	1
	9	Tuesday—shift 1	1	1	1	0	0	0
		Tuesday—shift 2	0	0	0	1	1	1
	10	Wednesday—shift 1	1	1	1	0	0	0
		Wednesday—shift 2	0	0	0	1	1	1
	11	Thursday—shift 1	1	0	1	0	0	0
		Thursday—shift 2	0	0	0	1	1	1
	12	Friday—shift 1	1	0	0	1	0	0
		Friday—shift 2	0	0	0	0	1	1
	13	Saturday	0	1	0	0	1	1
	14	Sunday	0	1	1	0	0	1

Note: 1 means possible availability and 0 means unavailable.

To give an idea about the availability matrix, an arbitrarily generated sample matrix for 2 weeks is shown in Table 10.4.

The above staff member availability matrix is a simple example of data preprocessing and the generated data/matrix clearly dictates how to model the problem. If the data preprocessing is not appropriately designed, the model solution would be different from the optimal solution. We have seen similar data preprocessing in Example 4.1 and other examples.

The problem can be complicated by imposing more constraints such as the following:

- Staff members must have at least 2 days break in succession.
- No one is expected to work more than 7 days in a row.
- The last shift before a break must be a morning or a weekend.
- The first shift after a break must be an evening or a weekend.

These constraints are not straightforward for mathematical modelling. Appropriately designed further preprocessing of data would reduce the complexity of modelling using these new constraints.

The preprocessing operations such as tightening bounds and removing redundant constraints, which are usually done after the formulation but before solving the model, are discussed in a later chapter.

10.6 Model-Driven Data versus Data-Driven Model

Consider the hierarchical production planning model discussed in an earlier chapter. An aggregate production planning model is an essential part of the hierarchical production planning process. As we know by now, it requires aggregate data to run this model. That means the model is dictating the data requirements.

In the above preprocessing example (Example 10.6), we have prepared the available data to formulate an easier but alternative model. That means, the data is dictating the formulation of a certain type of model.

10.7 Model Solving

To demonstrate the model-solving process, let us consider a simple product-mix problem as follows.

Example 10.7: A product mix problem

ABC Corporation produces two models of color television sets designated as Alfa and Beta. The company is in the market to maximize profit. The profit realized is $300 from set Alfa and $250 from set Beta. Obviously, the more sets produced and sold, the better. The trouble is that there are certain limitations that prevent ABC Corporation from producing and selling thousands of sets daily. The following are the limitations:

- Labor time availability in the production department per day is 40 h.
- Machine time availability per day is 45 h.
- No more than 12 sets of model Alfa can be sold in a day.

ABC's problem is to determine how many sets of each model to produce each day so that the total profit is as large as possible.

The variables of the problem can be defined as follows:

x_1 = number of sets of model Alfa (or identify as A) to be produced daily

x_2 = number of sets of model Beta (or identify as B) to be produced daily

	Model A	Model B		Objective Value		
Changing cells -->	0	0		0		
Coefficients				**Constraints**	LHS	RHS
Objective function	300	250		Labor avaliability	0	40
Constraint - Labor	2	1		Machine time	0	45
Constraint - Machine	1	3		Market restriction	0	12
Constraint - Market	1	0				

FIGURE 10.1
Solver's input as viewed in Excel sheet.

The corresponding linear programming model is

$$\begin{aligned}
&\text{Maximize } Z = 300x_1 + 250x_2 \\
&\text{Subject to} \\
&\quad 2x_1 + x_2 \le 40 \quad \text{Labor constraint} \\
&\quad x_1 + 3x_2 \le 45 \quad \text{Machine time constraint} \\
&\quad x_1 \le 12 \quad \text{Market constraint} \\
&\quad x_1, x_2 \ge 0 \quad \text{Nonnegativity constraint}
\end{aligned} \qquad \text{Model (10.1)}$$

In the following few subsections, we discuss the inputs required by Excel Solver, LINDO/LINGO, and MPL for the above model and the outputs provided by these software packages.

10.7.1 Excel Solver

The details of the Solver input preparation and solution generated can be found in the Appendix-9D. For the above model, the input can be prepared as Figure 10.1.

The objective function cell and the constraint LHS cells contain equations that are shown in Figure 10.2.

	Model A	Model B		Objective Value		
Changing cells -->	0	0		=Sumproduct (C7:D7,C4:D4)		
Coefficients				Constraints	LHS	RHS
Objective function	300	250		Labor avaliability	=Sumproduct (C8:D8,C4:D4)	40
Constraint - Labor	2	1		Machine time	=Sumproduct (C9:D9,C4:D4)	45
Constraint - Machine	1	3		Market restriction	=Sumproduct (C10:D10,C4:D4)	12
Constraint - Market	1	0				

FIGURE 10.2
Solver's input with Excel equations.

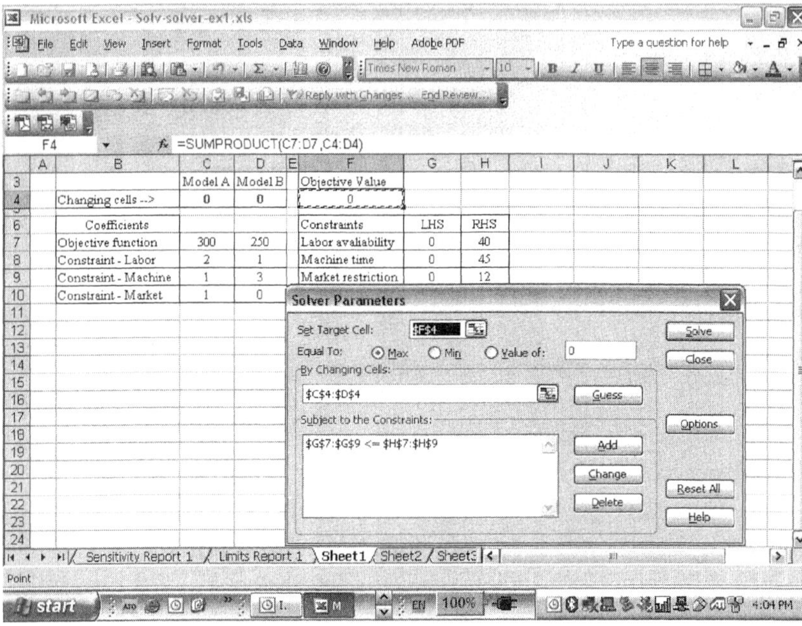

FIGURE 10.3
Input with solver's parameters.

The required solver parameters are shown in Figure 10.3. In the options to choose, we have selected linear model and nonnegative variables.

After solving the problem using Solver, the optimum values for the changing cells, the objective function cell, and the constraint LHS cells are recorded as shown in Figure 10.4.

As per the solution, the maximum possible profit that can be earned is \$6350. The number of models A and B that must be produced are 12 and 11, respectively. The answer report generated by Excel solver is presented in Figure 10.5.

	Model A	Model B	Objective Value		
Changing cells -->	12	11	6350		
Coefficients			**Constraints**	LHS	RHS
Objective function	300	250	Labor avaliability	35	40
Constraint - Labor	2	1	Machine time	45	45
Constraint - Machine	1	3	Market restriction	12	12
Constraint - Market	1	0			

FIGURE 10.4
Solver summary solution within input box.

Microsoft Excel 11.0 Answer Report
Worksheet: [Solv-solver-ex1.xls]Sheet1
Report Created: 18/03/2005 4:06:51 PM

Target Cell (Max)

Cell	Name	Original Value	Final Value
$F$4	Changing cells --> Objective Value	0	6350

Adjustable Cells

Cell	Name	Original Value	Final Value
$C$4	Changing cells --> Model A	0	12
$D$4	Changing cells --> Model B	0	11

Constraints

Cell	Name	Cell Value	Formula	Status	Slack
$G$7	Labor avaliability LHS	35	$G$7<=$H$7	Not Binding	5
$G$8	Machine time LHS	45	$G$8<=$H$8	Binding	0
$G$9	Market restriction LHS	12	$G$9<=$H$9	Binding	0

FIGURE 10.5
Solver answer report.

The answer report summarizes the solution to the problem and is fairly self-explanatory. The first section reports on the original and final (optimal) values of the objective function (also known as the target cell). The next section provides the original and final (optimal) values of the decision variables (also known as adjustable or changing cells). The final section presents information about the constraints. The Cell Value column shows the final value (optimal) assumed by each constraint cell (corresponds to each LHS formula). The Status column indicates which constraints are binding and which are nonbinding. A constraint is binding (or tight) if it is satisfied as a strict equality in the optimal solution; otherwise it is nonbinding (or loose). The last column represents the Slack of each constraint. Practically, slack means unutilized resource. In the solution, the slack indicates the difference between the LHS and RHS of each constraint. By definition, binding constraints have zero slack and nonbinding constraints have some positive level of slack. As provided in the solution above, 5 units of labor are slack (suggest not to use). Remember that all the constraints in our example problem were less than or equal to type. The difference between RHS and LHS for greater than or equal to constraints is known as surplus.

10.7.2 LINGO and MPL

The details of the input requirements of LINGO/LINDO and MPL are discussed in Appendices 9A and 9B. The input formats required by different old versions of LINDO/LINGO may vary slightly. However, they are

```
Model:
Max = 300 * x1 + 250 * x2;

!/*subject to*/;
2 * x1 + x2 <= 40;
x1 + 3 * x2 <= 45;
x1 <= 12;
End
```

FIGURE 10.6
Sample LINDO/LINGO inputs.

similar for all latest versions. An example of the input requirements for version 10.0 is shown in Figure 10.6.

The LINGO solver solution status is shown in Figure 10.7.

The solution report provided by LINGO is given in Figure 10.8.

The solution report in Figure 10.8 summarizes the solution provided by LINGO version 5.0. The first section reports on the parameters involved with the model such as number of constraints, number of variables, matrix density, smallest and largest elements in absolute value, type of objective, and type of constraints. The next section provides the number of iterations required to solve the problem and the optimal value of the objective function. The following section provides the optimal values of the decision variables and their reduced cost. The final section presents information about the constraints such as slack or surplus and dual price. Note that Row-1 represents the objective function as a constraint. The terms reduced cost, surplus variable, and dual price are discussed in the next chapter.

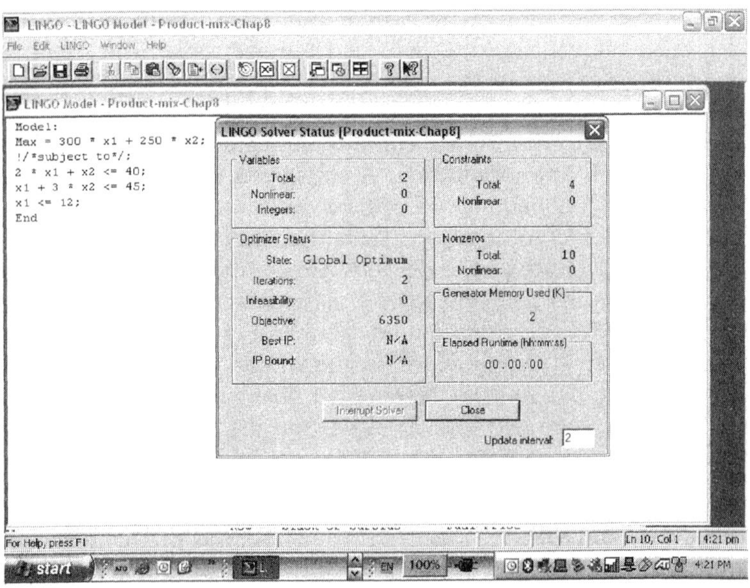

FIGURE 10.7
LINGO solver solution status.

```
Rows=     4 Vars=     2 No. integer vars=     0  ( all are linear)
Nonzeros=    10 Constraint nonz=    5(    3 are +- 1) Density=0.833
Smallest and largest elements in abs value=    1.00000      300.000
No. < :  3 No. =:  0 No. > :  0, Obj=MAX, GUBs <=  1
Single cols=   0

Global optimal solution found at step:              2
Objective value:                          6350.000

            Variable        Value        Reduced Cost
                  X1     12.00000           0.0000000
                  X2     11.00000           0.0000000

                 Row   Slack or Surplus     Dual Price
                   1       6350.000          1.000000
                   2       5.000000          0.000000
                   3       0.0000000         83.33334
                   4       0.0000000         216.6667
```

FIGURE 10.8
LINGO solutions.

LINGO 10.0 does not provide first section as part of its usual report. However, these information can be generated separately.

Although there are differences in format and information content in Solver and LINGO, the basic information provided are similar in all solution reports.

The input for MPL can be prepared as in Figure 10.9.

The solution status provided by the solution engine Conopt is Figure 10.10.

The solution produced by Conopt is provided in Figure 10.11.

The solution report in Figure 10.11 is very similar to Figure 10.8. However, the dual price is written as a shadow price.

Example 10.8: An integer programming model

Consider the model of Example 10.7. Now consider that the variables are to be integer and the machine time limitation is 40 h (instead of 45 h).

```
TITLE
   A_Product_Mix_Problem;

MAX
   Profit = 300 Model_A + 250 Model_B;

SUBJECT TO
   2 Model_A + Model_B <= 40;
   Model_A + 3 Mode_B <= 45;
   Model_A   <=  12;
END
```

FIGURE 10.9
MPL input.

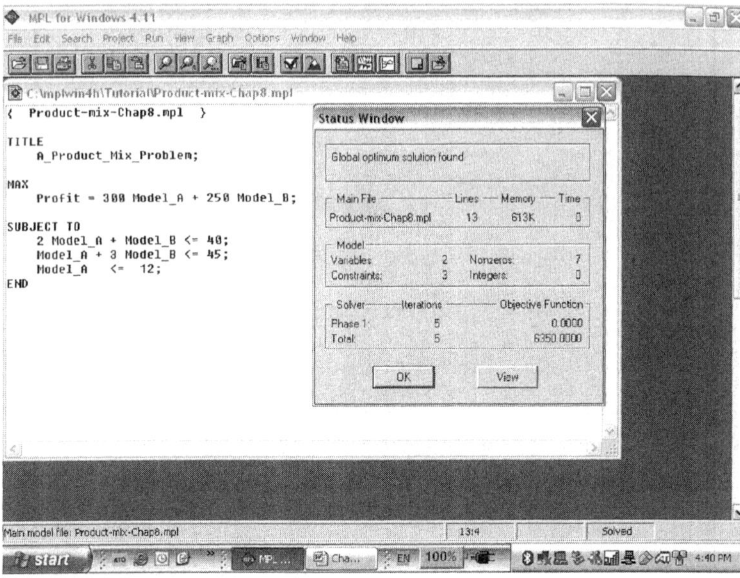

FIGURE 10.10
Solution status provided by Conopt.

$$
\begin{array}{lll}
\text{Maximize } Z = 300x_1 + 250x_2 & & \\
\text{Subject to} & & \\
 & 2x_1 + x_2 \leq 40 & \text{Labor constraint} \\
 & x_1 + 3x_2 \leq 40 & \text{Machine time constraint} \\
 & x_1 \leq 12 & \text{Market constraint} \\
 & x_1, x_2 \geq 0 \text{ and integer} & \text{Nonnegativity constraint}
\end{array}
$$

Model (10.2)

If you change the RHS value of constraint 2 in the previous Solver inputs, the optimum solution would be $x_1 = 12$, $x_2 = 9.3333$, and $Z = 5933.3333$. However, the values of x_1 and x_2 must be integers. In this case, one must set in the constraint section that they are to be integer as shown in Figure 10.12.

Now the optimum solution is $x_1 = 12$, $x_2 = 9$, and $Z = 5850$.

Example 10.9: A nonlinear model

Suppose the objective function of Example 10.8 is now nonlinear as given by the following equation:

$$\text{Maximize } Z = 300x_1^{0.5} + 250x_2^{0.9} \tag{10.22}$$

In solving this model, the objective function cell (in Figure 10.12) must contain the formula = C7 * (C4^0.5) + D7 * (D4^0.9). In addition, you

```
MPL Modelling System  -  Copyright (c) 1988-2000, Maximal Software, Inc.
--------------------------------------------------------------------------------

MODEL STATISTICS
Problem name:     A_Product_Mix_Problem
Filename:         Product-mix-Chap8.mpl
Date:             March 18, 2005
Time:             16:39
Parsing time:     0.00 sec
Solver:           Conopt

Objective value:  6350.00000000
Iterations:       5
Solution time:    0.05 sec
Constraints:      3
Variables:        2
Nonzeros:         5
Density:          83%

SOLUTION RESULT
Global optimum solution found

MAX Profit  =     6350.0000

DECISION VARIABLES
PLAIN VARIABLES

  Variable Name       Activity   Reduced Cost
  -------------------------------------------
  Model_A              12.0000      0.0000
  Model_B              11.0000      0.0000
  -------------------------------------------

CONSTRAINTS
PLAIN CONSTRAINTS

  Constraint Name    Slack    Shadow Price
  -------------------------------------------
  c1                5.0000     0.0000
  c2                0.0000     83.3334
  c3                0.0000     216.6667
  -------------------------------------------
END
```

FIGURE 10.11
Solution provided by Conopt.

must refrain from choosing Assume Linear Model under options. The additional options required for solving nonlinear programming model such as estimates, derivatives, and search are briefly discussed below:

Estimates

- Tangent: estimates values using linear extrapolation.
- Quadratic: estimates values using nonlinear/quadratic extrapolation.

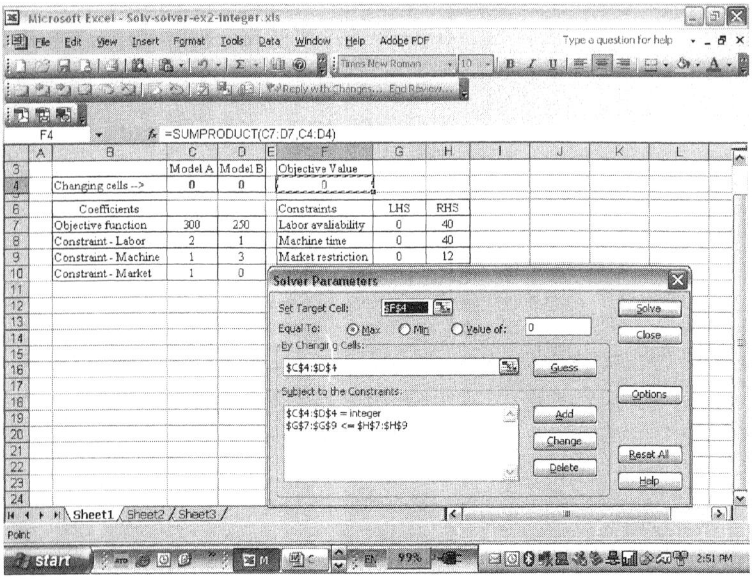

FIGURE 10.12
Solver input with integer restrictions.

Derivatives:

- Forward: estimates the first derivatives at a point by perturbing the point once in a forward direction and computes the rise over the run.
- Central: estimates the first derivatives by perturbing away from a point in both forward and backward directions and computes the rise over the run between the two points.

Search:

- Newton: uses Broyden-Fletcher-Goldfarb-Shanno quasi-Newton method to identify the search directions.
- Conjugate: uses the conjugate gradient method to find the search direction.

The default options are tangent, forward, and Newton. Using the default options, the Solver solution for this model is $x_1=7$, $x_2=11$, and $Z=2957.4074$.

Example 10.10: A two-stage transportation problem
Consider a two-stage transportation model as follows.

Variables and parameters:

$x1_{ij}$ = quantity transported from source i to transshipment point j
$x2_{jk}$ = quantity transported from transshipment point j to destination k
$C1_{ij}$ = unit transportation cost from source i to transshipment point j
$C2_{jk}$ = unit transportation cost from transshipment point j to destination k
S_i = total supply from source i
CP_j = capacity of transshipment point j
D_k = demand of destination k

The mathematical model of this problem can be formulated as follows:

$$\text{Minimize } Z = \sum_i \sum_j C1_{ij} x1_{ij} + \sum_j \sum_k C2_{jk} x2_{jk}$$

Subject to

$$
\begin{aligned}
&\sum_j x1_{ij} \le S_i && \forall i \\
&\sum_i x1_{ij} \le CP_j && \forall j \\
&\sum_j x2_{jk} \ge D_k && \forall k \\
&\sum_i x1_{ij} - \sum_k x2_{jk} \ge 0 && \forall j \\
&x1_{ij},\ x2_{jk} \ge 0 \text{ and integer} && \forall i, j, k
\end{aligned}
\qquad \text{Model (10.3)}
$$

The objective of this problem is to minimize the overall transshipment costs. The first, second, and third constraints ensure the supply limitation, transshipment point capacity and demand requirements, respectively. The fourth or the flow constraint indicates that the total supply (in the second stage) from a transshipment node j to all the destinations must be less than or equal to the supply (in the first stage) received by that node j from all sources. This is a standard integer programming model as the variables $x1_{ij}$ and $x2_{jk}$ are integer in nature. We assume there exist a number of feasible solutions.

If we consider 8 plants, 4 warehouses, and 8 distribution centers, the above model requires 64 variables and 24 constraints. It is not easy to organize the inputs for Excel Solver for such an instance of the problem. See Figure 10.13 the input requirements of LINGO for this problem.

Now suppose that there are 15 plants, 10 warehouses, and 15 distribution centers. In this case, the above model requires 300 variables and 50 constraints. The current instance contains five times the number of variables and twice the number of constraints; see Figure 10.14 how much it differs in terms of input requirements by LINGO. On the other hand, it would be unmanageable to organize these inputs for Excel Solver.

```
MODEL:
!8 PLANTS, 4 WAREHOUSE AND 8 DIST CENTER;
SETS:
PLANTS/ P1 P2 P3 P4 P5 P6 P7 P8 /: PLANT_CAPACITY;
WAREHOUSES/ WH1 WH2 WH3 WH4 /: WHOUSE_CAPACITY;
DISTCENTERS/ DC1 DC2 DC3 DC4 DC5 DC6 DC7 DC8 /: DC_CAPACITY;
LINKS1 (PLANTS, WAREHOUSES): COST1, VOLUME1;
LINKS2 (DISTCENTERS, WAREHOUSES): COST2, VOLUME2;
ENDSETS
!OBJECTIVE FUNCTION;
[OBJ] MIN = PRODUCTION + DISTRIBUTION;
PRODUCTION = @SUM ( LINKS1 (I, J): COST1 (I, J) * VOLUME1 (I,J));
DISTRIBUTION = @SUM (LINKS2 (I, J): COST2 (I,J) * VOLUME2 (I,J));
!CONSTRAINTS;
@FOR (WAREHOUSES(J):
      @SUM ( PLANTS(I): VOLUME1(I,J)) <= WHOUSE_CAPACITY(J));
@FOR (PLANTS (I):
      @SUM ( WAREHOUSES (J): VOLUME1(I,J)) <= PLANT_CAPACITY (I));
@FOR (DISTCENTERS(I):
      @SUM (WAREHOUSES (J): VOLUME2(I,J)) >= DC_CAPACITY(I));
@FOR (WAREHOUSES(J):
      @SUM ( PLANTS (I): VOLUME1(I,J)) - @SUM (DISTCENTERS (I):
VOLUME2(I,J)) >= 0);
@FOR (PLANTS(I):
      @FOR (WAREHOUSES(J):
            @ GIN(VOLUME1(I,J)))) ;
@FOR (DISTCENTERS(I):
      @FOR (WAREHOUSES(J):
            @ GIN(VOLUME2(I,J)))) ;
!DATA ARE BELOW;
DATA:
PLANT_CAPACITY = 1500.7 1400 1800.8 2200.6 1200 3200 4200 3800;
WHOUSE_CAPACITY = 4500 3000 5500 6800;
DC_CAPACITY = 2000 3000 1500 1800 2500 3700 4050 750;
COST1 =      2    3  3  5
             6    4  2  3
             8    6  7  9
             3    5  7  9
             1   12  3  7
             15   2  8  6
             5    7  3  9
             3    2  8  6;
COST2 =  7  3   5   5
         5  4  12   7
         4  5   2   9
         3  6   4  11
         5  7   6   9
         2  9   8   7
         1  12 10   5
         2  15  3  3;
ENDDATA
END
```

FIGURE 10.13
LINGO codes for 8–4–8 transportation problem.

```
MODEL:
!15 PLANTS, 10 WAREHOUSE AND 15 DIST CENTER ;

SETS:
PLANTS/ P1 P2 P3 P4 P5 P6 P7 P8 P9 P10 P11 P12 P13 P14 P15/:
PLANT_CAPACITY;
WAREHOUSES/ WH1 WH2 WH3 WH4 WH5 WH6 WH7 WH8 WH9 WH10/: WHOUSE_CAPACITY;
DISTCENTERS/ DC1 DC2 DC3 DC4 DC5 DC6 DC7 DC8 DC9 DC10 DC11 DC12 DC13
DC14 DC15/: DC_CAPACITY;
LINKS1 (PLANTS, WAREHOUSES): COST1, VOLUME1;
LINKS2 (DISTCENTERS, WAREHOUSES) : COST2, VOLUME2;
ENDSETS

!OBJECTIVE FUNCTION;
[OBJ] MIN = PRODUCTION + DISTRIBUTION;
PRODUCTION = @SUM ( LINKS1 (I, J): COST1 (I, J) * VOLUME1 (I,J));
DISTRIBUTION = @SUM (LINKS2 (I, J): COST2 (I,J) * VOLUME2 (I,J));

!CONSTRAINTS;
@FOR (WAREHOUSES(J):
      @SUM ( PLANTS(I): VOLUME1(I,J)) <= WHOUSE_CAPACITY(J));

@FOR (PLANTS (I):
      @SUM ( WAREHOUSES (J): VOLUME1(I,J)) <= PLANT_CAPACITY (I));

@FOR (DISTCENTERS(I):
      @SUM (WAREHOUSES (J): VOLUME2(I,J)) >= DC_CAPACITY(I));

@FOR (WAREHOUSES(J):
      @SUM ( PLANTS (I): VOLUME1(I,J)) - @SUM (DISTCENTERS (I):
VOLUME2(I,J)) >= 0);

@FOR (PLANTS(I):
      @FOR (WAREHOUSES(J):
            @ GIN(VOLUME1(I,J)))) ;

@FOR (DISTCENTERS(I):
      @FOR (WAREHOUSES(J):
            @ GIN(VOLUME2(I,J)))) ;

!DATA ARE BELOW;
DATA:
PLANT_CAPACITY = 1500.5 1400.5 1800.5 2200 1200 1700 1500 2000 2300
1950 2350 3700 1400 1600 2000;
WHOUSE_CAPACITY = 3500 2500 4000 2000 3000 4500 4500 1000 2000 2000 ;
DC_CAPACITY = 2000 2800 700 1400 1200 1600 1400 2100 2400 2050 2500
3450 1500 1700 1800 ;

COST1 =        2  3  3  5  7  8  9  2  3  4
               6  4  2  3  9  7  4  5  6  7
               8  6  7  9  3  3  2  1  6  9
               3  5  7  9 10 12 14 12 16 18
               1 12  3  7  7  5  7  5  3  1
              15  2  8  6  2  2  6  2  1  1
               5  7  3  9  8 15  1  2  7  8
               3  2  8  6 12  4  9  1  1  1
              13 12  1 10  5  2  1  3  5  7
               7  3  2 14  1  1  2 14  6  3
              17  2  5  7  9  1  3  5  9  1
              10  5  6  4  3  1  2  3  7  9
               5  6  7  9  3  1  2  5  9  4
               1  5  9  1  1  4  7 14  5  6
               1  3  4  7  9 12  1 11 12 13;
```

FIGURE 10.14
LINGO codes for 15–10–15 transportation problem.

(continued)

```
COST2 =          7  3  5  5  9  8  2  2  4  5
                 5  4 12  7  3  7  5  7  6  7
                 4  5  2  9  2  3  3  6 90  9
                 3  6  4 11  5  2  4  1  1  3
                 5  7  6  9  4  4  7 11 12  2
                 2  9  8  7  8  6  8 15 16  1
                 1 12 10  5 11  8  9  4  7  5
                 2 15  3  3 12  7  1 14  8  7
                 3  2  1  4  5  1 10  6  7  9
                 7  7  5  6  1  4  5  7  7  3
                 8  2  1  5  1  8 13 17  1  4
                 9  3  3  7  3  9  1  2  3  4
                 2  1 12  3  7  2  1 20  7  7
                 3  1  5  3  9  5  3  3  9  3
                 5  1  7  3  1  8  7  6  2  1;
ENDDATA
END
```

FIGURE 10.14 (continued)

As one can see, the only difference is in the data and data setting between the above two instances. If one now considers thousands of plants, warehouses, and distribution centers, the objective function and constraints section would remain the same.

10.8 Summary

One may get the feeling that one needs a significant effort (as discussed in this chapter) to prepare data for model solving. In fact, that reflects the difference between addressing textbook problems and real practical problems.

Data collection, data quality, data preservation, and data preparation are important tasks for optimization problem solving. In this chapter, we have discussed the data collection and data preparation methodologies from the point of view of inputs to optimization models. The influence of data preprocessing on the development of mathematical model has been analyzed. It is noted here that very little about the data requirement and preparation has been reported in the optimization literature. The input requirements for Excel Solver, LINGO/LINDO, and MPL-based solvers are also discussed with numerical examples. More problem-solving examples with LINGO can be found in the Appendix of this chapter.

Exercises

1. A small manufacturing company produces two products P1 and P2. Each product needs to be processed through two departments D1 and D2. Department D1 has three machines, L1, L2, and L3, while department

D2 has two machines M1 and M2. Product P1 must be processed on either machine L1 or L3 in department D1 first and then on either machine M1 and M2 in department D2. Product P2 can be processed on either machine L2 or L3 in department D1 and only on machine M1 in department D2. The time required by each product on each machine is given below.

Machine	Time for P1 (Minutes)	Time for P2 (Minutes)	Time Available (Minutes/Week)	Cost ($)/Week
L1	10	14	3000	1000
L2	—	10	2400	600
L3	8	—	2400	500
M1	12	9	3000	1200
M2	7	—	2400	800

Product P1 requires 1.1 kg of metal, whereas product P2 requires 1.5 kg of metal. The metal cost is $15 per kg. The selling prices for products P1 and P2 are $45 and $55, respectively.

Formulate the objective function for the problem in order to maximize the profit (ignore the constraints as this is mainly a data preparation exercise).

2. Consider the Problem 1 in Exercises. Suppose, there is an ordering cost of $300 per order of raw materials (metal). The company orders metal once a week. Write down your revised objective function.

3. Develop Excel Solver and LINGO inputs for the following mathematical models. Make sure these are free from errors.

(a) $\text{Max } Z = 24x_1 + 16x_2$

$\in$

$$50x_1 + 25x_2 \leq 4000$$
$$3x_1 + 9x_2 \leq 720$$
$$x_1 + x_2 \leq 100$$
$$x_1, x_2 \geq 0$$

(b) $\text{Min } Z = 1.2X_{12} + 0.1X_{13} + 1.2X_{22} + 0.1X_{23} + 3.7X_{24} + 1.3X_{25} + 0.2X_{26}$

$\in$

$$2X_{11} + 4X_{21} + 2X_{22} + 2X_{23} + X_{24} = 30$$
$$X_{12} + X_{22} + 2X_{24} + X_{25} = 50$$
$$X_{13} + X_{23} + X_{25} + 2X_{26} = 10$$
$$X_{11}, X_{12}, X_{13}, X_{21}, X_{22}, X_{23}, X_{24}, X_{25}, 2X_{26} \geq 0$$

(c) Maximize $Z = 50X_1 + 40X_2 + 70X_3$

$\in$

$$\begin{aligned}
6X_1 + 4X_2 + 9X_3 &\leq 5000\\
3X_1 + 7X_2 + 6X_3 &\leq 6000\\
X_1 + (1/2)X_2 + (1/3)X_3 &\leq 1600\\
X_1 &\geq 150\\
X_2 &\geq 250\\
X_3 &\geq 300\\
4X_1 - 3X_2 &= 0\\
5X_2 - 4X_3 &= 0\\
X_1, X_2, X_3 &\geq 0
\end{aligned}$$

(d) Maximize $Z = 0.09X_1 + 0.08X_2 + 0.065X_3 + 0.075X_4 + 0.10X_5$

$\in$

$$\begin{aligned}
X_1 + X_2 + X_3 + X_4 + X_5 &\leq 50\\
X_3 - 0.25X_0 &\leq 0\\
X_1 - X_4 &\leq 0\\
-0.7X_1 + 0.3X_2 - 0.7X_3 + 0.3X_4 - 0.7X_5 &\geq 0\\
X_2 - 3X_4 &\geq 0\\
X_i &\geq 0 \quad \forall i
\end{aligned}$$

(e) Maximize $Z = 4X_1 + 5X_2$

$\in$

$$\begin{aligned}
4X_1 + 3X_2 &\leq 480\\
3X_1 + 5X_2 &\leq 480\\
X_1 - 2X_2 - X_3 &\leq 0\\
2X_2 - X_1 - X_3 &\leq 0\\
X_3 &\leq 30\\
X_1, X_2, X_3 &\geq 0
\end{aligned}$$

(f) Maximize Y

$\in$

$$\begin{aligned}
6X_1 + 5X_2 + 7X_3 - 5Y &\geq 0\\
4X_1 + 8X_2 + 3X_3 - 4Y &\geq 0\\
7X_1 + 4X_2 + 2X_3 &\leq 120\\
5X_1 + 7X_2 + 9X_3 &\leq 240\\
X_1, X_2, X_3 &\geq 0
\end{aligned}$$

(g) $\text{Min } Z = 750y_A + 150y_B + 420y_C + 20x_A + 55x_B + 35x_C$

$$x_A + x_B + x_C \geq 800$$
$$x_A - 400y_A \leq 0$$
$$x_B - 700y_B \leq 0$$
$$x_C - 600y_C \leq 0$$
$$x_i \geq 0 \quad \text{for all } i$$
$$y_i = \text{either 0 or 1} \quad \text{for all } i$$

4. Consider the following vehicle mix model:

$$\text{Min } Z = \sum_i C_i \times P_i$$

Subject to

$$\sum_j \sum_t x_{ijt} \leq P_i \qquad \forall i$$
$$\sum_i \sum_t \text{CAP}_i \times x_{ijt} \geq Q_j \quad \forall j$$
$$x_{ijt} \geq 0 \text{ and integer}$$

where

x_{ijk} = number of vehicles of type i undertaking task j in period t
C_i = cost of purchasing a vehicle of type i
CAP_i = capacity of vehicle type i
Q_j = capacity required by task j
P_i = the number of vehicle to be purchased of type i

For this exercise, consider 3 vehicle types (heavy, medium, and small), 10 tasks in each period, and 10 periods problem with data as follows:

	Vehicle Type		
	Heavy	**Medium**	**Small**
Cost ($)	250,000	187,500	50,000
Capacity (tonne)	1,000	500	100

The tasks in each period contain 1000, 1000, 1000, 500, 500, 500, 100, 100, 100, and 100 tonnes.

Develop LINGO codes for this model.

References

Fadlalla, A., An experimental investigation of the impact of aggregation on the performance of data mining with logistic regression, *Information and Management*, 42, 695–707, 2005.

Garrett, T.A., Aggregated versus disaggregated data in regression analysis: Implications for inference, *Economics Letters*, 81, 61–65, 2003.

Neuman, W.L., *Social Research Methods: Qualitative and Quantitative Approaches*, Allyn and Bacon, Boston, MA, 2003.

Turban, E. and Meredith, J., *Fundamentals of Management Science*, 6th ed., Irwin McGraw-Hill, Boston, MA, 1994.

Appendix-10A Additional Problem-Solving Using LINGO

10A.1 Example 4.6 (Model 4.7)

$$\text{Maximize } Z = 450(X_{1A} + X_{2A} + X_{3A}) + 350(X_{1B} + X_{2B} + X_{3B}) + 150(X_{1C} + X_{2C} + X_{3C})$$

$\in$

$$5X_{1A} + 4X_{1B} + 3X_{1C} \leq 1600$$
$$5X_{2A} + 4X_{2B} + 3X_{2C} \leq 2000$$
$$5X_{3A} + 4X_{3B} + 3X_{3C} \leq 1000$$
$$X_{1A} + X_{1B} + X_{1C} \leq 500$$
$$X_{2A} + X_{2B} + X_{2C} \leq 600$$
$$X_{3A} + X_{3B} + X_{3C} \leq 400$$
$$X_{1A} + X_{2A} + X_{3A} \leq 600$$
$$X_{1B} + X_{2B} + X_{3B} \leq 700$$
$$X_{1C} + X_{2C} + X_{3C} \leq 300$$
$$6(X_{1A} + X_{1B} + X_{1C}) - 5(X_{2A} + X_{2B} + X_{2C}) = 0$$
$$4(X_{2A} + X_{2B} + X_{2C}) - 6(X_{3A} + X_{3B} + X_{3C}) = 0$$
$$X_{ij} \geq 0, \quad i = 1, 2, 3, \quad j = \text{A, B, C}$$

10A.1.1 LINGO Code

```
Model:
!Objective function;
Max = 450* (X1A + X2A + X3A) + 350* (X1B + X2B + X3B) + 150*
     (X1C + X2C + X3C);
!Constraint;
5*X1A + 4*X1B + 3*X1C <= 1600;
5*X2A + 4*X2B + 3*X2C <= 2000;
5*X3A + 4*X3B + 3*X3C <= 1000;
```

```
X1A + X1B + X1C <= 500;
X2A + X2B + X2C <= 600;
X3A + X3B + X3C <= 400;
X1A + X2A + X3A <= 600;
X1B + X2B + X3B <= 700;
X1C + X2C + X3C <= 300;
6* (X1A + X1B + X1C) - 5* (X2A + X2B + X2C) = 0;
4* (X2A + X2B + X2C) - 6* (X3A + X3B + X3C) = 0;
End
```

10A.1.2 LINGO Solution

```
Global optimal solution found.
Objective value:        399772.7

Variable        Value          Reduced Cost
   X1A         236.3636          0.000000
   X2A         363.6364          0.000000
   X3A         0.000000          153.4091
   X1B         104.5455          0.000000
   X2B         45.45455          0.000000
   X3B         181.8182          0.000000
   X1C         0.000000          153.4091
   X2C         0.000000          153.4091
   X3C         90.90909          0.000000

Row      Slack or Surplus       Dual Price
 1          399772.7             1.000000
 2          0.000000             46.59091
 3          0.000000             46.59091
 4          0.000000             200.0000
 5          159.0909             0.000000
 6          190.9091             0.000000
 7          127.2727             0.000000
 8          0.000000             53.40909
 9          368.1818             0.000000
10          209.0909             0.000000
11          0.000000             27.27273
12          0.000000             75.00000
```

10A.2 A Transportation Model

Consider a transportation model as shown below, which is similar to Example 4.13 (Model 4.14):

$$\text{Minimize } Z = 1X_{AR} + 2X_{AS} + 3X_{AT} + 4X_{BR} + 1X_{BS} + 5X_{BT}$$

Subject to

$$X_{AR} + X_{AS} + X_{AT} = 100$$
$$X_{BR} + X_{BS} + X_{BT} = 120$$
$$X_{AR} + X_{BR} \leq 80$$
$$X_{AS} + X_{BS} \leq 120$$
$$X_{AT} + X_{BT} \leq 60$$
$$X_{ij} \geq 0 \quad \forall i, j$$

10A.2.1 LINGO in Algebraic Form

```
Model:
!Objective function;
Min = 1*XAR + 2*XAS + 3*XAT + 4*XBR + 1*XBS + 5*XBT;
!Constraint;
XAR + XAS + XAT = 100;
XBR + XBS + XBT = 120;
XAR + XBR <= 80;
XAS + XBS <= 120;
XAT + XBT <= 60;
End
```

10A.2.2 LINGO Solution Report

```
Global optimal solution found.
Objective value:              260.0000
Total solver iterations: 4

Variable        Value         Reduced Cost
   XAR        80.00000          0.000000
   XAS        0.000000          0.000000
   XAT        20.00000          0.000000
   XBR        0.000000          4.000000
   XBS        120.0000          0.000000
   XBT        0.000000          3.000000

Row      Slack or Surplus      Dual Price
 1           260.0000          -1.000000
 2           0.000000          -3.000000
 3           0.000000          -2.000000
 4           0.000000           2.000000
 5           0.000000           1.000000
 6           40.00000           0.000000
```

An alternative formulation for this problem is shown in Section 6.2 (Model 6.1). LINGO allows the coding of such a compact model as shown below.

10A.2.3 LINGO Codes (Alternative)

```
MODEL:
! Flexible LINGO coding for the Transportation Problem;
SETS:
! Declaring the sources - we have two sources here and
capacity data are given for each source;
  SOURCES /A B/: CAPACITY;
! Declaring the destinations - we have three destinations
here and the demand is given for each destination;
  DESTINATIONS /R S T/: DEMAND;
! Identifying the links between each source to each
destination - cost for each link is given and volume is the
variable representing the transportation in the link;
  LINKS (SOURCES, DESTINATIONS) : COST, VOLUME;
ENDSETS

!Objective function;
  MIN=@SUM(LINKS(I,J):
    COST(I,J) * VOLUME(I,J));
!Capacity constraints;
  @FOR(SOURCES(I):
    @SUM(DESTINATIONS(J): VOLUME(I,J)) =CAPACITY(I));
! Demand constraints;
  @FOR(DESTINATIONS(J):
    @SUM(SOURCES(I): VOLUME(I,J))<= DEMAND(J));
! Here is the data;
DATA:
  CAPACITY=100  120;
  DEMAND=80  120  60;
  COST=1  2  3
       4  1  5;
ENDDATA
END
```

10A.2.4 LINGO Solution Report (Using Alternative Codes)

```
Global optimal solution found.
Objective value:            260.0000
Total solver iterations: 4
```

Variable	Value	Reduced Cost
CAPACITY(A)	100.0000	0.000000
CAPACITY(B)	120.0000	0.000000
DEMAND(R)	80.00000	0.000000
DEMAND(S)	120.0000	0.000000
DEMAND(T)	60.00000	0.000000
COST(A, R)	1.000000	0.000000
COST(A, S)	2.000000	0.000000
COST(A, T)	3.000000	0.000000
COST(B, R)	4.000000	0.000000
COST(B, S)	1.000000	0.000000
COST(B, T)	5.000000	0.000000
VOLUME(A, R)	80.00000	0.000000
VOLUME(A, S)	0.000000	0.000000
VOLUME(A, T)	20.00000	0.000000
VOLUME(B, R)	0.000000	4.000000
VOLUME(B, S)	120.0000	0.000000
VOLUME(B, T)	0.000000	3.000000

Row	Slack or Surplus	Dual Price
1	260.0000	-1.000000
2	0.000000	-3.000000
3	0.000000	-2.000000
4	0.000000	2.000000
5	0.000000	1.000000
6	40.00000	0.000000

10A.2.5 A Modified Transportation Model

Suppose there is no link from source A to destination T in the above problem. This situation can be handled by either putting a big cost on that link or forcing zero shipment through that link. For the latter one see LINGO codes below:

```
MODEL:
!Flexible LINGO coding for the Transportation Problem;
SETS:
!Declaring the sources - we have two sources here and
capacity data are given for each source;
  SOURCES /A B/: CAPACITY;
!Declaring the destinations - we have three destinations
here and the demand is given for each destination;
  DESTINATIONS /R S T/: DEMAND;
!Identifying the links between each source to each
destination - cost for each link is given and volume is the
variable representing the transportation in the link;
  LINKS(SOURCES, DESTINATIONS): COST, VOLUME;
ENDSETS
```

```
!Objective function;
  MIN=@SUM(LINKS(I,J):
    COST(I,J) * VOLUME(I,J));
!Forcing zero shipment from A to T;
  @FOR(SOURCES(I):
    @SUM(DESTINATIONS(J)| I #EQ# 1 #AND# J #EQ# 3:
                                  VOLUME(I,J)) =0);
!Capacity constraints;
  @FOR(SOURCES(I):
    @SUM(DESTINATIONS(J): VOLUME(I,J)) =CAPACITY(I));
!Demand constraints;
  @FOR(DESTINATIONS(J):
    @SUM(SOURCES(I): VOLUME(I,J)) <=DEMAND(J));
!Here is the data;
DATA:
  CAPACITY=100  120;
  DEMAND=80  120  60;
  COST=1  2  3
       4  1  5;
ENDDATA
END
```

10A.2.6 LINGO Solution Report (with Restricted Path)

```
Global optimal solution found.
Objective value:          320.0000
Total solver iterations: 3
```

Variable	Value	Reduced Cost
CAPACITY(A)	100.0000	0.000000
CAPACITY(B)	120.0000	0.000000
DEMAND(R)	80.00000	0.000000
DEMAND(S)	120.0000	0.000000
DEMAND(T)	60.00000	0.000000
COST(A, R)	1.000000	0.000000
COST(A, S)	2.000000	0.000000
COST(A, T)	3.000000	0.000000
COST(B, R)	4.000000	0.000000
COST(B, S)	1.000000	0.000000
COST(B, T)	5.000000	0.000000
VOLUME(A, R)	80.00000	0.000000
VOLUME(A, S)	20.00000	0.000000
VOLUME(A, T)	0.000000	0.000000
VOLUME(B, R)	0.000000	4.000000
VOLUME(B, S)	100.0000	0.000000
VOLUME(B, T)	20.00000	0.000000

```
Row     Slack or Surplus     Dual Price
 1         320.0000          -1.000000
 2         0.000000           3.000000
 3         0.000000           0.000000
 4         0.000000          -6.000000
 5         0.000000          -5.000000
 6         0.000000           5.000000
 7         0.000000           4.000000
 8         40.00000           0.000000
```

10A.3 Example 4.14 (Model 4.15)

$$\text{Minimize } Z = 25X_{11} + 20X_{12} + 30X_{13} + 20X_{21} + 15X_{22} + 35X_{23} + 18X_{31} + 19X_{32} + 28X_{33}$$

Subject to

$$X_{11} + X_{12} + X_{13} = 1$$
$$X_{21} + X_{22} + X_{23} = 1$$
$$X_{31} + X_{32} + X_{33} = 1$$
$$X_{11} + X_{21} + X_{31} = 1$$
$$X_{12} + X_{22} + X_{32} = 1$$
$$X_{13} + X_{23} + X_{33} = 1$$
$$X_{ij} \geq 0 \text{ (or } X_{ij} \in 0, 1) \quad \forall i, j$$

10A.3.1 LINGO in Algebraic Form

```
Model:
!Objective function;
Min=25*X11 + 20*X12 + 30*X13 + 20*X21 + 15*X22 + 35*X23
     + 18*X31 + 19*X32 + 28*X33;
!Constraint;
X11 + X12 + X13=1;
X21 + X22 + X23=1;
X31 + X32 + X33=1;
X11 + X21 + X31=1;
X12 + X22 + X32=1;
X13 + X23 + X33=1;
End
```

10A.3.2 LINGO Solution Report

```
Global optimal solution found.
Objective value:              63.00000
Total solver iterations:  5

Variable        Value          Reduced Cost
   X11        0.000000          5.000000
   X12        0.000000          0.000000
   X13        1.000000          0.000000
   X21        0.000000          5.000000
   X22        1.000000          0.000000
   X23        0.000000          10.00000
   X31        1.000000          0.000000
   X32        0.000000          1.000000
   X33        0.000000          0.000000

Row      Slack or Surplus      Dual Price
 1           63.00000          -1.000000
 2           0.000000          -20.00000
 3           0.000000          -15.00000
 4           0.000000          -18.00000
 5           0.000000           0.000000
 6           0.000000           0.000000
 7           0.000000          -10.00000
```

10A.3.3 LINGO Codes (Alternative Form)

```
MODEL:
!Flexible LINGO coding for the Assignment Problem;
SETS:
!Declaring the service team - we have three teams here;
  TEAM /S1  S2  S3/;
!Declaring the locations - we have three locations here;
  LOCATION /L1  L2  L3/;
!Identifying the assignment links between each service
team to each location - cost for each link is given and
ASSIGN is the variable representing the assignment;
  LINKS (TEAM, LOCATION) : COST, ASSIGN;
ENDSETS

!Objective function - total cost of assignment;
  MIN = @SUM (LINKS (I,J) :
    COST (I,J) * ASSIGN (I,J));
!Team assignment constraint - one team will be assigned in
one of three locations;
  @FOR (TEAM (I) :
    @SUM (LOCATION (J) : ASSIGN (I,J)) = 1);
!Location constraints - each location will receive only one
service team;
```

```
   @FOR(LOCATION(J):
     @SUM(TEAM(I): ASSIGN(I,J)) = 1);
 !Here is the data;
 DATA:
   COST = 25  20  30
          20  15  35
          18  19  28;
 ENDDATA
 END
```

10A.3.4 LINGO Solution Report (for Alternative Codes)

```
Global optimal solution found.
Objective value:             63.00000
Total solver iterations:  5

   Variable            Value        Reduced Cost
COST(S1, L1)         25.00000        0.000000
COST(S1, L2)         20.00000        0.000000
COST(S1, L3)         30.00000        0.000000
COST(S2, L1)         20.00000        0.000000
COST(S2, L2)         15.00000        0.000000
COST(S2, L3)         35.00000        0.000000
COST(S3, L1)         18.00000        0.000000
COST(S3, L2)         19.00000        0.000000
COST(S3, L3)         28.00000        0.000000
ASSIGN(S1, L1)       0.000000        5.000000
ASSIGN(S1, L2)       0.000000        0.000000
ASSIGN(S1, L3)       1.000000        0.000000
ASSIGN(S2, L1)       0.000000        5.000000
ASSIGN(S2, L2)       1.000000        0.000000
ASSIGN(S2, L3)       0.000000        10.00000
ASSIGN(S3, L1)       1.000000        0.000000
ASSIGN(S3, L2)       0.000000        1.000000
ASSIGN(S3, L3)       0.000000        0.000000

Row     Slack or Surplus     Dual Price
 1          63.00000         -1.000000
 2          0.000000         -20.00000
 3          0.000000         -15.00000
 4          0.000000         -18.00000
 5          0.000000          0.000000
 6          0.000000          0.000000
 7          0.000000         -10.00000
```

Suppose that service team 3 cannot be assigned to location 3 for political reasons. How can we incorporate this information into the model? We can either (1) use cost equal to big-M for this assignment or (2) force X33 to be zero. If we use the former one, LINGO function BIGMVL for big-M would not work here (as it must be linked to a constraint) although setting M = 1000 would work nicely.

10A.4 Example 3.6 (Model 4.1)

$$\text{Max } Z = 18x_1 + 16x_2 + 12x_3 + 25x_4 + 14x_5$$

Subject to

$$6x_1 + 18x_2 + 10x_3 + 9x_4 + 4x_5 \leq 30$$

$$x_i \text{ either 1 or 0} \quad \forall i$$

10A.4.1 LINGO in Algebraic Form

```
Model:
!Objective function;
Max = 18*x1 + 16*x2 + 12*x3 + 25*x4 + 14*x5;
!Constraint;
6*x1 + 18*x2 + 10*x3 + 9*x4 + 4*x5 <= 30;
!Binary declaration;
@BIN (x1); @BIN (x2); @BIN (x3); @BIN (x4); @BIN (x5);
End
```

10A.4.2 LINGO Model Statistics

```
Rows = 2 Vars = 5 No. integer vars = 5 (all are linear)
Nonzeros = 11 Const nonz = 5 (0 are + - 1) Density = 0.917
Smallest and largest elements in abs value = 4.00000
30.0000
No. <: 1 No. =: 0 No. >:0, Obj = MAX, GUBs <= 1
Single cols = 0
```

10A.4.3 LINGO Solution

```
Global optimal solution found.
Objective value: 69.00000

Variable        Value        Reduced Cost
   X1         1.000000        -18.00000
   X2         0.000000        -16.00000
   X3         1.000000        -12.00000
```

```
   X4          1.000000        -25.00000
   X5          1.000000        -14.00000

 Row      Slack or Surplus      Dual Price
  1           69.00000           1.000000
  2           1.000000           0.000000
```

10A.4.4 LINGO Codes (Alternative Form)

```
Model:
!Defining Sets: project is a set of projects and each
project has a binary variable X, utility and cost;
SETS:
PROJECT: X, UTILITY, COST;
ENDSETS

!Objective function;
Max = @SUM (PROJECT(I): X(I)*UTILITY(I));
!Constraint;
@SUM (PROJECT(I): X(I)*COST(I)) <= BUDGET;

!Binary declaration;
@FOR (PROJECT(I): @BIN (X(I)));


DATA:
!Number of project with an identification code;
PROJECT = P1  P2  P3  P4  P5;
!Utility of each project;
UTILITY = 18  16  12  25  14;
!Cost of each project;
COST = 6  18  10  9  4;
!Budget available;
BUDGET = 30;
ENDDATA
End
```

10A.4.5 LINGO Solution for Alternative Codes

```
Global optimal solution found.
Objective value: 69.00000

 Variable           Value        Reduced Cost
BUDGET            30.00000         0.000000
X(P1)             1.000000       -18.00000
X(P2)             0.000000       -16.00000
X(P3)             1.000000       -12.00000
```

```
X(P4)            1.000000      -25.00000
X(P5)            1.000000      -14.00000
UTILITY(P1)      18.00000       0.000000
UTILITY(P2)      16.00000       0.000000
UTILITY(P3)      12.00000       0.000000
UTILITY(P4)      25.00000       0.000000
UTILITY(P5)      14.00000       0.000000
COST(P1)         6.000000       0.000000
COST(P2)         18.00000       0.000000
COST(P3)         10.00000       0.000000
COST(P4)         9.000000       0.000000
COST(P5)         4.000000       0.000000

Row     Slack or Surplus     Dual Price
 1          69.00000          1.000000
 2          1.000000          0.000000
```

10A.5 Example 5.3 (Model 5.2)

$$\text{Maximize } Z = 4x_1 + 7x_2$$

Subject to

$$2x_1 + x_2 \leq 6{,}000 + M(1 - y_1)$$
$$3x_1 + 7x_2 \leq 13{,}000 + M(1 - y_2)$$
$$5x_1 + 6x_2 \leq 12{,}500 + M(1 - y_3)$$
$$y_1 + y_2 + y_3 = 1$$
$$x_1, x_2 \geq 0$$
$$\text{and } y_i = 0, 1 \quad \text{for all } i$$

10A.5.1 LINGO Codes

```
Model:
!Objective function;
Max = 4*x1 + 7*x2;
!Constraint;
!Assuming M = 100,000;
2*x1 + x2 <= 6000 + 100000*(1 - y1);
3*x1 + 7*x2 <= 13000 + 100000*(1 - y2);
5*x1 + 6*x2 <= 12500 + 100000*(1 - y3);
y1 + y2 + y3 = 1;
!Binary declaration;
@BIN (y1); @BIN (y2); @BIN (y3);
End
```

10A.5.2 LINGO Solution Report

```
Global optimal solution found.
Objective value:            42000.00
Extended solver steps:      3
Total solver iterations:    15

Variable        Value        Reduced Cost
   X1         0.000000        10.00000
   X2         6000.000        0.000000
   Y1         1.000000        700000.0
   Y2         0.000000        0.000000
   Y3         0.000000        0.000000

Row     Slack or Surplus      Dual Price
 1          42000.00          1.000000
 2          0.000000          7.000000
 3          71000.00          0.000000
 4          76500.00          0.000000
 5          0.000000          0.000000
```

10A.5.3 LINGO Alternative Codes

```
Model:
!Objective function;
Max = 4*x1 + 7*x2;
!Constraint;
!Using LINGO Big-M value function;
2*x1 + x2 <= 6000 + BIGMVL*(1 - y1);
3*x1 + 7*x2 <= 13000 + BIGMVL*(1 - y2);
5*x1 + 6*x2 <= 12500 + BIGMVL*(1 - y3);
y1 + y2 + y3 = 1;
!Binary declaration;
@BIN (y1); @BIN (y2); @BIN (y3);
End
```

10A.5.4 LINGO Solution Report (for Alternative Codes)

```
Linearization components added:
Constraints:                12
Variables:                  3
Global optimal solution found.
Objective value:            42000.00
Extended solver steps:      3
Total solver iterations:    21
```

```
Variable        Value          Reduced Cost
 X1            0.000000        10.00000
 X2            6000.000        0.000000
 BIGMVL        29000.00        0.000000
 Y1            1.000000        700000.0
 Y2            0.000000        0.000000
 Y3            0.000000        0.000000

Row      Slack or Surplus      Dual Price
 1          42000.00            1.000000
 2          0.000000            7.000000
 3          0.000000            0.000000
 4          5500.000            0.000000
 5          0.000000            0.000000
```

10A.6 Example 5.16

$$\text{Minimize } Z = 3X_1 + 2X_2 + 2X_3$$
Subject to
$$X_1 + 4X_2 + X_3 \geq 7$$
$$2X_1 + X_2 + X_4 \geq 10$$
$$X_1, X_2, X_3 \geq 0$$
$$X_4 \text{ unrestricted in sign}$$

10A.6.1 LINGO Codes

```
Model:
!Objective function;
Min=3*X1 + 2*X2 + 2*X3;
!Constraint;
X1 + 4*X2 + X3 >=7;
2*X1 + X2 + X4 >=10;
!Declaration for unrestriction in sign;
@FREE (x4);
End
```

10A.6.2 LINGO Solution Report

```
Global optimal solution found.
Objective value:             3.500000
Total solver iterations:     0
```

```
Variable        Value         Reduced Cost
   X1         0.000000          2.500000
   X2         1.750000          0.000000
   X3         0.000000          1.500000
   X4         8.250000          0.000000

Row      Slack or Surplus      Dual Price
 1          3.500000           -1.000000
 2          0.000000           -0.5000000
 3          0.000000            0.000000
```

10A.7 Example 4.11 (Model 4.12)

Minimize $Z = P_1d_1^- + P_2d_2^- + P_3d_3^-$

Subject to

$$x_1 + x_2 + x_3 + 0.333x_4 \leq 100$$

$$x_2 + x_3 + 0.143x_4 \leq 100$$

$$x_3 + x_4 \leq 150$$

$$168x_1 + 288x_2 + 288x_3 + 391x_4 + d_1^- - d_1^+ = 72{,}000$$

$$150x_1 + 80x_2 + 86x_3 + 70x_4 + d_2^- - d_2^+ = 16{,}000$$

$$112x_1 + 162x_2 + 192x_3 + 89x_4 + d_3^- - d_3^+ = 23{,}000$$

$$x_i,\ d_i^-,\ d_i^+ \geq 0 \quad \forall i$$

10A.7.1 LINGO Codes

```
Model:
!Objective function;
!Assuming P1=1000, P2=100 and P3=1;
Min=1000*DNEG1 + 100*DNEG2 + DNEG3;
!Hard constraint;
x1 + x2 + x3 + 0.333*x4 <=100;
x2 + x3 + 0.143*x4 <=100;
x3 + x4 <=150;
!Goal /Soft constraint;
168*x1 + 288*x2 + 288*x3 + 391*x4 + DNEG1 - DPOS1=72000;
150*x1 + 80*x2 + 86*x3 + 70*x4 + DNEG2 - DPOS2=16000;
112*x1 + 162*x2 + 192*x3 + 89*x4 + DNEG3 - DPOS3=23000;
End
```

10A.7.2 LINGO Solution Report

```
Global optimal solution found.
Objective value:              89495.40
Total solver iterations:      7

Variable         Value         Reduced Cost
  DNEG1        0.000000          942.0833
  DNEG2        875.1000          0.000000
  DNEG3        1985.400          0.000000
     X1        8.870000          0.000000
     X2        41.18000          0.000000
     X3        0.000000          20832.03
     X4        150.0000          0.000000
  DPOS1        0.000000          57.91667
  DPOS2        0.000000          100.0000
  DPOS3        0.000000          1.000000

Row      Slack or Surplus       Dual Price
 1           89495.40           -1.000000
 2           0.000000            24842.00
 3           37.37000            0.000000
 4           0.000000            21462.03
 5           0.000000           -57.91667
 6           0.000000           -100.0000
 7           0.000000           -1.000000
```

10A.8 Example 5.10 (Model 5.7)

$$\text{Minimize } Z = (0.85)^{X_{H1}} \times (0.70)^{X_{H2}} \times (0.75)^{X_{H3}} \times (0.65)^{X_{H4}} \times (0.90)^{X_{M1}} \times (0.85)^{X_{M2}} \times (0.88)^{X_{M3}} \times (0.80)^{X_{M4}}$$

$\in$

$$450X_{H1} + 500X_{H2} + 550X_{H3} + 650X_{H4} + 333.33X_{M1} + 366.67X_{M2} + 400X_{M3} + 466.67X_{M4} \leq 30{,}000.$$

$$X_{H1} + X_{H2} + X_{H3} + X_{H4} \leq 30.$$

$$X_{M1} + X_{M2} + X_{M3} + X_{M4} \leq 25.$$

$$X_{ij} \geq 0 \quad \text{for all } i \text{ and } j.$$

10A.8.1 LINGO Codes

```
Model:
!Objective function;
Min=085^XH1+0.70^XH2+0.75^XH3+0.65^XH4+0.90^XM1
+0.85^XM2+0.88^XM3+0.80^XM4;
!Constraint;
```

```
450*XH1 + 500*XH2 + 550*XH3 + 650*XH4 + 333.33*XM1
+ 366.67*XM2 + 400*XM3 + 466.67*XM4 <= 30000;
XH1 + XH2 + XH3 + XH4 <= 30;
XM1 + XM2 + XM3 + XM4 <= 25;
End
```

10A.8.2 LINGO Solution Report

```
Local optimal solution found.
Objective value:            2.667184
Total solver iterations:    34

Variable         Value         Reduced Cost
   XH1         0.000000          4.453168
   XH2         9.879749          0.000000
   XH3         11.50190          0.000000
   XH4         8.618351          0.000000
   XM1         5.928616          0.000000
   XM2         6.510309          0.000000
   XM3         6.398812          0.000000
   XM4         6.162263          0.000000

Row      Slack or Surplus        Dual Price
 1          2.667184            -1.000000
 2          3333.564             0.000000
 3          0.000000             0.1051671E-01
 4          0.000000             0.5641561E-01
```

11

Output Analysis and Practical Issues

11.1 Introduction

After solving a mathematical model, the next stages of the decision process (as of Figure 2.1) are to perform a sensitivity analysis and then implement the solutions. Before implementation, however, the solutions must be analyzed and interpreted in a nontechnical manner and in nonmathematical terms so that nonexpert users can understand, accept, and implement the solutions. Otherwise, the whole effort would be to no avail, which frequently occurs in practice.

In this chapter, we will discuss the solutions of models and the outputs provided by different software packages, and some practical issues faced by the users when dealing with optimization models and software packages. A number of examples to demonstrate these issues will be provided.

11.2 Solutions and Reports

When model solving, we usually expect to find the values of the decision variables and the objective function. Depending on the model type, it is possible to generate much more information, which may assist in arriving at better decision alternatives. For example, information derived from *sensitivity analysis* and *parametric programming* in linear programming (LP). In addition to computationally more efficient algorithms, the ability of the algorithm to perform and report upon associated sensitivity analysis has made LP the most popular optimization tool. However, due to the discrete nature of the solution space, sensitivity analysis is not automatic for integer programming models and limited sensitivity information can be generated for nonlinear models.

In solving any LP model, Excel Solver provides three report options: Answer, Sensitivity, and Limits. Any or all of these reports can be saved after a model has been solved. LINDO/LINGO and MPL/Conopt

Microsoft Excel 11.0 Answer Report
Worksheet: [Solv-solver-ex1.xls]Sheet1
Report Created: 18/03/2005 4:06:51 PM

Target Cell (Max)

Cell	Name	Original Value	Final Value
$F$4	Changing cells --> Objective Value	0	6350

Adjustable Cells

Cell	Name	Original Value	Final Value
$C$4	Changing cells --> Model A	0	12
$D$4	Changing cells --> Model B	0	11

Constraints

Cell	Name	Cell Value	Formula	Status	Slack
$G$7	Labor availability LHS	35	$G$7<=$H$7	Not Binding	5
$G$8	Machine time LHS	45	$G$8<=$H$8	Binding	0
$G$9	Market restriction LHS	12	$G$9<=$H$9	Binding	0

FIGURE 11.1
Solver answer report.

provide one report that contains all the necessary solutions and sensitivity information.

Consider Example 10.7. Although the solver answer report has already been presented in Chapter 10, the sensitivity report (Figure 11.2) together with the answer report (Figure 11.1) is presented here for ease of explanation.

In the first section of the sensitivity report, reduced costs are provided together with the allowable increases and decreases in the original objective

Microsoft Excel 11.0 Sensitivity Report
Worksheet: [Solv-solver-ex1.xls]Sheet1
Report Created: 18/03/2005 3:48:14 PM
Adjustable Cells

Cell	Name	Final Value	Reduced Cost	Objective Coefficient	Allowable Increase	Allowable Decrease
$C$4	Changing cells --> Model A	12	0	300	1E+30	216.6666667
$D$4	Changing cells --> Model B	11	0	250	650	250

Constraints

Cell	Name	Final Value	Shadow Price	Constraint R.H. Side	Allowable Increase	Allowable Decrease
$G$7	Labor availability LHS	35	0	40	1E+30	5
$G$8	Machine time LHS	45	83.33333333	45	15	33
$G$9	Market restriction LHS	12	216.6666667	12	3	12

FIGURE 11.2
Solver sensitivity report.

function coefficients. In the next section, the report presents shadow prices and the allowable increases and decreases in the original constraints' right-hand side (RHS). To understand the sensitivity report, let us define shadow price and reduced cost.

11.2.1 Shadow Price

The shadow price for a constraint indicates the amount by which the objective function value changes given a unit increase in the RHS value of the constraint, assuming all other coefficients remain constant. If a shadow price is positive (negative), a unit increase in the RHS value of the associated constraint results in an increase (decrease) in the optimal objective function value. When a unit is decreased in the RHS value, a positive shadow price would decrease the optimal objective function value. The shadow price values apply, provided that the increase or decrease in the RHS value falls within the allowable increase or allowable decrease limits in the sensitivity report of each constraint.

For example, the shadow price for the machine time constraint is \$83.333. Therefore, if the number of available machine hours is increased by any amount in the range 0–15 h, the optimum objective function value changes (increases) by \$83.333 for each additional machine hour. If the number of available machine hours is decreased by any amount in the range 0–33 h, the optimum objective function value changes (decreases) by –\$83.333 for each machine hour decreased. A similar interpretation holds for the other constraints.

The shadow price for a nonbinding constraint is always zero. For example, the shadow price for the labor constraint is zero with an allowable increase of infinity (1E + 30 in the report) and allowable decrease of 5 h. Therefore, if the RHS value for the labor constraint increases by any amount, the objective function value does not change. This result is not surprising, because the optimal solution to the problem leaves 5 h (slack) of labor unused. Interestingly, we can reduce the RHS of the labor constraint by 5 h without affecting the optimal solution. For a given constraint, there is a relationship between shadow price and slack amount as shown in Table 11.1.

TABLE 11.1

The Relationship between Shadow Price and Slack or Surplus Amounts

	Shadow Price	Slack or Surplus
Condition-1	Zero	Nonnegative
Condition-2	Nonzero	Zero

TABLE 11.2

The Relationship between Shadow Price and Objective Function Value

Right-Hand Side	Shadow Price	Objective Function Value
Add one unit	Positive	Increases by the amount of shadow price
	Negative	Decreases by the amount of shadow price
	Zero	Remains unchanged
Delete one unit	Positive	Decreases by the amount of shadow price
	Negative	Increases by the amount of shadow price
	Zero	Remains unchanged

The relationship between the shadow price and the objective function value can be shown as of Table 11.2.

11.2.2 Reduced Cost

By definition, all the variables are assumed nonnegative ($x_i \geq 0$) as is the case in this book. The reduced cost for a variable indicates the amount by which the objective function value changes given a *unit increase* (from zero) in the variable value. This is similar to the shadow price concept but for nonnegative constraints. The reduced cost for each variable is equal to the per-unit amount the variable contributes to the objective function minus the per-unit value of the resources it consumes (where the consumed resources are priced at their shadow price). A reduced cost value is valid only within its allowable increase and allowable decrease as per the sensitivity report.

In our example, Model-A consumes 2 units of labor hours, 1 unit of machine hours, and 1 unit of market limitation. The shadow prices for labor hours, machine hours, and market limitation are \$0, \$83.333, and \$216.667, respectively. The profit per unit of Model-A is \$300. If we increase the production by one unit of Model-A, the cost of resources (as per shadow prices) $= 2 \times \$0 + 1 \times \$83.333 + 1 \times \$216.667 = \$300$. So the reduced cost $=$ (unit contribution − unit price) $= \$300 - \$300 = \$0$. There is a relationship between the optimal value of the decision variable and the optimal value of reduced cost (provided by Solver) as shown in Table 11.3.

TABLE 11.3

Nature of Optimal Reduced Cost Values

Type of Objective	Optimal Value of Decision Variable	Optimal Value of Reduced Cost
Maximization	At simple lower bound	$\leq$
	Between lower and upper bounds	$=$
	At simple upper bound	$\geq$
Minimization	At simple lower bound	$\geq$
	Between lower and upper bounds	$=$
	At simple upper bound	$\leq$

11.3 Sensitivity Analysis

The optimal solution to an LP problem is based on a set of assumptions and on the forecasting of future data such as prices and market demand. In a deterministic model, there is no provision to incorporate risk or uncertainty. Therefore, it is important for the management to know what will happen to the optimal solution if changes occur in the input data on which the LP model is based. Because such analysis is done after the optimal solution is found, this approach is also known as *post-optimality* analysis. However, sensitivity analysis is the most common type of post-optimality analysis.

Sensitivity analysis is the investigation of the effect on optimal solutions related to changes in the model parameters after solving the model under consideration. If you want to determine the effect of some change in the model, the simplest approach is to change the model and re-solve it. This approach is suitable if the model does not take an excessive amount of time to change or solve. If you are interested in studying the consequences of changing a number of coefficients one at a time or simultaneously changing several coefficients in the model, then the sensitivity analysis may be the only practical approach. Most optimization software (if simplex-based) provide some sensitivity information after solving an LP model. The information is used for sensitivity analysis without re-solving the model for every change. Sensitivity analysis can help answer a number of practical managerial questions that might arise about the solution to an LP model. In this section, we would demonstrate sensitivity analysis with some examples. The parameters that can be changed in any LP model are the coefficients of the objective function, the constraint RHS values, and the coefficients of the constraints. In addition, we may introduce a new product or variable after solving an LP problem. In the following few subsections, we will briefly discuss these changes.

11.3.1 Changes in the Objective Coefficients

After solving an LP problem you might want to investigate the effect of objective coefficient changes on the optimal solution as the coefficients may either have been predicted values or they may vary due to the changing environment. For example, Model-A sales prices may decrease due to a new competitor coming into the market with a lower price and a better after-sales service offer.

In a previous problem, the objective function coefficients for Model-A and Model-B were set to \$300 and \$250, respectively. The LP solution suggests that it is optimal to produce 12 units of Model-A and 11 units of Model-B in order to maximize the profit. In Table 11.2, the allowable increases and decreases of the objective function coefficients are provided. For example, the objective function coefficient associated with Model-B can be increased by as much as \$650 or decreased by \$250 without changing the optimal solution, assuming all other coefficients remain constant. A similar conclusion can be drawn for Model-A.

11.3.2 Changes in the RHS Values

After solving an LP problem you might want to determine how much better or worse the solution would be if we had more or less of a given resource available. For example, one might wonder how much more profit could be earned if additional machine hours were available.

We must mention here that the RHS values of nonbinding constraints do not, within their allowable increase and decrease, improve the objective function value. On the other hand, the RHS values of binding constraints are very important as they prevent us from further improving (maximizing or minimizing) the objective function value. We have already learnt that shadow price information is very useful in determining the level of possible resource addition.

The machine time and market restriction constraints are binding in our example. Let us now suppose that the RHS value of the machine time constraint in our example increases by 9 h (from 45 to 54) due to the addition of a new machine. Because this increase is within the allowable increase listed for the machine time constraint in Table 11.2, you might expect that the optimal objective function value would increase by $83.333 × 9 = $750.00. That is, the new objective function value would be ($6350.00 + $750.00 = $7100.00). After resolving the model using a machine time of 54 h, the new objective function value turns out to be $7100.00 as expected. However, the new solution involves producing 12 units of Model-A and 14 units of Model-B with a profit of $7100.00. The optimal solution is different from the solution to the original problem because of the additional resources.

As discussed above, an additional 9 h of machine time would increase the profit by $750.00. A question might then arise as to how much we should be willing to pay to acquire these additional 9 h of machine time. This is up to management as the profit is determined on the basis of the cost of regular machine time. However, to maintain a profit of $6350.00 or more, the management can pay up to $750.00 extra to acquire the additional 9 h of machine time.

11.3.3 Changes in the Constraint Coefficients

Using shadow price and reduced cost information, one can analyze how changes in some constraint coefficients affect the optimal solution to an LP problem. As discussed earlier, the reduced cost for each variable is equal to the per-unit profit the product contributes to the overall profit (objective function value) minus the per-unit value of the resources it consumes. In that case, the consumed resources are priced at their shadow prices. For example, the reduced cost for Model-A can be calculated as follows:

Reduced cost of Model-A = $300 − 2 × $0 − 1 × $83.333 − 1 × $216.667 = $0

In our example, Model-A consumes 2 units of labor hours, 1 unit of machine hours, and 1 unit of market limitation. The shadow prices for labor hours, machine hours, and market limitation are \$0, \$83.333, and \$216.667, respectively. The profit per unit of Model-A is \$300.

If the resource requirements for Model-A are increased, due to improved quality demand, the reduced cost for Model-A would be negative. For example, the reduced cost would be –\$8.3333 for a 10% increase in machine time per-unit product as calculated below:

$$\begin{aligned}\text{Reduced cost of Model-A} &= \$300 - 2 \times \$0 - 1.10 \times \$83.333 - 1 \times \$216.667 \\ &= -\$8.3333\end{aligned}$$

It indicates that the increased resource requirements would decrease the profitability of Model-A by \$8.3333 per unit. For 12 units of Model-A, the total decrease in profitability would be $12 \times \$8.3333 = \$100.00$. So the new objective value would be $\$6350 - \$100 = \$6250$. Because of the increase in resource requirements by one product, the new optimal solution will be different.

11.3.4 Addition of New Product or Variable

Suppose we now consider introducing a new product called Model-C. Each unit of Model-C requires 1.5 h of labor and 2.5 h of machine time and the product can be sold to generate a marginal profit of \$225. Would the production of this new model be profitable?

The cost of resources required to produce one unit of Model-C can be calculated as follows:

$$\begin{aligned}\text{Cost of resources for Model-C} &= 1.5 \times \$0 + 2.5 \times \$83.333 + 0 \times \$216.667 \\ &= \$208.33\end{aligned}$$

The reduced cost for Model-C = \$225.00 – \$208.33 = +\$16.67. This indicates that the profitability would be increased by \$16.67 for each unit of Model-C produced. Therefore, the production of Model-C would be profitable. In this case, the solution we obtained in Tables 11.1 and 11.2 would no longer be optimal.

11.3.5 Sensitivity Analysis for Integer and Nonlinear Models

As discussed earlier, the sensitivity report is not meaningful for integer models. However, the answer report for Example 10.8 is reported in Figure 11.3.

The integer programming answer report is similar to the LP answer report except it includes integer variables under the constraint section.

Microsoft Excel 11.0 Answer Report
Worksheet: [Solv-solver-ex2-integer.xls]Sheet1
Report Created: 12/04/2005 4:28:42 PM

Target Cell (Max)

Cell	Name	Original Value	Final Value
$F$4	Changing cells --> Objective Value	0	5850

Adjustable Cells

Cell	Name	Original Value	Final Value
$C$4	Changing cells --> Model A	0	12
$D$4	Changing cells --> Model B	0	9

Constraints

Cell	Name	Cell Value	Formula	Status	Slack
$G$7	Labor availability LHS	33	$G$7<=$H$7	Not Binding	7
$G$8	Machine time LHS	39	$G$8<=$H$8	Not Binding	1
$G$9	Market restriction LHS	12	$G$9<=$H$9	Binding	0
$C$4	Changing cells --> Model A	12	$C$4=integer	Binding	0
$D$4	Changing cells --> Model B	9	$D$4=integer	Binding	0

FIGURE 11.3
Answer report for an integer model.

The answer report, sensitivity report, and limits report for a problem with nonlinear objective and linear constraints are presented below (see Figures 11.4 through 11.6). The problem is similar to our Example 10.9 except that the variables are considered as continuous.

Microsoft Excel 11.0 Answer Report
Worksheet: [Solv-solver-ex3-nonlinear.xls]Sheet1
Report Created: 12/04/2005 4:37:32 PM

Target Cell (Max)

Cell	Name	Original Value	Final Value
$F$4	Changing cells --> Objective Value	0	2958.018204

Adjustable Cells

Cell	Name	Original Value	Final Value
$C$4	Changing cells --> Model A	0	6.481746338
$D$4	Changing cells --> Model B	0	11.17275122

Constraints

Cell	Name	Cell Value	Formula	Status	Slack
$G$7	Labor availability LHS	24.1362439	$G$7<=$H$7	Not Binding	15.8637561
$G$8	Machine time LHS	40	$G$8<=$H$8	Binding	0
$G$9	Market restriction LHS	6.481746338	$G$9<=$H$9	Not Binding	5.518253662

FIGURE 11.4
Answer report for a nonlinear model.

Microsoft Excel 11.0 Sensitivity Report
Worksheet: [Solv-solver-ex3-nonlinear.xls]Sheet1
Report Created: 12/04/2005 4:37:32 PM

Adjustable Cells

Cell	Name	Final Value	Reduced Gradient
$C$4	Changing cells --> Model A	6.481746338	0
$D$4	Changing cells --> Model B	11.17275122	0

Constraints

Cell	Name	Final Value	Lagrange Multiplier
$G$7	Labor availability LHS	24.1362439	0
$G$8	Machine time LHS	40	58.91762288
$G$9	Market restriction LHS	6.481746338	0

FIGURE 11.5
Sensitivity report for a nonlinear model.

The answer report is similar to the LP answer report.

The sensitivity report provides the variables with their reduced gradients and the constraints with their Lagrange multipliers. The reduced gradient approximately represents the reduced cost discussed in LP. Similarly, the Lagrange multiplier approximately substitutes for the shadow price. In nonlinear programming, there is no simple way to compute the allowable increase and allowable decrease over which the reduced gradient and Lagrange multiplier are valid. However, the reduced gradient and Lagrange multiplier information can still be used to estimate the approximate impact

Microsoft Excel 11.0 Limits Report
Worksheet: [Solv-solver-ex3-nonlinear.xls]Limits
Report 1
Report Created: 12/04/2005
4:37:32 PM

Cell	Target Name	Value
$F$4	Changing cells --> Objective Value	2958.018204

Cell	Adjustable Name	Value	Lower Limit	Target Result	Upper Limit	Target Result
$C$4	Changing cells --> Model A	6.481746338	0	2194.239983	6.481746338	2958.018204
$D$4	Changing cells --> Model B	11.17275122	0	763.7782207	11.17275122	2958.018204

FIGURE 11.6
Limits report for a nonlinear model.

on the objective function for a small change in the value of a given variable and in a constraint's RHS, respectively.

11.4 Practical Issues and Tips

A number of practical issues frequently faced by practitioners are discussed here and some tips are provided to overcome them.

11.4.1 Solutions to Goal Programming Problems

Although goal programming handles multiple objectives as a single composite objective function, it does not optimize all the objectives simultaneously. The solutions to goal programming problems are considered as good feasible compromises and workable solutions rather than true optimal solutions.

11.4.2 Multi-Objective Optimization

Many real world problems involve simultaneous optimization of several incommensurable and often competing objectives. In such problems, there is no single optimal solution, but rather a set of alternative solutions. These solutions are optimal in the wider sense in that no other solutions in the search space are superior to them when all objectives are simultaneously considered. These solutions are known as *Pareto-optimal* solutions. However, the decision maker usually requires only one solution from the whole set of solutions for implementation. Theoretically, the solutions in the set are *indifferent* (in some sense!), which makes it very difficult to choose one specific solution.

11.4.3 Reduction of Variables and Constraints

To reduce the size of a model, one must not include unnecessary variables and constraints; they should be excluded from the model. Reduction in size means reduction in computational complexity and time. In some cases where more than one variable is to have the same value in the final solution, such variables can be represented by one common variable. As an example, consider the crop planning model of Section 15.4. The variables for double and triple cropped land can be reduced as demonstrated in Section 15.4.1.

In many modelling situations, extra constraints may be included to find additional information from the modelling solution. If this information could have been derived after the model had been solved, then to reduce the model size, one should exclude such constraints. Consider the model of Example 10.7 as presented below again.

Variables:

x_1 = Number of sets of model A to be produced daily
x_2 = Number of sets of model B to be produced daily

Linear programming model is

$$\begin{aligned}
&\text{Maximize } Z = 300x_1 + 250x_2 \\
&\text{Subject to} \\
&\quad 2x_1 + x_2 \le 40 \quad \text{Labor constraint} \\
&\quad x_1 + 3x_2 \le 45 \quad \text{Machine time constraint} \\
&\quad x_1 \le 12 \quad \text{Market constraint} \\
&\quad x_1, x_2 \ge 0 \quad \text{Nonnegativity constraint}
\end{aligned} \qquad \text{Model (11.1)}$$

In this model, additional constraints can be incorporated to calculate total labor and machine hours used as follows:

$$2x_1 + x_2 - L = 0 \text{ labor hours calculation} \tag{11.1}$$

$$x_1 + 3x_2 - M = 0 \text{ machine hours calculation} \tag{11.2}$$

Where, L and M are the variables to represent the total labor hours and machine hours used, respectively. However, if the model is solved without these additional constraints, one can still find the values of L and M using the Equations 11.1 and 11.2 and model solutions.

By removing redundant constraints one can reduce the problem size, hence the computational time; however, it is not easy to identify redundant constraints when a model involves a large number of constraints and variables. By removing redundant constraints, however, the feasible solution space is unaffected. It is expected that if redundant constraints have been identified then they would be excluded from the model. The issue of identifying redundant constraints before solving a model is an interesting research topic.

11.4.4 Solutions and Number of Basic Variables

The number of basic variables is equal to the number of functional constraints in an LP model's solution. Suppose you have modelled a problem that contains 15 variables (each variable represents one product) but 6 key constraints, then only a maximum of 6 variables will have values different than zero in the solution. This solution may not sit well with management. The management will not be convinced to implement your solution as they are currently producing all 15 products (representing 15 variables) and making reasonable profits from their sales, and they may be interested to continue producing all 15 products to keep the company's market share steady and face the competitors. What is wrong with your solution? To answer this question, see the example in Section 15.2.

11.4.5 Variables with No Restriction in Sign

Many optimization techniques (for example, LP) assume that the values of variables must be nonnegative, for example, production of cars, manpower required, or machine hours used. However, there are variables that can be positive, negative, or both (+ve, zero, or –ve), for example, temperature (minus, zero, or plus), return on investment (profit, zero, or loss), or performance measuring scale/indicators. To handle variables with negative or unrestricted signs, one has to express each variable as a difference between two nonnegative variables. Suppose, a variable x is unrestricted in sign, then in a model, x can be replaced by $(x' - x'')$ where both x' and $x'' \geq 0$.

$$x = (x' - x'') \tag{11.3}$$

If $x' = 10$ and $x'' = 5$ (both positive), then $x = 5$ (positive value).

If $x' = 5$ and $x'' = 10$ (both positive), then $x = -5$ (negative value).

If you want to ensure that x will take either a negative value or zero (no positive value), then a restriction like $x' \leq x''$ must be imposed.

It is appropriate to mention here that Excel Solver and some optimization packages can directly handle variables that are unrestricted in sign.

11.4.6 Negative RHS

It is not unusual to have negative figures in the RHS of a functional constraint. However, most optimization techniques and software packages require a positive value or entry for the constraint RHS when inputting the model. Consider the following constraint:

$$2x - 3y \leq -5 \tag{11.4}$$

Because of the solution approach or some software package requirements, the constraint must be rewritten as follows:

$$-2x + 3y \geq +5 \tag{11.5}$$

It is a convention in the optimization domain to identify a model with (one or more) negative (–ve) constraint RHS as infeasible although the original problem is feasible. Some optimization techniques, such as simplex method of LP, cannot be applied to solve such a model. However, there are also techniques such as the dual simplex method that will remove such a technical infeasibility. It is noted that Excel Solver and some optimization packages can handle negative RHS.

11.4.7 Scaling Factors in Modelling

It is encouraged to model any problem with similar orders of magnitude. If the largest number in the model is greater than 1000 times the smallest number in the model, optimization software packages may encounter problems when solving the model. As numbers become extremely large or small, computers often run into storage or representation problems that force them to use approximations (by rounding) of the actual numbers. This opens the door for problems to occur in the accuracy of the results and, in some cases, can prevent the computer from solving the problem at all. So if some coefficients in the initial model are extremely large or extremely small, it is recommended to rescale the problem so that all the coefficients are of a similar magnitude.

For example, consider a financial problem with equations expressing an interest rate of 9.0% (.09) and budget constraints of $12,850,000. The difference in magnitude between these numbers is of the order of 10^9 (1/100th compared to 10,000,000). A difference of 10^4 or less between the largest and smallest units would be preferable. In this case, the budget could be expressed in units of millions of dollars. That is, $12.85 would be used to represent $12,850,000. This lowers the difference in magnitude between the numbers of the different units to 10^4.

Note that Excel Solver attempts to rescale the data automatically before solving the model. However, it is advisable that all the scaling is done during model formulation (that is before solving).

11.4.8 Linear vs. Nonlinear Relationships

When possible, one is encouraged to use linear rather than nonlinear relationships. Some nonlinear expressions can be reformulated in a linear manner. A simple example is a constraint involving the ratio of two variables as shown in the following equation:

$$x/y \leq 20 \tag{11.6}$$

This constraint is nonlinear because we are dividing by y. To linearize the constraint, one can multiply both sides by y. The equivalent, linear constraint becomes as given in the following equation:

$$x \leq 20y \quad \text{or} \quad x - 20y \leq 0 \tag{11.7}$$

11.4.9 Non-Smooth Relationships

It is appropriate to avoid non-smooth relationships when possible. Models with non-smooth constraints are generally much more difficult to solve. Where possible, approximate the non-smooth relationship with smooth expressions and, perhaps, use integer variables.

11.4.10 Linear vs. Integer and Nonlinear Models

Compared to linear models, it is usually more difficult to solve integer and nonlinear models. The solving of such models is usually done by relaxing the conditions and approximating the functions. Even if the alternative linear model involves more variables and constraints, it is preferable to use such models because of their (1) computationally efficient solution approaches, (2) easily available software packages, and (3) capability of software packages in generating useful sensitivity reports.

11.4.11 Rounding for Integer Solutions

Relaxation of integer restrictions can drastically reduce the solution time. In instances involving large numbers of variables, you may find that solving the model without integer restrictions and then rounding yields acceptable answers in a fraction of the time required by the integer model. Be forewarned, however, that rounding a solution will not necessarily yield a feasible or an optimal solution.

11.4.12 Improved Initial Solutions

The initial values you provide for the variables in a model can affect the *path* your solution approach or software package takes to obtain the optimal solution. Starting with values close to the optimal solution can noticeably reduce the solution time. In many situations, you may not know what the good initial values are. However, when you do know some reasonable ones, it may be to your benefit to use them as the starting point values.

11.4.13 Variable Bounds

Most LP software packages use a module called upper bounding technique, which helps to avoid treating variable bounds as constraints. That means it removes all single variable constraints from the constraint count.

The use of appropriate upper and lower bounds on variables can help make the solution search as efficient as possible. Using good bounds can save computational time as the algorithm is restricted to search a smaller solution space. For example, suppose you know that, even though the feasible value for a particular variable is between 10 and 200, it is highly improbable the optimal value is outside the range of 60–80; then in this case, the use of a lower bound of 60 and an upper bound of 80 could significantly reduce the solution time.

Bounding can also help keep the solution search clear of mathematically troublesome areas like undefined regions. For example, if you have a constraint with the term $5/x$, it may be helpful to add a lower bound on x so it does not get close to zero.

Consider the network flow problem of Example 4.15 as presented again below:

$$\text{Minimize } Z = 30X_{12} + 40X_{24} + 35X_{42} + 50X_{34} + 45X_{43} + 120X_{36} + 100X_{51} + 45X_{54} + 110X_{56}$$

Subject to

$$
\begin{aligned}
& X_{51} - X_{12} \geq 100 \\
& X_{12} + X_{42} - X_{24} \geq 150 \\
& -X_{43} + X_{34} + X_{36} \leq 300 \\
& X_{24} + X_{34} + X_{54} - X_{42} - X_{43} \geq 70 \\
& X_{51} + X_{54} + X_{56} \leq 200 \\
& X_{56} + X_{36} \geq 80 \\
& X_{ij} \geq 0 \quad \forall i, j
\end{aligned}
\qquad \text{Model (11.2)}
$$

The upper limit of X_{ij} is unlimited because the arcs do not have any capacity constraints. It is not difficult to find an appropriate bound for each flow variable. In theory, the tighter bounds mean a smaller feasible solution space.

11.4.14 Management Issues in Solution Implementation

The implementation issues are discussed in Section 2.9. In this section, we discuss the management issues for solution implementation. In some organizations, senior management appear to be obstacles when implementing new and sophisticated plans and technologies. This may be for a number of reasons as discussed below:

- They are unaware of the optimization tools and techniques and their capability. As a result, they cannot appreciate the new solutions and are afraid of the outcomes. However, without their acceptance, the solution cannot be implemented.
- Sometimes senior management do not want change if the current method is workable. Because any failure due to the suggested change may cause them to possibly lose their jobs. So who will take the risk?
- Some senior management cannot accept the good work carried out by junior colleagues as their own capability will be under question and their position may be under threat. Sometimes they are simply jealous of talented people. So they would not approve such work and allow the suggested implementation.
- In some instances senior management do not want to hastily implement junior colleagues' suggestions as that may not leave anything left for demonstrating future improvement. If no improvement can be shown for a long time, senior level jobs could be under risk.

It is not an easy task to solve the problems discussed above. There is no known model or solution approach for such problems. However, a few points can be made, which may help in solving such problems in some cases. These are as follows:

- Involve senior management at every stage of the decision process. To encourage their participation or sponsorship, the level of language in explaining the terms and processes must be in terms of their knowledge to ensure their understanding of the approach and solution technique.
- Seek their advice—by asking questions such as what can be done and how things can be improved? Allow them to feel that their participation and contribution to the process is valuable.
- Do not try to educate them directly or force them to do or say something they do not feel comfortable with at that stage. You need to make them become aware slowly of the development of the model and the possible solutions and convince them gradually of the appropriateness of the solutions.
- Do not try to demonstrate that you are more knowledgeable and capable than they are as it may backfire. However, on the other hand, creating a positive impression about your capability is necessary so that they can rely on you and appreciate your contributions.

As you can see, it could be a very long process. Remember that if you cannot convince them that your solution is the appropriate one then your whole effort will be discredited and wasted.

11.4.15 Gap between Solutions and Outcomes

Do not forget the assumptions that were made in the modelling and data preparation phases, the limitations of the solution approaches, and possible problems that may occur in the implementation phase. These factors may contribute to a possible gap between the model solution and the company goals after implementation. So it can be risky unless senior management is aware of the whole decision process.

11.4.16 Nontechnical Report

As we indicated earlier, the solutions must be interpreted in an easy and nontechnical form so that the users (usually officers and workers with different skills and knowledge levels) understand and find them easy for use. Otherwise, they may oppose any implementation of the derived solution.

11.4.17 Special Cases in LP Models

A number of special cases can arise in LP modelling such as alternative optimal solutions, redundant constraints, unbounded solutions, and infeasibility. The first two conditions do not prevent the solution of an LP model and are not really problems, although it may indicate that the formulation of the problem is not complete or the problem has not been correctly identified. However, the last two conditions prevent us from solving an LP model. In these situations, the optimization software packages cannot complete the solution of the problem and will indicate whether the problem is unbounded or infeasible. This may indicate that the problem has not been correctly identified or defined.

Although it is not unusual to encounter an unbounded solution when solving an LP model, such a solution indicates that there is something wrong with the formulation. For example, one or more constraints were omitted from the formulation, or a constraint was erroneously entered (such as a less than or equal to constraint being entered as a greater than or equal to constraint). In such a case, it is recommended that one revisits the formulation.

Infeasibility can occur in LP models, perhaps due to an error in the formulation of the model. For example, unintentionally making a "less than or equal to" constraint to "a greater than or equal to constraint," or there might not be a way to satisfy all the constraints in the model. In this case, constraints will have to be eliminated or loosened in order to obtain a feasible solution of the model.

11.5 Risk Analysis

Managing risk is a very important duty placed on senior management. Senior management must assess possible risks that are associated with the implementation of their decisions. Since operations research is involved in supporting the decision maker, by providing analysis and model building in support of the selection of appropriate decision alternatives, it has risk associated with it. Therefore, it is important that the analyst assists the decision maker in identifying the risks associated with complex problem solving and ways of minimizing those risks.

In some respects, the operations researcher does carry out some aspects of risk assessment when he or she undertakes sensitivity analysis. However, that is not the whole risk management story. In order to carry out a complete risk assessment one needs to undertake all of the following stages:

- Context identification
- Risk identification
- Risk analysis
- Risk evaluation
- Monitor and review the risk management issues
- Communicate the issues with stakeholders

Since the operations researcher, when conducting a study, starts with an identification and definition of the problem, he or she should have considered the environment and context within which the problem exists. With that knowledge he or she should be able to identify the possible risks that may eventuate from the decision solution or alternative that he or she arrives at through their study considerations. A sensitivity analysis will assist in analyzing the impact of those risks and evaluate the consequences given that the events associated with that risk occur. The operations researcher must consider carrying out a risk assessment based on their findings and communicate them to the decision maker, so that the decision maker can consider these factors when making his or her decision. Sensitivity analysis as discussed in an earlier chapter is therefore an important aspect of modelling.

Some real problems that have been resolved using optimization techniques have had major impacts on either individuals or organizations and have had significant risks associated with them.

11.6 Summary

In this chapter, we have discussed the issues of problem solving using different computer packages and the outputs provided by them. The sensitivity analysis is presented with numerical examples. The practical issues faced by the users when dealing with optimization models and software packages are briefly analyzed. The problem that arose in solution implementation, particularly when dealing with senior management, is briefly discussed. The risk-associated implementation of solution is also discussed.

Exercises

1. The ABC corporation manufactures two special-sized electrical devices for ADF: generators and alternators. Both of these products require wiring and testing during the assembly process. Each generator requires 3 h of wiring and 1 h of testing and can be sold for a profit of $250. Each alternator requires 1 h of wiring and 2 h of testing and can be sold for a profit of $180. There are 240 h of wiring time and 160 h of testing time available in the next production period.
   (a) Formulate the problem as a linear programming (LP) model.
   (b) Do you think the problem defined is realistic or complete? If not, what information do you think are missing?
   (c) Create a spreadsheet model for the LP developed in part (a).

(d) If the demand for generator and alternator is limited to 65 and 45, respectively, what would be the optimal product-mix?

(e) Due to excessive demand of alternators, the management decided to increase the price (considering the model in (a)). As a result, the new profit per alternator is \$260. Comment on your new optimal solution.

(f) If the management decide to increase the working hours (either wiring or testing time, as of model in (a)), which one they should increase and why?

2. Consider the LP model developed in 1(a) of Exercises.

(a) Solve the model using Solver and create reports on sensitivity and limits.

(b) Use your solutions in (a) to answer the following:

(i) What range of values can the objective function coefficient for variable X_1 (=number of generators) assume without changing the optimal solution?

(ii) Is the optimal solution to this problem unique, or are there alternate optimal solutions?

(iii) What is the optimal objective function value if X_2 (=number of alternators) is assumed to be at least 1?

(iv) What is the optimal objective function value if the testing time changes from 160 to 170?

(v) Is the current solution still optimal if the coefficient for X_2 (=number of alternators) in the testing time constraint changes from 2 to 1? Explain.

3. Implement the following LP model in a spreadsheet. Use Solver to solve the problem and create a sensitivity report. Use this information to answer the following questions:

$$\text{Maximize } 3X_1 + 2X_2$$

Subject to

$$2X_1 + 4X_2 \leq 20$$
$$3X_1 + 5X_2 \leq 15$$
$$X_1, X_2 \geq 0$$

(a) What range of values can the objective function coefficient for variable X_1 assume without changing the optimal solution?

(b) Is the optimal solution to this problem unique, or are there alternate optimal solutions?

(c) How much does the objective function coefficient for variable X_2 have to increase before it enters the optimal solution at a strictly positive level?

(d) What is the optimal objective function value if X_2 equals 1?

(e) What is the optimal objective function value if the RHS value for the second constraint changes from 15 to 25?

(f) Is the current solution still optimal if the coefficient for X_2 in the second constraint changes from 5 to 1? Explain.

12

Basic Optimization Techniques

12.1 Introduction

The purpose of this book is to cover the basic modelling techniques of optimization problem solving, discuss the widely used theoretical optimization models and some practical optimization problems, and demonstrate the use of available software packages for optimization problem solving. Although the interpretation of solutions and sensitivity analysis are also provided with sufficient detail, an understanding of the basic optimization techniques would help to gain more insight about the problem solution space and the decision to be made. In addition, it will provide greater confidence of the quality of solutions to be obtained and decision to be made. With these views in mind, we present a few basic techniques briefly in this chapter as follows:

- Graphical solution method for linear programming (LP)
- Simplex method for LP
- Branch-and-bound technique for integer programming

12.2 Graphical Method

The graphical method is used to solve optimization models involving two variables and a few constraints, and the method helps to gain insights about the feasible solution space, optimality, and the interaction between different model parameters. The educational software Win QSB, discussed in Chapter 9, is capable of producing graphical solutions. In this section, we solve a small LP problem using Win QSB and then discuss how the method works. To demonstrate the method, let us consider an example.

Example 12.1: A simple example

Consider Example 3.1 with different right-hand sides (RHSs) for the first two constraints ($RHS_1 = 190$ and $RHS_2 = 180$). Let us here define the decision variables as follows:

X_1 = the number of tables to be produced

X_2 = the number of chairs to be produced

The corresponding LP model is

$$\begin{aligned} &\text{Maximize } Z = 100X_1 + 60X_2 \\ &\text{Subject to} \\ &\qquad 5X_1 + 2X_2 \leq 190 \quad \text{Machining} \\ &\qquad 4X_1 + 3X_2 \leq 180 \quad \text{Sanding} \\ &\qquad 3X_1 + 4X_2 \leq 200 \quad \text{Assembly} \\ &\qquad X_1 \geq 0 \text{ and } X_2 \geq 0 \quad \text{Nonnegativity} \end{aligned} \qquad \text{Model (12.1)}$$

The Win QSB input and graphical solution are provided in Figures 12.1 and 12.2, respectively.

As per the solution in Figure 12.2, $Z = 4200.00$, $X_1 = 30.00$, and $X_2 = 20.00$. However, a natural question that would arise is how did Win QSB produce the solution? To answer this question, the steps involved in the graphical method are discussed below:

- Graphing the feasible region
  - Draw lines for X (horizontal axis) and Y (vertical axis) axes
  - We need to only consider the first quadrant because of the nonnegativity constraints
  - X-axis represents one variable and Y-axis the other
  - Choose a suitable range and scale for each axis
  - Graph the constraints: Draw a straight line for each constraint. Constraint with an equal to sign is simply an equation of a

Variable -->	X1	X2	Direction	R. H. S.
Maximize	100	60		
C1	5	2	<=	190
C2	4	3	<=	180
C3	3	4	<=	200
LowerBound	0	0		
UpperBound	M	M		
VariableType	Continuous	Continuous		

FIGURE 12.1
Win QSB input for Model 12.1.

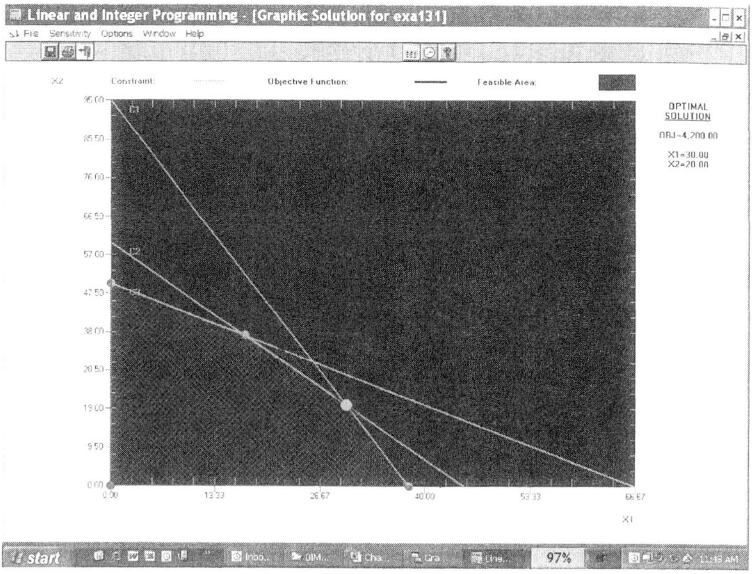

FIGURE 12.2
Win QSB graphical solution for Model 12.1.

straight line. For an inequality constraint, first replace the inequality sign by an equal to sign and draw the straight line. Then for the inequality constraint, identify the area that is satisfied by the constraint.

- Identify the area that is simultaneously satisfied by all the constraints of the problem. This area is known as the feasible area/region.

- Identifying the optimal solution
    - Once the feasible solution space is drawn, identify the best solution point in the feasible region either by using the iso-profit line approach or analyzing the corner points of the feasible region. These methods will be demonstrated later.

To understand the method clearly, let us graph the constraints one by one. We assume that the horizontal axis represents variable X_1 and the vertical axis represents X_2. To graph all the constraints, we need up to 66.67 units on the horizontal axis and 95 on the vertical axis. For the convenience of scaling, we choose up to 80 on the X-axis and 100 on the Y-axis.

Let us draw the machining constraint first. For the time being, we replace the inequality sign of the constraint by the equal to sign so that it can be treated as an equation for a straight line as follows:

$$5X_1 + 2X_2 = 190 \tag{12.1}$$

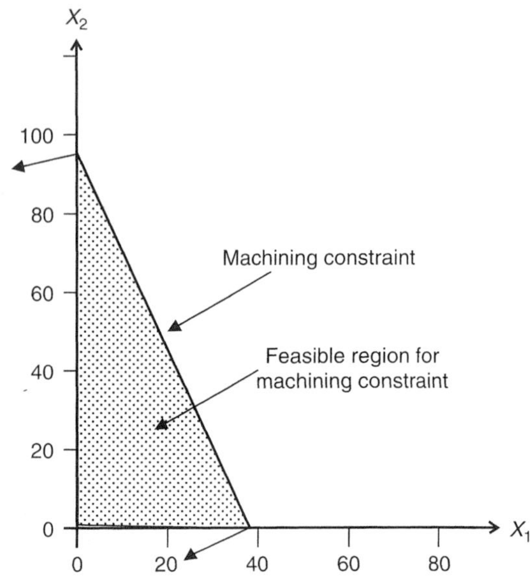

FIGURE 12.3
Graphing machining constraint.

If $X_1 = 0$, then from the equation $2X_2 = 190$ or $X_2 = 95$

If $X_2 = 0$, then $5X_1 = 190$ or $X_1 = 38$

Draw a straight line joining the points (0, 95) and (38, 0) as shown in Figure 12.3. Now the constraint ($5X_1 + 2X_2 \leq 190$) represents the area on the left-hand side (LHS) of this line drawn. The highlighted triangular area, in Figure 12.3, is bounded by the machining constraint and two nonnegativity constraints. This area is recognized as the feasible region/area for the machining constraint.

We can draw the sanding constraint in a similar way as shown in Figure 12.4. The highlighted area is the feasible region for both machining and sanding constraints. In Figure 12.5, the assembly constraint is drawn and the feasible region, which is satisfied by all the constraints of the model, is highlighted. One can now compare the feasible region in Figure 12.5 with the Win QSB solution (shown in Figure 12.2).

Any point taken from the feasible region (as shown in Figure 12.5) would satisfy all the constraints of the model (Model 12.1). One can test that by taking points such as (20, 20), (30, 15), and any other point from the feasible region. The best points in the feasible region would represent the optimal solution of the model (Model 12.1). However, there are an infinite number of points in the feasible region. For the convenience of analysis, we have redrawn a magnified version of the feasible region in Figure 12.6.

To identify the optimal point in the above feasible region, we will demonstrate the iso-profit line approach first. In this approach, we assume a value for the objective function, which helps to draw a straight line inside the feasible region. For example, if we assume $Z = 2400$, the corresponding iso-profit line will be $100X_1 + 60X_2 = 2400$ as shown in Figure 12.7.

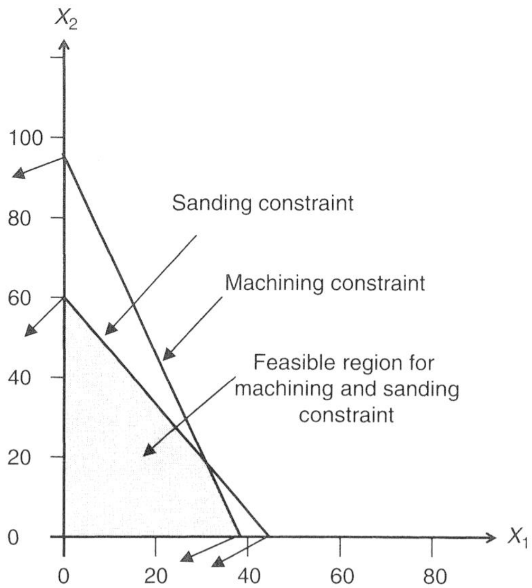

FIGURE 12.4
Graphing sanding constraint.

Any point taken on this line will have the same Z value of 2400. However, the values of X_1 and X_2 will vary depending on the location of the point chosen. We can now draw other iso-profit lines with higher Z values as shown in Figure 12.8. In other words, we draw parallel lines to the iso-profit line already drawn, in the direction of objective function increase, until we reach the last possible point of the feasible region. The last intersecting point is the optimal point of the model.

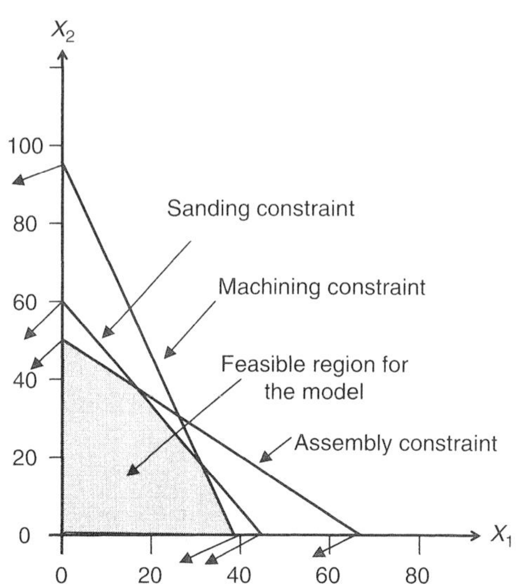

FIGURE 12.5
Graphing assembly constraint.

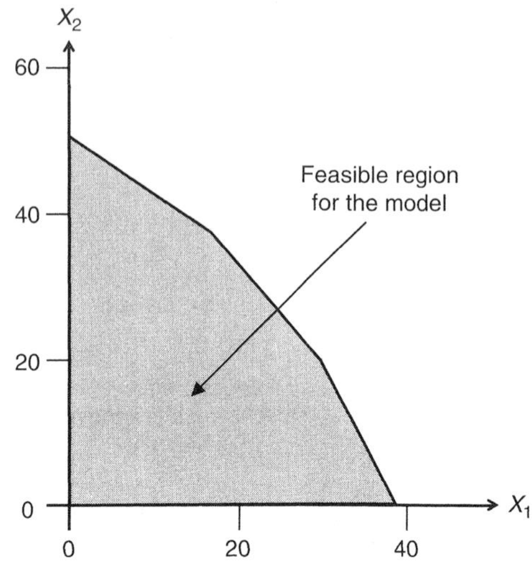

FIGURE 12.6
Feasible region of the model.

The direction of the objective function increase can easily be found using the gradient of the objective function. For example, as the gradient of our objective function is (100, 60), the direction of the objective function increase would be in the direction from (0, 0) to (100, 60). The iso-profit lines drawn in Figure 12.8 would be perpendicular to the direction of gradient (see Figure 12.7).

As shown in Figure 12.9, the optimal point is the intersection of the lines determined by the machining and sanding constraints. So the exact

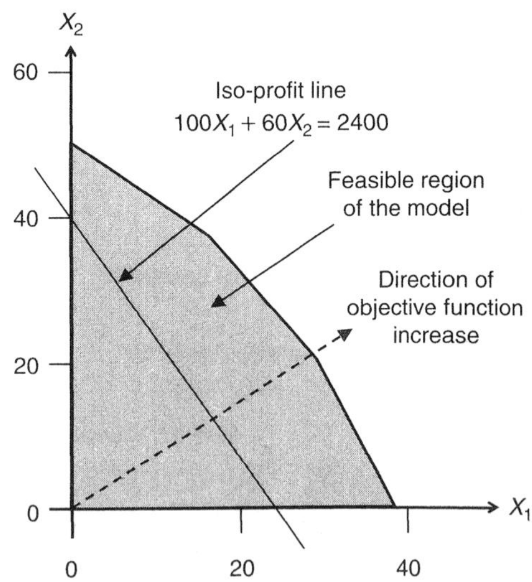

FIGURE 12.7
Iso-profit line and direction of profit increase.

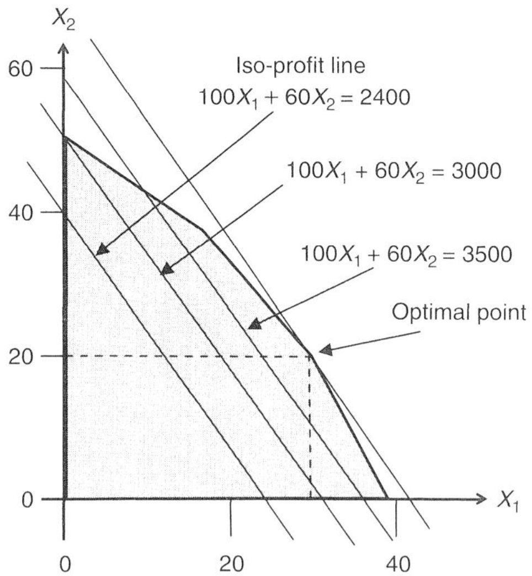

FIGURE 12.8
Finding the location optimal point.

coordinate of the optimal point can be found by determining the intersection of these two boundary lines as follows:

$$\text{Machining} \quad 5X_1 + 2X_2 = 190 \tag{12.2}$$

$$\text{Sanding} \quad 4X_1 + 3X_2 = 180 \tag{12.3}$$

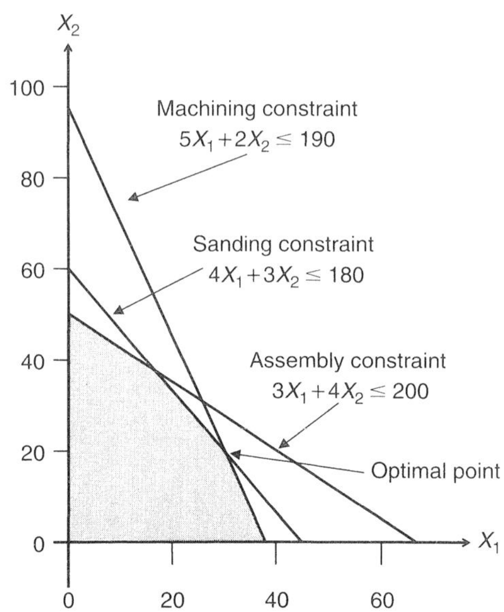

FIGURE 12.9
Finding optimal point.

From Equation 12.2

$$X_1 = (190 - 2X_2)/5 = 38 - 0.4X_2 \tag{12.4}$$

Substituting Equation 12.4 in Equation 12.3, we get

$$\begin{aligned} 4(38 - 0.4X_2) + 3X_2 &= 180 \\ \text{or} \quad 152 - 1.6X_2 + 3X_2 &= 180 \\ \text{or} \quad 1.4X_2 &= 28 \\ \text{or} \quad X_2 &= 20 \end{aligned} \tag{12.5}$$

Substituting $X_2 = 20$ in Equation 12.2, we get

$$\begin{aligned} 5X_1 + 2(20) &= 190 \\ \text{or} \quad X_1 &= 30 \end{aligned} \tag{12.6}$$

For $X_1 = 30$ and $X_2 = 20$, the objective function value $Z = 100X_1 + 60X_2 = 100 \times 30 + 60 \times 20 = \$4200$. So, the optimal solution suggests that producing 30 tables and 20 chairs would generate a profit of \$4200.

It is well known that the optimal solution point in LP exists in one (or more for some cases) of the corner points of the feasible region. Therefore, the optimal point can be found by simply analyzing the corner points. The exact coordinates of corner points can be calculated by finding the intersection of two straight lines forming the corner. The coordinates of the corner points of the feasible region are shown in Figure 12.10. The corner points are used to find the best objective function value as shown in Table 12.1.

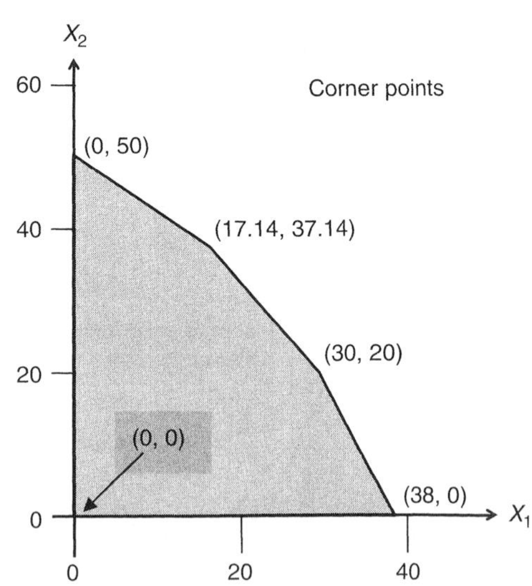

FIGURE 12.10
Finding optimal point using corner points.

TABLE 12.1

Corner Points and Their Corresponding Objective Values

Corner Point (X_1, X_2)	$Z = 100X_1 + 60X_2$	Z Value ($)	Remark
(0, 0)	$100 \times 0 + 60 \times 0$	0	
(38, 0)	$100 \times 30 + 60 \times 0$	3000	
(30, 20)	$100 \times 30 + 60 \times 20$	4200	Maximum
(17.14, 37.14)	$100 \times 17.14 + 60 \times 37.14$	3942.4	
(0, 50)	$100 \times 0 + 60 \times 50$	3000	

As per the above corner point analysis, the corner point (30, 20) provides the maximum possible profit of $4200. This solution is the same as the iso-profit approach presented earlier.

Example 12.2

Consider Example 3.1 once again with the variable definition as follows:

X_1 = number of tables to be produced

X_2 = number of chairs to be produced

$$
\begin{aligned}
&\text{Maximize } Z = 100X_1 + 60X_2\\
&\text{Subject to}\\
&\qquad 5X_1 + 2X_2 \le 270 \quad \text{Machining}\\
&\qquad 4X_1 + 3X_2 \le 250 \quad \text{Sanding}\\
&\qquad 3X_1 + 4X_2 \le 200 \quad \text{Assembly}\\
&\qquad X_1 \ge 0 \text{ and } X_2 \ge 0 \quad \text{Nonnegativity}
\end{aligned}
\qquad \text{Model (12.2)}
$$

The feasible region of this model is presented in Figure 12.11. The sanding constraint is recognized here as a redundant constraint as it has no effect on the feasible region of the model. The redundant constraints can be excluded when solving a model.

12.3 Simplex Method

We can solve LP problems, using graphical method, only when there are two decision variables. Most real-world LP models have more than two variables. As we discussed in earlier chapters, we solve LP models in this book using software packages such as LINDO/LINGO, GAMS, Excel Solver, and Win QSB. One may be interested to know what methodology/algorithm these software packages implement to solve LP models. As discussed in the following few sections, the models are solved through a

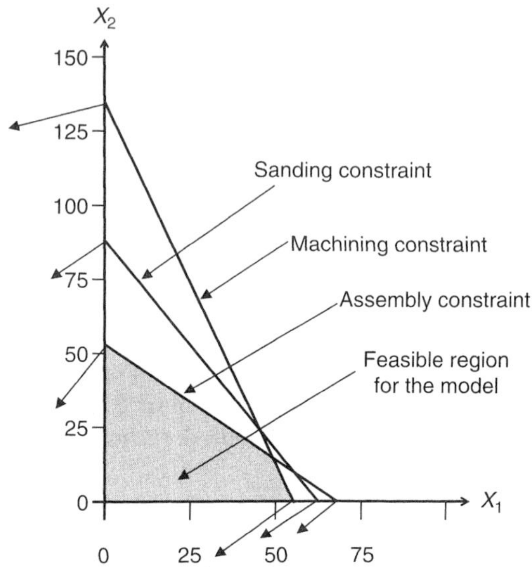

FIGURE 12.11
Feasible region for Example 3.1.

number of iterations. Excel Solver has an option to provide solution summary at each iteration. Win QSB also provides iteration details. The iteration-wise solutions for Example 12.1 using both Win QSB and Excel Solver are provided in the next few tables. Both software packages require only two iterations to solve the model. However, three tables for each software package represent the initial, iteration-1, and iteration-2 solutions.

From Tables 12.2 and 12.3, we get $X_1 = 0$ and $X_2 = 0$ with $Z = 0$. This is the point (0, 0) in Figure 12.10 (graphical solution).

As per Tables 12.4 and 12.5, we get $X_1 = 38$ and $X_2 = 0$ with $Z = 3800$ after iteration 1. This is the point (38, 0) in Figure 12.10 (graphical solution).

As per Tables 12.6 and 12.7, we get $X_1 = 30$ and $X_2 = 20$ with $Z = 4200$ after iteration 2. This is the point (30, 20) in Figure 12.10 (graphical solution). As indicated by both the software packages, this is the optimal solution. If we look at the solutions presented in Tables 12.2 through 12.7, as per graphical solution in Figure 12.10, it started from a corner point (0, 0) and then moved to an adjacent corner point (38, 0) in iteration 1 when the choices were (0, 50)

TABLE 12.2
Win QSB Initial Solution

		X1	X2	Slack_C1	Slack_C2	Slack_C3		
Basis	C(i)	100.0000	60.0000	0	0	0	R. H. S.	Ratio
Slack_C1	0	5.0000	2.0000	1.0000	0	0	190.0000	38.0000
Slack_C2	0	4.0000	3.0000	0	1.0000	0	180.0000	45.0000
Slack_C3	0	3.0000	4.0000	0	0	1.0000	200.0000	66.6667
	C(i)-Z(i)	100.0000	60.0000	0	0	0	0	

TABLE 12.3

Excel Solver Initial Solution

	Table	Chair			Objectove function
Number to produce	0	0			0
Profit per unit	100	60			
Constraints:			LHS	RHS	
Machining	5	2	0	190	
Sanding	4	3	0	180	
Assembly	3	4	0	200	

and (38, 0). In iteration 2, it moved from (38, 0) to another corner point (30, 20) and then stopped as it reached the optimal point. The process has not explored the corner points (17.14, 37.14) and (0, 50) at all.

We know the solutions presented in Tables 12.2 through 12.7 and the points discussed above will create many questions in your mind. To clarify some of these questions, we now explain how the software packages come up with such numbers and solutions in the above tables.

For LP problems with two or more variables, an analytical method known as the simplex algorithm is widely used. As discussed in the graphical solution section, we know that the optimal solutions of any LP exist at the corner points of the feasible solution space. The simplex method is an iterative method that finds the optimal solution by searching only some of the corner points. The process usually starts at the origin and then moves to an adjacent corner point that provides better objective-function value. The process of moving to adjacent corner points continues until no further improvement is possible. The basic steps of the algorithm can be expressed briefly as follows:

Step 1: Standardize the problem into an LP tableau.

Step 2: Generate an initial feasible solution, called a basis.

Step 3: Test the solution for optimality. If not optimal, improve it (following Step 4); otherwise go to Step 6.

Step 4: Generate an improved solution by identifying the leaving and entering variables to the basis and updating the tableau.

TABLE 12.4

Win QSB Iteration 1 Solution

		X1	X2	Slack_C1	Slack_C2	Slack_C3		
Basis	C(j)	100.0000	60.0000	0	0	0	R. H. S.	Ratio
X1	100.0000	1.0000	0.4000	0.2000	0	0	38.0000	95.0000
Slack_C2	0	0	1.4000	-0.8000	1.0000	0	28.0000	20.0000
Slack_C3	0	0	2.8000	-0.6000	0	1.0000	86.0000	30.7143
	C(j)-Z(j)	0	20.0000	-20.0000	0	0	3,800.0000	

TABLE 12.5

Excel-Solver Iteration-1 Solution

	Table	Chair			Objectove function
Number to produce	38	0			3800
Profit per unit	100	60			
Constraints:			LHS	RHS	
Machining	5	2	190	190	
Sanding	4	3	152	180	
Assembly	3	4	114	200	

Step 5: Check for optimality (as in Step 3). If not optimal, repeat Steps 4 and 5. If optimal, go to Step 6.

Step 6: Stop.

The method can be implemented using either a tabular or matrix form. For ease of understanding and explanation, we will demonstrate the method using the tabular form. First of all, let us find out how Win QSB generated the initial solution as of Table 12.2. To do that, recall Model 12.1.

$$\begin{aligned} &\text{Maximize } Z = 100X_1 + 60X_2 \\ &\text{Subject to} \\ &\qquad 5X_1 + 2X_2 \leq 190 \quad \text{Machining} \\ &\qquad 4X_1 + 3X_2 \leq 180 \quad \text{Sanding} \\ &\qquad 3X_1 + 4X_2 \leq 200 \quad \text{Assembly} \\ &\qquad X_1 \geq 0 \text{ and } X_2 \geq 0 \quad \text{Nonnegativity} \end{aligned} \qquad \text{Model (12.3)}$$

The inequalities in the model (Model 12.3) can be converted to equalities by adding one extra variable in the LHS of each constraint as follows:

$$5X_1 + 2X_2 + S_1 = 190 \tag{12.7}$$

$$4X_1 + 3X_2 + S_2 = 180 \tag{12.8}$$

$$3X_1 + 4X_2 + S_3 = 200 \tag{12.9}$$

TABLE 12.6

Win QSB Iteration 2 (also final) Solution

		X1	X2	Slack_C1	Slack_C2	Slack_C3		
Basis	C(i)	100.0000	60.0000	0	0	0	R. H. S.	Ratio
X1	100.0000	1.0000	0.0000	0.4286	-0.2857	0	30.0000	
X2	60.0000	0.0000	1.0000	-0.5714	0.7143	0	20.0000	
Slack_C3	0	0	0	1.0000	-2.0000	1.0000	30.0000	
	C(i)-Z(i)	0	0	-8.5714	-14.2857	0	4,200.0000	

TABLE 12.7

Excel-Solver Iteration-2 (also Final) Solution

	Table	Chair			Objectove function
Number to produce	30	20			4200
Profit per unit	100	60			
Constraints:			LHS	RHS	
Machining	5	2	190	190	
Sanding	4	3	180	180	
Assembly	3	4	170	200	

S_1, S_2, and S_3 are known as slack variables. By definition, the slacks are greater than or equal to zero and have zero contribution to the objective function. After inclusion of the objective function, the revised model is as follows:

$$\text{Maximize } Z = 100X_1 + 60X_2 + 0S_1 + 0S_2 + 0S_3$$

Subject to

$$\begin{aligned} &5X_1 + 2X_2 + S_1 + 0S_2 + 0S_3 = 190 \\ &4X_1 + 3X_2 + 0S_1 + S_2 + 0S_3 = 180 \\ &3X_1 + 4X_2 + 0S_1 + 0S_2 + S_3 = 200 \\ &X_1, X_2, S_1, S_2, S_3 \geq 0 \end{aligned} \qquad \text{Model (12.4)}$$

We now present this model in tabular form.

If we exclude column 1, the Table 12.8 is the middle part of the Win QSB initial solution as presented in Table 12.2. We now add the other columns and rows as of Table 12.2. However, we will add one extra row (second last row in Table 12.9) for convenience of explanation and analysis.

We now fill in the first and second columns of Table 12.9. The first column can be generated in two ways. The first approach is to write down the slack variables in order of their constraints as shown in Table 12.10.

As you can see in Table 12.10, there is an identity matrix under the slack variables. The second approach is to identify the entry "1" for each column of slack variables (that is the columns of the identity matrix) and then write the slack variable in the corresponding row cell of the first column as shown in Table 12.10. The middle three cells in second column contain the objective

TABLE 12.8

Tabular Representation of Model 12.4

	X_1	X_2	S_1	S_2	S_3	
Objective function	100	60	0	0	0	**RHS**
Constraint 1	5	2	1	0	0	190
Constraint 2	4	3	0	1	0	180
Constraint 3	3	4	0	0	1	200

TABLE 12.9

Model 12.4 with Additional Columns and Rows

		X_1	X_2	S_1	S_2	S_3		
Basis	$C(j)$	100	60	0	0	0	**RHS**	Ratio
		5	2	1	0	0	190	
		4	3	0	1	0	180	
		3	4	0	0	1	200	
	$Z(j)$							
	$C(j) - Z(j)$							

function coefficients of the variables written in the first column. In the present case, they are all zero.

For easy identification of the constraints, we have added a new column (first one) to Table 12.11. Let us now fill in the bottom two rows in Table 12.10 or Table 12.9.

To find the value of the bottom cell of column X_1, we use the entries in the constraint rows for $C(j)$ and X_1 columns. For each constraint row, we multiply the cell of $C(j)$ column with the cell of X_1 column and then sum them all. At this stage, their sum is zero. This value is recognized as $Z(j)$ and it is then subtracted from the coefficient of X_1 in the objective function (as shown in second row under X_1), which is the value of the bottom cell in X_1 column. That is $C(j) - Z(j) = (100 - 0) = 100$ here. The detailed calculations are shown below:

	$C(j)$	X_1	**Multiplication**
C1	0	5	$0 \times 5 = 0$
C2	0	4	$0 \times 4 = 0$
C3	0	3	$0 \times 3 = 0$

$Z(j) = 0 + 0 + 0 = 0$
$C(j) - Z(j) = (100 - 0) = 100$

The value of the bottom cell of column X_2 can be found using columns $C(j)$ and X_2 as $[60 - (0 \times 2 + 0 \times 3 + 0 \times 4) =]$ 60. In a similar way, we can

TABLE 12.10

Filling First Column of Table 12.9

		X_1	X_2	S_1	S_2	S_3		
Basis	$C(j)$	100	60	0	0	0	**RHS**	Ratio
S_1	0	5	2	**1**	**0**	**0**	190	
S_2	0	4	3	**0**	**1**	**0**	180	
S_3	0	3	4	**0**	**0**	**1**	200	
	$Z(j)$							
	$C(j) - Z(j)$							

TABLE 12.11

Filling the Bottom Row of Table 12.10

			X_1	X_2	S_1	S_2	S_3		
	Basis	$C(j)$	100	60	0	0	0	**RHS**	Ratio
C1	S_1	**0**	**5**	2	1	0	0	190	
C2	S_2	**0**	**4**	3	0	1	0	180	
C3	S_3	**0**	**3**	4	0	0	1	200	
		$Z(j)$	0	0	0	0	0	0	
		$C(j) - Z(j)$	**100**	**60**	**0**	**0**	**0**		

find for S_1, S_2, S_3, and RHS columns as presented in Table 12.11. This table represents the initial feasible solution. The $C(j) - Z(j)$ row is known as the simplex criterion (evaluator/net evaluation) row. The $C(j) - Z(j)$ value indicates the opportunity cost of not having one unit of the corresponding value in the solution. However, it may not be found to be meaningful at the initial solution stage.

The variables under the basis column are known as basic variables. The values of the basic variables can be read directly from the RHS column. The variables not in the basis are known as nonbasic variables. The nonbasic variables have zero values. At this stage, the values of the decision variables are $X_1 = 0$ and $X_2 = 0$.

To start the simplex iteration, we need to identify the entering variable and the leaving variable in Table 12.11. The entering variable is a nonbasic variable with the largest $C(j) - Z(j)$ value that is the largest opportunity cost. As per Table 12.11, X_1 is the entering variable with $C(j) - Z(j) = 100$. The column of the entering variable is recognized as the pivot column. To find the leaving variable, we take the ratio of RHS and entry of pivot column for each constraint (see the last column of Table 12.12). Table 12.12 is similar to the Win QSB initial solution presented in Table 12.2.

The leaving variable is the one with the minimum ratio as shown in the last column of Table 12.12. As the minimum ratio lies with constraint 1 (or C1), the leaving variable is S_1. The row of the leaving variable is recognized

TABLE 12.12

Filling the Last Column of Table 12.11

			X_1	X_2	S_1	S_2	S_3		
	Basis	$C(j)$	**100**	60	0	0	0	**RHS**	Ratio
C1	S_1	0	**5**	**2**	**1**	**0**	**0**	**190**	**38.00**
C2	S_2	0	**4**	3	0	1	0	180	45.00
C3	S_3	0	**3**	4	0	0	1	200	66.67
		$Z(j)$	**0**	0	0	0	0	0	
		$C(j) - Z(j)$	**100**	60	0	0	0		

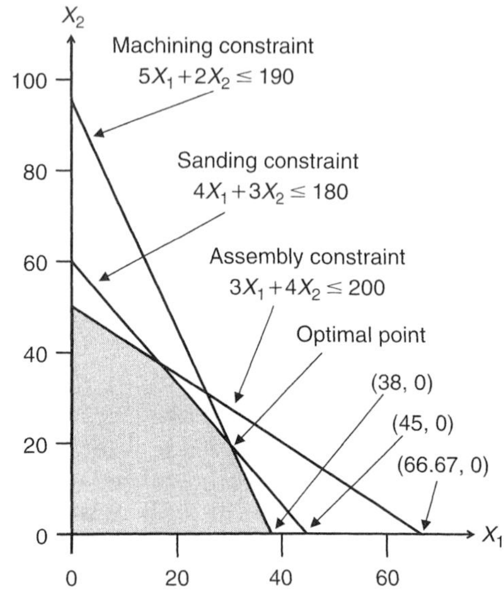

FIGURE 12.12
Feasible region.

as the pivot row. The element at the intersection of the pivot row and the pivot column is known as the pivot element.

Let us recall the graphical solution in Figure 12.9, which is redrawn here as Figure 12.12 for ease of explanation. As you can see in Figure 12.12, the points (38, 0), (45, 0), and (66.67, 0) are the intersection points of the boundary lines of three constraints with X-axis. The point $X_1 = 38$ is the only point that ensures feasibility. That is why we consider the minimum ratio for selection of the leaving variable.

In the first iteration, we are required to replace the leaving variable S_1 in the basis by the entering variable X_1, update the $C(j)$ entry, and convert the pivot column as a column of the identity matrix where the pivot element must be one. To change the pivot column, it is necessary to perform a matrix operation as follows:

- Divide row C1 by 5. The resulting row is shown in Table 12.13.
- Multiply the modified C1 (as shown in Table 12.13) by 4 and subtract from C2. The resulting C2 row is shown in Table 12.14.
- Multiply the modified C1 (as shown in Table 12.13) by 3 and subtract from C3. The resulting C3 row is shown in Table 12.15.

The $Z(j)$ and $C(j) - Z(j)$ rows must now be updated. The $Z(j)$ element under the X_1 column is $(100 \times 1 + 0 \times 0 + 0 \times 0 =)$ 100 and under the X_2 column is $(100 \times 0.4 + 0 \times 1.4 + 0 \times 2.8 =)$ 40. So the corresponding elements for the $C(j) - Z(j)$ rows are $(100 - 100 =)$ 0 and $(60 - 40 =)$ 20, respectively are shown in Table 12.16. Similarly, one can find the elements of the $Z(j)$ and $C(j) - Z(j)$ rows under the slack variables and $Z(j)$ for RHS.

TABLE 12.13

First Simplex Iteration—Changing Pivot Row

	Basis	$C(j)$	X_1	X_2	S_1	S_2	S_3	RHS	Ratio
			100	60	0	0	0		
C1	X_1	**100**	**1**	**0.4**	**0.2**	**0**	**0**	**38**	
C2	S_2	0	**4**	3	0	1	0	180	
C3	S_3	0	**3**	4	0	0	1	200	
		$Z(j)$	**0**	0	0	0	0	0	
		$C(j) - Z(j)$	**100**	60	0	0	0		

TABLE 12.14

First Simplex Iteration—Changing C2 Row

	Basis	$C(j)$	X_1	X_2	S_1	S_2	S_3	RHS	Ratio
			100	60	0	0	0		
C1	X_1	100	**1**	0.4	0.2	0	0	38	
C2	S_2	**0**	**0**	**1.4**	**−0.8**	**1**	**0**	**28**	
C3	S_3	0	**3**	4	0	0	1	200	
		$Z(j)$	**0**	0	0	0	0	0	
		$C(j) - Z(j)$	**100**	60	0	0	0		

TABLE 12.15

First Simplex Iteration—Changing C3 Row

	Basis	$C(j)$	X_1	X_2	S_1	S_2	S_3	RHS	Ratio
			100	60	0	0	0		
C1	X_1	100	**1**	0.4	0.2	0	0	38	
C2	S_2	0	**0**	1.4	−0.8	1	0	28	
C3	S_3	**0**	**0**	**2.8**	**−0.6**	**0**	**1**	**86**	
		Z(j)	**0**	0	0	0	0	0	
		$C(j) - Z(j)$	**100**	60	0	0	0		

TABLE 12.16

First Simplex Iteration—Changing Bottom Two Rows

	Basis	$C(j)$	X_1	X_2	S_1	S_2	S_3	RHS	Ratio
			100	60	0	0	0		
C1	X_1	100	1	0.4	0.2	0	0	38	
C2	S_2	0	0	1.4	−0.8	1	0	28	
C3	S_3	0	0	2.8	−0.6	0	1	86	
		$Z(j)$	**100**	**40**	**20**	**0**	**0**	**3800**	
		$C(j) - Z(j)$	**0**	**20**	**−20**	**0**	**0**		

TABLE 12.17

Simplex Solution after First Iteration

			X_1	X_2	S_1	S_2	S_3		
	Basis	$C(j)$	100	60	0	0	0	**RHS**	Ratio
C1	X_1	100	1	0.4	0.2	0	0	38	
C2	S_2	0	0	1.4	−0.8	1	0	28	
C3	S_3	0	0	2.8	−0.6	0	1	86	
		$Z(j)$	100	40	20	0	0	3800	
		$C(j) - Z(j)$	0	20	−20	0	0		

After the first iteration, the solution is $X_1 = 38$ and $X_2 = 0$ with $Z = 3800$, and the corresponding simplex tableau looks like Table 12.17. This is a corner point (38, 0) as per the graphical solution. Is this solution optimum? No—as there is a positive entry in the $C(j) - Z(j)$ row. So, one is required to perform a further iteration. The new entering variable will now be X_2, with the highest $C(j) - Z(j)$ entry, for the next iteration.

For the second iteration, the new pivot column is highlighted in Table 12.18 and the ratios are calculated in order to identify the leaving variable. Table 12.18 is similar to the Win QSB iteration 1 solution as presented in Table 12.4.

As per the minimum ratio, S_2 is the leaving variable. The iteration is performed as follows:

- Replace S_2 by X_2 in the basis.
- Update $C(j)$ for C2 (for X_2).
- Divide row C2 of Table 12.18 by 1.4. The resulting row is shown in Table 12.19.
- Multiply the modified C2 (as shown in Table 12.19) by 0.4 and subtract from C1 of Table 12.18. The resulting C1 row is shown in Table 12.19.

TABLE 12.18

Simplex Method—Second Iteration

			X_1	X_2	S_1	S_2	S_3		
	Basis	$C(j)$	100	**60**	0	0	0	**RHS**	Ratio
C1	X_1	100	1	**0.4**	0.2	0	0	38	95.00
C2	S_2	**0**	**0**	**1.4**	**−0.8**	**1**	**0**	**28**	**20.00**
C3	S_3	0	0	**2.8**	−0.6	0	1	86	30.71
		$Z(j)$	100	**40**	20	0	0	3800	
		$C(j) - Z(j)$	0	**20**	−20	0	0		

TABLE 12.19

Simplex Solution after Second Iteration

			X_1	X_2	S_1	S_2	S_3		
	Basis	$C(j)$	100	60	0	0	0	**RHS**	Ratio
C1	X_1	100	1	0	0.429	−0.289	0	**30**	
C2	X_2	60	0	1	−0.571	0.714	0	**20**	
C3	S_3	0	0	0	1	−2.00	1	**30**	
		$Z(j)$	100	60	8.57	14.29	0	**4200**	
		$C(j) - Z(j)$	0	0	−8.57	−14.29	0		

- Multiply the modified C2 (as shown in Table 12.19) by 2.8 and subtract from C3 of Table 12.18. The resulting C3 row is shown in Table 12.19.
- Update the $Z(j)$ and $C(j) - Z(j)$ rows.

The above table is same as the Win QSB solution presented in Table 12.6 and the solution is optimum because there is no positive entry remaining for nonbasic variables in the $C(j) - Z(j)$ row.

We believe it is now clear to the reader how Win QSB and Excel Solver came up with the optimal solutions for the model solved. However one must remember that

- a minimization problem must be changed into a maximization problem when using this approach, and
- any zero and negative ratio must be ignored when identifying leaving variables.

The above model involves less than equal to ($\leq$) constraints only. To handle equal to ($=$) and greater than equal to ($\geq$) constraints, we need artificial variables and surplus variables to generate initial feasible solutions. For examples on solving models with equal to and greater than equal to constraints and for further details on the simplex method, see one of the operations research books listed in the bibliography section.

12.4 Branch-and-Bound Method

The branch-and-bound (B&B) method is widely used for solving integer programming models. To demonstrate the method here, let us consider

Model 12.1 once again but with different RHS_1, say 200 instead of 190. We identify this model as the following model:

$$\begin{aligned}&\text{Maximize } Z = 100X_1 + 60X_2\\&\text{Subject to}\\&\qquad 5X_1 + 2X_2 \leq 200 \quad \text{Machining}\\&\qquad 4X_1 + 3X_2 \leq 180 \quad \text{Sanding}\\&\qquad 3X_1 + 4X_2 \leq 200 \quad \text{Assembly}\\&\qquad X_1 \geq 0 \text{ and } X_2 \geq 0 \quad \text{Nonnegativity}\end{aligned} \qquad \text{Model (12.5)}$$

The LP solution using Win QSB, Solver, or LINGO provided $X_1 = 34.29$, $X_2 = 14.29$ with $Z = 4285.71$. Obviously, this solution is not acceptable as the number of tables and chairs is non-integer. After solving the model as integer programming as shown in the following model 12.6, we get $X_1 = 33$, $X_2 = 16$ with $Z = 4260$.

$$\begin{aligned}&\text{Maximize } Z = 100X_1 + 60X_2\\&\text{Subject to}\\&\qquad 5X_1 + 2X_2 \leq 200 \quad \text{Machining}\\&\qquad 4X_1 + 3X_2 \leq 180 \quad \text{Sanding}\\&\qquad 3X_1 + 4X_2 \leq 200 \quad \text{Assembly}\\&\qquad X_1, X_2 \geq 0 \text{ and integer} \quad \text{Nonnegativity}\end{aligned} \qquad \text{Model (12.6)}$$

Now we will describe how the software packages found this integer solution. First, let us present the iteration-wise solutions of both Win QSB and Excel Solver.

	Excel Solver Solution			**Win QSB Solution**		
Iteration	X_1	X_2	Z	X_1	X_2	Z
1	40	0	4000	34.29	14.29	4285.71
2	34.29	14.29	4285.71	35	12.5	4250
3	34	0	3400	—	—	Infeasible
4	34	14.67	4280	35.20	12	4240
5	40	0	4000	36	10	4200
6	35	12.5	4250	35	12	4220
7	34	0	3400	34	14.67	4280
8	34	14	4240	33.75	15	4275
9	33.75	15	4275	—	—	Infeasible
10	33	15	4200	**33**	**16**	**4260**
11	**33**	**16**	**4260**	34	14	4240

The B&B algorithm can be described as follows:

- First solve the problem as a continuous problem that is an LP problem. If the solutions satisfy integrality conditions, then stop.
- Assume that a variable x is required to be an integer, but is found to be fractional. The portion of the feasible space indicated by $\lfloor x \rfloor < x < \lceil x \rceil$ cannot include any feasible integer solutions of x. Therefore, a feasible integer value of x must satisfy one of the two conditions: either $x \leq \lfloor x \rfloor$ or $x \geq \lceil x \rceil$.
- When one applies these two conditions to the continuous LP, the search space is divided into two mutually exclusive subproblems.
- The original problem has now been branched or divided into two subproblems.
- One then proceeds by solving each of the subproblems as a continuous LP problem. Each subproblem contains the original model plus an additional constraint for the above condition.
- If its optimum solution satisfies the integrality conditions then stop. Otherwise, each subproblem must be divided further into two subproblems by introducing an integer condition on one of its integer variables that currently has a fractional value.
- Repeat the branching of subproblems until either all subproblems have been explored or stopping criteria have been met. Note that as feasible integer solutions are found they should be compared with the current best integer solution, which is then updated if necessary.

The efficiency of the computations can be considerably enhanced by introducing the principle of bounding or fathoming. In the case when the continuous optimum solution of a subproblem yields a worse objective value than the one associated with the best available integer solution of the parent problem, it is not necessary to explore that subproblem any further, because the integer solution will always be inferior or equal to the continuous solution. Once a feasible integer solution is found, its associated objective value can be used as an upper bound (for minimization) or lower bound (for maximization) to discard inferior subproblems.

To demonstrate how the software packages solve integer programming models (using the B&B algorithm) we consider Model 12.6 presented earlier. This model provides non-integer solutions, $X_1 = 34.29$, $X_2 - 14.29$ with $Z = 4285.71$ when solving the problem as an LP model. Let us now branch it using variable X_1. For X_1, there is no integer value between 34 and 35. That means it will be either less than or equal to 34 or greater than or equal to 35. So, the following are the subproblems:

Subproblem A

$$\text{Maximize } Z = 100X_1 + 60X_2$$

Subject to

$$
\begin{aligned}
5X_1 + 2X_2 &\leq 200 \\
4X_1 + 3X_2 &\leq 180 \\
3X_1 + 4X_2 &\leq 200 \\
X_1 &\geq 35 \\
X_1 \geq 0 \text{ and } X_2 &\geq 0
\end{aligned}
\qquad \text{Model (12.7)}
$$

Solution: $X_1 = 35$, $X_2 = 12.50$, and $Z = 4250$

Subproblem B

$$\text{Maximize } Z = 100X_1 + 60X_2$$

Subject to

$$
\begin{aligned}
5X_1 + 2X_2 &\leq 200 \\
4X_1 + 3X_2 &\leq 180 \\
3X_1 + 4X_2 &\leq 200 \\
X_1 &\leq 34 \\
X_1 \geq 0 \text{ and } X_2 &\geq 0
\end{aligned}
\qquad \text{Model (12.8)}
$$

Solution: $X_1 = 34$, $X_2 = 14.67$, and $Z = 4280$

These two subproblems associated with the original model are shown in Figure 12.13 as branching.

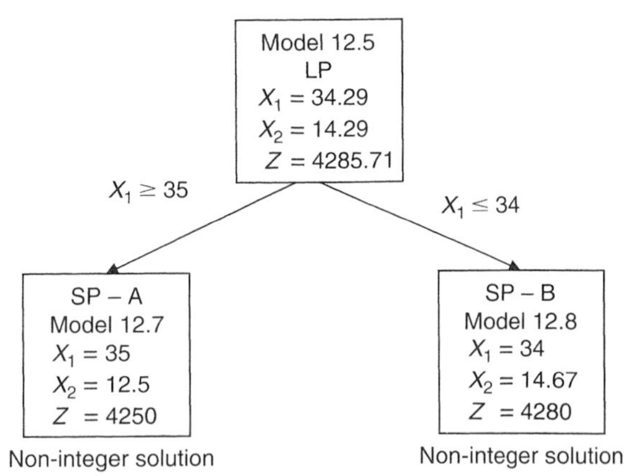

FIGURE 12.13
First branching in B&B approach.

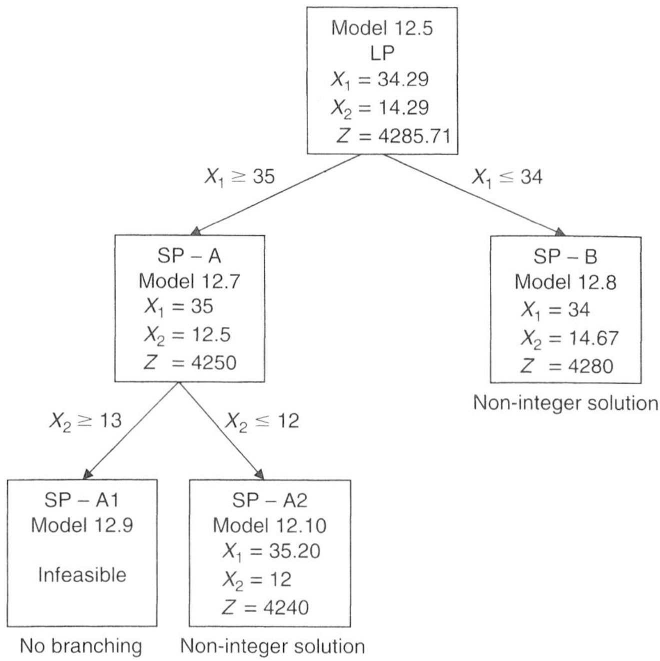

FIGURE 12.14
Second level branching for Subproblem A.

As no subproblem (A and B in Figure 12.13) provided integer solution, we would further branch them. First, we divide Subproblem A into Subproblem A1 and Subproblem A2 and the resulting tree is shown in Figure 12.14.

Subproblem A1

$$\text{Maximize } Z = 100X_1 + 60X_2$$

Subject to

$$\begin{aligned} 5X_1 + 2X_2 &\leq 200 \\ 4X_1 + 3X_2 &\leq 180 \\ 3X_1 + 4X_2 &\leq 200 \\ X_1 &\geq 35 \\ X_2 &\geq 13 \\ X_1 \geq 0 \text{ and } X_2 &\geq 0 \end{aligned} \qquad \text{Model (12.9)}$$

Solution: Infeasible

Remarks: A1 cannot be branched as it is infeasible.

Subproblem A2

$$\begin{aligned}
&\text{Maximize } Z = 100X_1 + 60X_2 \\
&\text{Subject to} \\
&\qquad 5X_1 + 2X_2 \le 200 \\
&\qquad 4X_1 + 3X_2 \le 180 \\
&\qquad 3X_1 + 4X_2 \le 200 \\
&\qquad X_1 \ge 35 \\
&\qquad X_2 \le 12 \\
&\qquad X_1 \ge 0 \text{ and } X_2 \ge 0
\end{aligned} \qquad \text{Model (12.10)}$$

Solution: $X_1 = 35.20$, $X_2 = 12.00$, and $Z = 4240$

Remarks: Can be branched further using X_1

As the Subproblem A1 is infeasible, it cannot be explored further. The Subproblem A2 provided non-integer solution. So it will be branched further, if necessary. Let us finish second level branching by dividing the Subproblem B into Subproblems B1 and B2.

Subproblem B1

$$\begin{aligned}
&\text{Maximize } Z = 100X_1 + 60X_2 \\
&\text{Subject to} \\
&\qquad 5X_1 + 2X_2 \le 200 \\
&\qquad 4X_1 + 3X_2 \le 180 \\
&\qquad 3X_1 + 4X_2 \le 200 \\
&\qquad X_1 \le 34 \\
&\qquad X_2 \ge 15 \\
&\qquad X_1 \ge 0 \text{ and } X_2 \ge 0
\end{aligned} \qquad \text{Model (12.11)}$$

Solution: $X_1 = 33.75$, $X_2 = 15$, and $Z = 4275$

Remarks: Can be branched further using X_1

Subproblem B2

$$\begin{aligned}
&\text{Maximize } Z = 100X_1 + 60X_2 \\
&\text{Subject to} \\
&\qquad 5X_1 + 2X_2 \le 200 \\
&\qquad 4X_1 + 3X_2 \le 180 \\
&\qquad 3X_1 + 4X_2 \le 200 \\
&\qquad X_1 \le 34 \\
&\qquad X_2 \le 14 \\
&\qquad X_1 \ge 0 \text{ and } X_2 \ge 0
\end{aligned} \qquad \text{Model (12.12)}$$

Solution: $X_1 = 34$, $X_2 = 14$, and $Z = 4240$

Remarks: Integer solution, no further branching

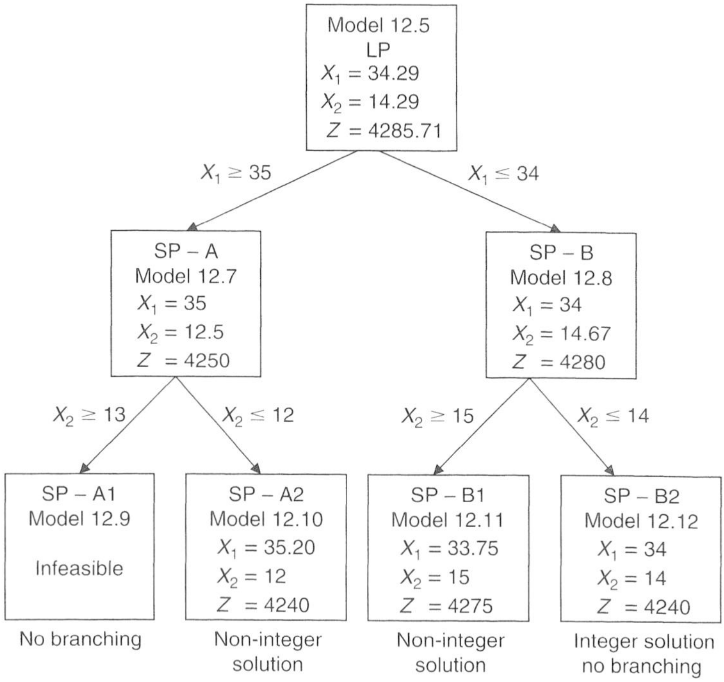

FIGURE 12.15
Second level branching for Subproblem B.

The Subproblem B1 must be further explored as it provided a non-integer solution with a Z value greater than 4240 and it is not possible to branch Subproblem B2 since it provided an integer solution (see Figure 12.15). At this stage, we will divide Subproblem A2 into Subproblems A2A and A2B as follows:

Subproblem A2A

Maximize $Z = 100X_1 + 60X_2$

Subject to

$$\begin{aligned} 5X_1 + 2X_2 &\leq 200 \\ 4X_1 + 3X_2 &\leq 180 \\ 3X_1 + 4X_2 &\leq 200 \\ X_1 &\geq 35 \\ X_2 &\leq 12 \\ X_1 &\geq 36 \\ X_1 \geq 0 \text{ and } X_2 &\geq 0 \end{aligned} \qquad \text{Model (12.13)}$$

Solution: $X_1 = 36$, $X_2 = 10$, and $Z = 4200$

Remarks: Integer solution, no further branching

Subproblem A2B

$$\text{Maximize } Z = 100X_1 + 60X_2$$

Subject to

$$5X_1 + 2X_2 \leq 200$$
$$4X_1 + 3X_2 \leq 180$$
$$3X_1 + 4X_2 \leq 200$$
$$X_1 \geq 35 \qquad \text{Model (12.14)}$$
$$X_2 \leq 12$$
$$X_1 \leq 35$$
$$X_1 \geq 0 \text{ and } X_2 \geq 0$$

Solution: $X_1 = 35$, $X_2 = 12$, and $Z = 4220$

Remarks: Integer solution, no further branching

Both Subproblems A1A and A2B produced integer solutions. So, they cannot be further explored (see Figure 12.16). Now we branch Subproblem B1 into Subproblems B1A and B1B as follows:

Subproblem B1A

$$\text{Maximize } Z = 100X_1 + 60X_2$$

Subject to

$$5X_1 + 2X_2 \leq 200$$
$$4X_1 + 3X_2 \leq 180$$
$$3X_1 + 4X_2 \leq 200$$
$$X_1 \leq 34 \qquad \text{Model (12.15)}$$
$$X_2 \geq 15$$
$$X_1 \geq 34$$
$$X_1 \geq 0 \text{ and } X_2 \geq 0$$

Solution: Infeasible

Remarks: Cannot be branched further

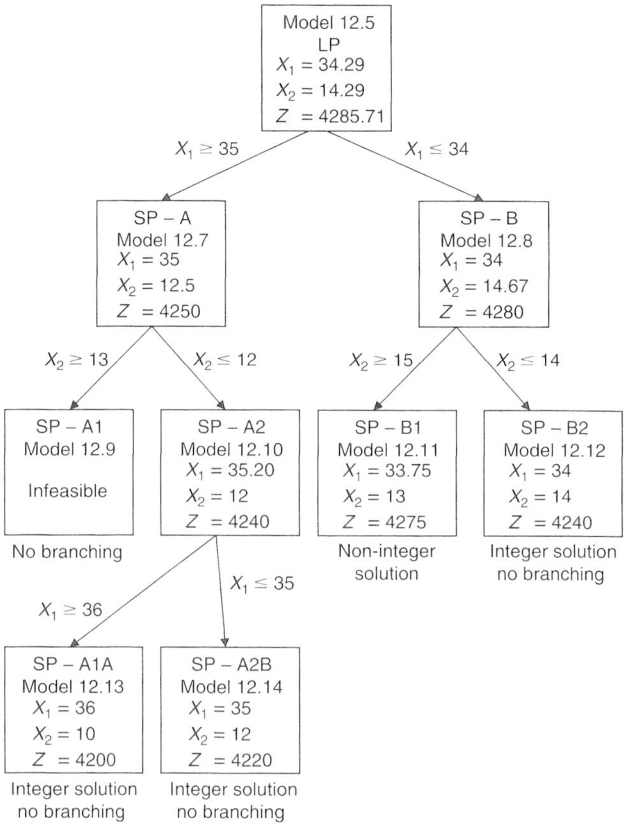

FIGURE 12.16
Branching for Subproblem A2.

Subproblem B1B

Maximize $Z = 100X_1 + 60X_2$

Subject to

$$
\begin{aligned}
5X_1 + 2X_2 &\leq 200 \\
4X_1 + 3X_2 &\leq 180 \\
3X_1 + 4X_2 &\leq 200 \\
X_1 &\leq 34 \\
X_2 &\geq 15 \\
X_1 &\leq 33 \\
X_1 \geq 0 \text{ and } X_2 &\geq 0
\end{aligned}
\qquad \text{Model (12.16)}
$$

Solution: $X_1 = 33$, $X_2 = 16$, and $Z = 4260$

Remarks: Integer solution, no further branching

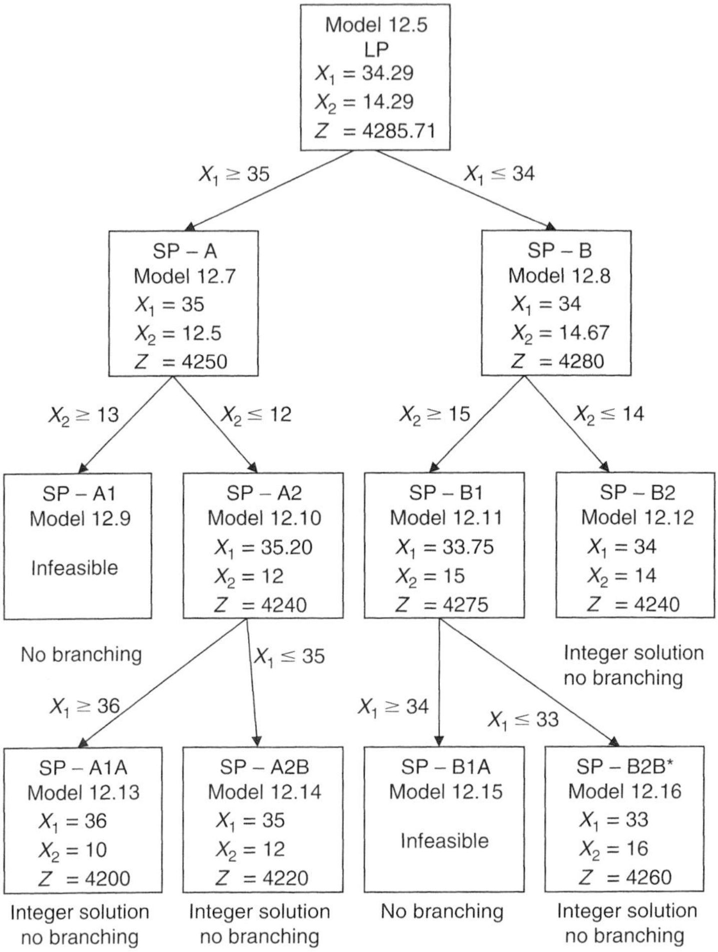

FIGURE 12.17
Complete branching.

Subproblem B1A is infeasible. The solution of Subproblem B1B is the best integer solution among all the subproblem integer solutions, which is also same as the Win QSB's solution (see Figure 12.17).

12.5 Summary

In this chapter, we have described three basic optimization techniques briefly illustrated by numerical examples. Although the book is not intended to cover the optimization algorithms in depth, these basic techniques will provide readers with a view of how the software packages were developed to solve the optimization problems. As you are aware, most operations research and optimization books cover solution techniques in depth; so, one can consult such a book if one is interested in delving more deeply into optimization algorithmic approaches.

Exercises

1. Solve graphically the following LP models:

   (a) Maximize $Z = 9X_1 + 3X_2$

   Subject to

   $$2X_1 + 3X_2 \leq 13$$
   $$2X_1 + X_2 \leq 5$$
   $$X_1, X_2 \geq 0$$

   (b) Maximize $Z = 2X_1 + 5X_2$

   Subject to

   $$X_1 + 4X_2 \leq 24$$
   $$3X_1 + X_2 \leq 21$$
   $$X_1 + X_2 \leq 9$$
   $$X_1, X_2 \geq 0$$

   (c) Maximize $Z = 9X_1 + 10X_2$

   Subject to

   $$11X_1 + 9X_2 \leq 9900$$
   $$7X_1 + 12X_2 \leq 8400$$
   $$6X_1 + 16X_2 \leq 9600$$
   $$X_1, X_2 \geq 0$$

2. Solve the following LP models using simplex method:

   (a) Maximize $Z = 9X_1 + 3X_2$

   Subject to

   $$2X_1 + 3X_2 \leq 13$$
   $$2X_1 + X_2 \leq 5$$
   $$X_1, X_2 \geq 0$$

(b) Maximize $Z = 3X_1 + 4X_2$

Subject to

$$2X_1 + 3X_2 \leq 6$$
$$X_1 + 2X_2 \leq 20$$
$$X_1, X_2 \geq 0$$

(c) Maximize $Z = 3X_1 + 2X_2 + 3X_3$

Subject to

$$X_1 + 2X_2 + 3X_3 \leq 18$$
$$3X_1 + 4X_2 + X_3 \leq 20$$
$$X_1, X_2, X_3 \geq 0$$

3. Solve the following integer programming models using B&B method. Hint: Use Excel Solver, LINGO, or any other optimization packages for solving the initial linear program and the subsequent linear sub-problems.

(a) Maximize $Z = 9X_1 + 3X_2$

Subject to

$$2X_1 + 3X_2 \leq 13$$
$$2X_1 + X_2 \leq 5$$
$$X_1, X_2 \geq 0 \text{ and integer}$$

(b) Maximize $Z = 2X_1 + 5X_2$

Subject to

$$X_1 + 4X_2 \leq 24$$
$$3X_1 + X_2 \leq 21$$
$$X_1 + X_2 \leq 9$$
$$X_1, X_2 \geq 0 \text{ and integer}$$

(c) Maximize $Z = 3X_1 + 2X_2 + 3X_3$

Subject to

$$X_1 + 2X_2 + 3X_3 \leq 18$$
$$3X_1 + 4X_2 + X_3 \leq 20$$
$$X_1, X_2, X_3 \geq 0 \text{ and integer}$$

Bibliography

Bazaraa, M., Jarvis, J., and Sherali, H., *Linear Programming and Network Flows*, John Wiley & Sons, New York, 1990.

Bradley, S.P., Hax, A.C., and Magnanti, T.L., *Applied Mathematical Programming*, Addison-Wesley Publishing Company, Readings, MA, 1977.

Daellenbach, H.G., George, J.A., and McNickle, D.C., *Introduction to Operations Research Techniques*, Allyn & Bacon, Boston, MA, 1983.

Eiselt, H.A., Pederzoli, G., and Sandblom, C.-L., *Continuous Optimization Models*, Walter deGruyter, Berlin, 1987.

Hillier, F. and Lieberman, G., *Introduction to Operations Research*, 8th ed., McGraw-Hill, Boston, MA, 2005.

Morse, P.M. and Kimball, G.E., *Methods of Operations Research*, John Wiley & Sons, New York, 1951.

Render, B. and Stair, R., *Quantitative Analysis for Management*, Prentice Hall, Englewood Cliffs, NJ, 1994.

Sierksma, G., *Linear and Integer Programming: Theory and Practice*, Marcel Dekker, New York, 1996.

Taha, H.A., *Operations Research: An Introduction*, 4th ed., MacMillan Publishing Company, New York, 1989.

Turban, E. and Meredith, J., *Fundamentals of Management Science*, 6th ed., Irwin McGraw-Hill, Boston, MA, 1994.

Section IV

Practical Problems

13

Models for Practical Problems I

13.1 Introduction

Many mathematical models used for practical optimization problems have been reported in the literature. The articles describing these models have been published in different journals, conference proceedings, reports, and books under different disciplines and organizations. Although it is infeasible to compile a compendium of all the models and present them in one or two chapters, we present a few representative problems and their models referring to the published materials.

In this chapter, we briefly present a number of practical problems where optimization models were developed for solving them. The problems discussed are crop planning, power generation planning, water supply management, supply chain, coal upgrading and marketing, and general blending. We also discuss their mathematical models and related issues.

13.2 A Crop Planning Problem

We present here a crop planning problem that is similar to one presented by Sarker and Quaddus (2002) and Sarker et al. (1997). Assume that a country grows a wide variety of crops in different seasons, and it has different types of soils. The yield rate, the cost of crop cultivation, and the return from the crops are functions of soil characteristics, regional aspects, the crop being produced, cropping pattern, and method (crop being produced and their sequence, irrigation, nonirrigation, etc.). We assume that the planning period is 1 year long. For a single-cropped area, there are a number of alternative crops (or crop groups) from which the crops to be cultivated may be chosen. Similarly, there are many different combinations of crops for double-cropped (two crops in a period) and triple-cropped (three crops in a period) areas. Different combinations produce different outputs. The utilization of the land for appropriate crops is the key issue for the crop planning problem. So the problem is to determine an annual crop production plan that would provide the area to be used for different crops while fulfilling the demand, land, capital, import, and region limitations.

Sarker et al. (1997) formulated the problem as a single-objective linear program (LP) with an objective of contribution maximization. Later, Sarker and Quaddus (2002) reformulated this problem as a multi-objective problem and solved it using the goal programming (GP) technique. In this section, we discuss both the single- and multi-objective crop planning models.

13.2.1 Linear Programming Model

The problem is formulated here as a linear programming model.* As per our problem description above, we have a number of alternative crops for single-cropped areas from which we can cultivate one crop in a year. Similarly, we have many combinations of crops for double- and triple-cropped areas. From the above considerations, the following variables, coefficients, and data can be identified and defined:

Parameters:

n_1 = number of alternative crops for single-cropped areas

n_2 = number of crop combinations for double-cropped areas

n_3 = number of crop combinations for triple-cropped areas

$N1_j$ = a crop in each j for single-cropped area (j is an index crop combination)

$N2_j$ = a set of crops (two crops) in each j for double-cropped area

$N3_j$ = a set of crops (three crops) in each j for triple-cropped area

YR_{ijk} = yield rate that is the amount of production per unit area for crop i of crop combination j in soil type k

CP_{ijk} = cost of production per unit area for crop i of crop combination j in soil type k

B_{ijk} = contribution that is the benefit that can be obtained per unit area of land from crop i of crop combination j in soil type $k = (P_i \times A_{ijk} - \mathrm{CP}_{ijk})$

P_i = market price of crop i per metric tonne

IC_i = contribution from import of crop i (= market revenue − import cost) for crop i

D_i = yearly demand of crop i

L_k = available area of soil type k

LT_k = land type coefficient for soil type k

---

* Produced as a slightly modified version of Sarker, R., Talukdar, S., and Haque, A., *Appl. Math. Model.*, 21, 621, 1997. With permission from Elsevier.

C_a = capital available, this indicates the total amount of money that can be invested for cropping

CE = set of crops that can be imported

CL_1 = set of crops having area limitation

CL_2 = set of crops having import limitation

A = area suitable and available for crop when $k=1$

IL = upper limit of total crop import

Variables:

X_{ijk} = area of land to be cultivated for crop i of crop combination j in soil type k

I_i = amount of crop i that should be imported

The LP model for the crop planning problem can then be developed as follows:

Objective function:
The objective is to maximize the total contribution (from cultivated plus imported crops) that can be obtained from cropping in a single-crop year.

$$\text{Maximize } Z = \sum_{j=1}^{n_1} \sum_{i \in N1_j} B_{ij(k=1)} X_{ij(k=1)} + \sum_{j=1}^{n_2} \sum_{i \in N2_j} B_{ij(k=2)} X_{ij(k=2)} + \sum_{j=1}^{n_3} \sum_{i \in N3_j} B_{ij(k=3)} X_{ij(k=3)} + \sum_{i \in \text{CE}} \text{IC}_i I_i \tag{13.1}$$

The first, second, third, and fourth terms represent the contribution from single-crop land, double-crop land, triple-crop land, and imported crop, respectively. Note that there is only one crop for each j in single-crop land, two crops in double-crop land, and three crops in triple-crop land.

Constraints:
Demand constraint: The sum of local production and the imported quantity of crop i in a year must be greater than or equal to the total requirements in the country.

$$\sum_j \sum_k \text{YR}_{ijk} X_{ijk} + I_{i \in \text{CE}} \geq D_i \quad \forall i \tag{13.2}$$

Soil constraint: The total land used for a given type of soil must be less than or equal to the total available land of that type.

$$\sum_i \sum_j \text{LT}_k X_{ijk} \leq L_k \quad \forall k \tag{13.3}$$

Here, for $k = 1, 2$, and 3, the coefficients (LT_k) are 1, ½, and ⅓, respectively. If a piece of land is used by two crops (in a double-cropped land) one after another (consecutive production) in a given year, it is assumed equivalent to the use of half the land for one of these two crops in a year—that is $LT_k = ½$. This assumption makes the constraint (Equation 13.3) simpler and it is required only for land constraint.

Capital constraint: The total amount of money that can be spent for crop production must be less than or equal to the capital available.

$$\sum_i \sum_j \sum_k \mathrm{CP}_{ijk} X_{ijk} \leq C_a \tag{13.4}$$

Contingent constraint: The area used for any crop under a crop combination for double- or triple-cropped land must be equal for every crop. For example, in a double-cropped land, the area used by two crops belonging to any crop combination must be equal.

$$X_{(i_1 \in N2_j)\,j(k=2)} - X_{(i_2 \in N2_j)\,j(k=2)} = 0 \quad \forall j \tag{13.5}$$

In double-cropped land, for a given crop combination j there are only two crops i_1 and i_2, where i_1 is the first crop and i_2 is the second crop in the combination. Both crops use the same area of land but one after another. In a triple-cropped land, the area used by three crops belonging to any crop combination must be equal.

$$X_{(i_1 \in N3_j)\,j(k=3)} - X_{(i_2 \in N3_j)\,j(k=3)} = 0 \quad \forall j \tag{13.6}$$

$$X_{(i_2 \in N3_j)\,j(k=3)} - X_{(i_3 \in N3_j)\,j(k=3)} = 0 \quad \forall j \tag{13.7}$$

Here, i_1 is the first crop, i_2 is the second crop, and i_3 is the third crop for combination j.

Area and import bound constraint: Because of soil characteristics and regional aspects, in some regions, the amount of area to be used for certain crops is restricted. For example, the unsuitability of certain soils for fruit cultivation needs to set an area limit for fruit production. Similarly, a constraint needs to be set for import restrictions as there is an upper limit on importation of some crops.

$$\text{Area bound: } \sum_{i \in \mathrm{CL}_1} X_{ijk} \leq A \quad \forall j,k \tag{13.8}$$

$$\text{Import bound: } \sum_{i \in \mathrm{CL}_2} I_i \leq \mathrm{IL} \tag{13.9}$$

Nonnegativity constraint: The decision variables must be greater than or equal to zero.

$$\begin{aligned} &X_{ijk} \geq 0 \quad \forall i,j,k \quad \text{and} \\ &I_i \geq 0 \quad \forall i \end{aligned} \tag{13.10}$$

13.2.2 Goal Programming (GP) Model

In this section, we have attempted to formulate a GP model for the above problem, as a GP model may represent the problem better. To formulate a GP model, we need to convert the goals into goal constraints. The following three goals are considered in this problem where the import reduction is chosen as the first goal and the capital requirements as the second and the contribution as the third.

1. Import restriction: minimize the dependency on import of basic foods like cereal
2. Capital restriction: minimize the investment required for cultivation
3. Contribution: maximize the return from cultivated land

The goal constraints of the case problem are

Import goal constraint

$$\sum_{i \in \mathrm{CL2}} I_i + d_1^- - d_1^+ = \mathrm{IL} \tag{13.11}$$

Capital goal constraint

$$\sum_i \sum_j \sum_k \mathrm{CP}_{ijk} X_{ijk} + d_2^- - d_2^+ = C_a \tag{13.12}$$

Contribution goal constraint

$$\sum_{j=1}^{n_1} \sum_{i \in N1_j} B_{ij(k=1)} X_{ij(k=1)} + \sum_{j=1}^{n_2} \sum_{i \in N2_j} B_{ij(k=2)} X_{ij(k=2)} + \sum_{j=1}^{n_3} \sum_{i \in N3_j} B_{ij(k=3)} X_{ij(k=3)} + \sum_{i \in \mathrm{CE}} \mathrm{IC}_i I_i + d_3^- - d_3^+ = \mathrm{ECon} \tag{13.13}$$

where

ECon = expected contribution per year

d_i^- and d_i^+ ($i = 1$, 2, and 3) are the underachievement and overachievement deviations from the goals

In GP, it is required to set a target for each goal. These target values are used as the right-hand side (RHS) value of the goal constraints. The RHS values for import goal restrictions and capital goal restrictions would be same as the corresponding LP value. However, we need to decide/determine a target for the RHS (defined as ECon) for the contribution goal restriction.

The objective of the goal program is to minimize the weighted sum of all undesirable deviations. In our case, following are the undesirable deviations:

- Excess import than the set target or overachievement of import expectation d_1^+
- Overuse of capital or overachievement of capital availability d_2^+
- Less contribution or underachievement of contribution target d_3^-

So the objective function of the GP is

$$\text{Minimize } Z_1 = w_1 d_1^+ + w_2 d_2^+ + w_3 d_3^- \tag{13.14}$$

Where, w_1, w_2, and w_3 are the weights assigned to the undesirable deviations.

So the resulting GP formulation is

Minimize $Z_1 = w_1 d_1^+ + w_2 d_2^+ + w_3 d_3^-$

Subject to

$$\sum_{i \in \text{CL2}} I_i + d_1^- - d_1^+ = b$$

$$\sum_i \sum_j \sum_k \text{CP}_{ijk} X_{ijk} + d_2^- - d_2^+ = C_a$$

$$\sum_{j=1}^{n_1} \sum_{i \in N1_j} B_{ij(k=1)} X_{ij(k=1)} + \sum_{j=1}^{n_2} \sum_{i \in N2_j} B_{ij(k=2)} X_{ij(k=2)}$$

$$+ \sum_{j=1}^{n_3} \sum_{i \in N3_j} B_{ij(k=3)} X_{ij(k=3)} + \sum_{i \in \text{CE}} \text{IC}_i I_i + d_3^- - d_3^+ = \text{ECon} \quad \text{Model (13.1)}$$

$$\sum_j \sum_k \text{YR}_{ijk} X_{ijk} + I_{i \in \text{CE}} \geq D_i \quad \forall i$$

$$\sum_i \sum_j \text{LT}_k X_{ijk} \leq L_k \quad \forall k$$

$$X_{(i_1 \in N2_j)j(k=2)} - X_{(i_2 \in N2_j)j(k=2)} = 0 \quad \forall j$$

$$X_{(i_1 \in N3_j)j(k=3)} - X_{(i_2 \in N3_j)j(k=3)} = 0 \quad \forall j$$

$$X_{(i_2 \in N3_j)j(k=3)} - X_{(i_3 \in N3_j)j(k=3)} = 0 \quad \forall j$$

$$\sum_{i \in \text{CL}_1} X_{ijk} \leq A \quad \forall j,k$$

$$\sum_{i \in \text{CL}_2} I_i \leq b$$

$$X_{ijk} \geq 0,\ I_i \geq 0,\ d_i^+ \geq 0, \text{ and } d_i^- \geq 0 \quad \forall i,j,k$$

13.3 Power Generation Planning

A long-term power generation plan considers the operational schedule of existing power-generating units and the installation schedule of new plants for meeting electricity demand, for a number of future years, at minimum possible cost. The electricity demand is described by a load duration curve (LDC) as is shown in Figure 13.1. The area under the LDC represents the requirements of electricity production in megawatt-hours. In order to keep the model simple such as linear, the LDC is approximated by a piecewise-constant function as shown as a histogram (or step function) in Figure 13.1. Bar 1 refers to the minimum load, bar 4 to peak load, and two intermediate bars represent middle-load demand.

In any company/country, a number of different types of technology, such as steam turbines, gas turbines, hydro, combined cycle and diesel cycle, are used for electricity generation. These power-generating units require different inputs, installation costs, and variable costs. In addition, they have different technical lives and follow different maintenance policies. The planning problem is to find the operations schedule of existing units, the new generation units to be installed in future, and the installation schedule of new units by minimizing the sum of installation and operating costs for a given number of years while not violating the technical and financial constraints. An LP model for such a problem is formulated below.

13.3.1 Linear Programming Model

The parameters and decision variables are defined as follows:

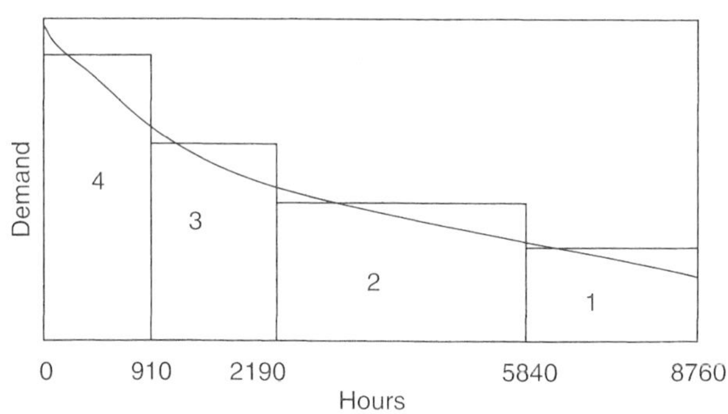

FIGURE 13.1
The load duration curve (LDC).

Parameters:

CE_{it} = capacity (MW) of existing units type i in period t
TD_p = duration of load block p in hours
AF_{it} = availability factor for type i in period t
PD_{pt} = maximum power demand in MW in a load block p in a period t
VC_{it} = variable cost (fuel cost plus other variable (\$/MW-h) for plant type i in period t) for both existing and new
FC_{it} = fixed cost (annual investment cost plus annual operating and maintenance costs, \$/MW-h) for plant type i in period t (for newly proposed)

Index:

i = plant type as discussed above, $i = 1, I$
t = time period, $t = 1, T$
p = load duration block, $p = 1, P$, in a given period

Variables:

X_{it} = capacity required (MW) for new units of type i in period t
CX_i = maximum capacity (MW) of type i required to install during the entire planning horizon
OX_{ipt} = output used (MW) from a new unit type i in load block p and during period t
OE_{ipt} = output used (MW) from an existing type i in load block p and during period t

Objective Function:
The objective is to minimize the total fixed costs plus the variable costs for the entire planning horizon.

$$\text{Minimize } Z = \sum_{p=1}^{P} TD_p \sum_{i=1}^{I} \sum_{t=1}^{T} VC_{it}(OX_{ipt} + OE_{ipt}) + \sum_{i=1}^{I} \sum_{t=1}^{T} FC_{it} X_{it} \tag{13.15}$$

The first term represents the variable costs of existing units plus newly proposed plants. The second term represents the fixed cost of new plants to be installed. The fixed cost for the existing plants is ignored here since it has no influence on the optimization process under consideration.

Constraints:
Capacity constraint: The output of each type of power-generation unit cannot exceed the total capacity of the existing or planned units of this type, multiplied by the corresponding availability factor in each load block of a given year t:

$$\text{Existing plants: } \text{OE}_{ipt} \leq \text{AF}_{it}\text{CE}_{it} \quad \text{for all } i, p, \text{ and } t \tag{13.16}$$

$$\text{New plants: } \text{OX}_{ipt} - \text{AF}_{it}X_{it} \leq 0 \quad \text{for all } i, p, \text{ and } t \tag{13.17}$$

Demand satisfaction: The power demand at each load block, in any period, must be satisfied from the existing and new plant generation.

$$\sum_{i=1}^{I} (\text{OX}_{ipt} + \text{OE}_{ipt}) \geq \text{PD}_{pt} \quad \text{for all } p \text{ and } t \tag{13.18}$$

New plant constraint: Once a certain type of plants with a given capacity is added, that plant will be retained throughout the planning horizon. In addition, further capacity may be added in a future period. For example, suppose the model has decided to install a new plant of type $i=1$ with 50 MW capacity in period 1, it will exist in the system until the technical life of this unit is expired. To ensure this, the following constraints are necessary:

$$X_{it+1} \geq X_{it}$$

or

$$X_{it+1} - X_{it} \geq 0 \quad \text{for all } i \text{ and } t \tag{13.19}$$

Note that the length of the planning horizon, considered in this problem, is less than the technical life of any new power-generating units available in the market.

Management preference for new plants: Management may prefer certain types of technology over others or like to have all the technologies in certain ratios.

The maximum capacity required from each new type i can be found using the following constraints:

$$\text{CX}_i \geq X_{it} \quad \text{for all } i \text{ and } t$$

or

$$\text{CX}_i - X_{it} \geq 0 \quad \text{for all } i \text{ and } t \tag{13.20}$$

Then the total capacity required from all new types of plants is $\sum_{i=1}^{I} \text{CX}_i$.

Now the management preference constraint can be written as follows:

$$CX_i \geq W_i \sum_{i=1}^{I} CX_i \text{ for some } i'\text{s}$$

or

$$CX_i - W_i \sum_{i=1}^{I} CX_i \geq 0 \text{ for some } i'\text{s} \tag{13.21}$$

where W_i is a fraction set by the management to their preferences. For example, if the management wants 50% of new capacity should be combined cycles ($i=4$), then set $W_4=0.50$. Note that $\sum_i W_i$ must be equal to 1.0.

The final LP model is

$$\text{Minimize } Z = \sum_{p=1}^{P} TD_p \sum_{i=1}^{I} \sum_{t=1}^{T} VC_{it}\left(OX_{ipt} + OE_{ipt}\right) + \sum_{i=1}^{I} \sum_{t=1}^{T} FC_{it}X_{it}$$

Subject to

$$OE_{ipt} \leq AF_{it}CE_{it}, \quad \text{for all } i, p, \text{ and } t$$

$$OX_{ipt} - AF_{it}X_{it} \leq 0 \quad \text{for all } i, p, \text{ and } t$$

$$\sum_{i=1}^{I} \left(OX_{ipt} + OE_{ipt}\right) \geq PD_{pt} \quad \text{for all } p \text{ and } t$$

$$X_{it+1} - X_{it} \geq 0 \quad \text{for all } i \text{ and } t \qquad \text{Model (13.2)}$$

$$CX_i - X_{it} \geq 0 \quad \text{for all } i \text{ and } t$$

$$CX_i - W_i \sum_{i=1}^{I} CX_i \geq 0 \quad \text{for selected } i'\text{s}$$

$$OX_{ipt}, OE_{ipt}, X_{it}, CX_i \geq 0 \quad \forall i, p, t$$

Calculation of the number of constraints (considering $I=5$, $P=4$, and $T=6$):

- Capacity $= 2 \times I \times P \times T = 2 \times 5 \times 4 \times 6 = 240$
- Demand satisfaction $= P \times T = 4 \times 6 = 24$
- New plant $= I \times T = 5 \times 6 = 30$
- New plant capacity required $= I \times T = 5 \times 6 = 30$
- Management preference $= I = 5$

Total number of constraints $= 240 + 24 + 30 + 30 + 5 = 329$.

Total variables required $= (2 \times I \times P \times T) + (I \times T) + I = I(1+T+2PT)$

With I = 5, $P = 4$, and $T = 6$, there are 120 OX_{ipt} variables, 120 OE_{ipt} variables, 30 X_{it} variables, and 5 CX_i variables, which make a total of 275 variables.

13.3.2 Multi-Objective Optimization Model

Mavrotas et al. (1999) considered a bi-objective model for power generation planning. The second objective in their model was to minimize the total amount of SO_2 (sulfur dioxide) emissions (in tonnes) produced annually from the electricity generation sector. These emissions are responsible for serious impacts on human health and the natural environment. In addition to SO_2, we believe it is important to minimize the total amount of CO_2 (carbon dioxide) emissions, which is a major concern for the greenhouse gas effect. So the above model can be reformulated with three objectives as follows.

The first objective is to minimize the total fixed costs plus the variable costs for the entire planning horizon.

$$\text{Minimize } Z_1 = \sum_{p=1}^{P} TD_p \sum_{i=1}^{I} \sum_{t=1}^{T} VC_{it}(OX_{ipt} + OE_{ipt}) + \sum_{i=1}^{I} \sum_{t=1}^{T} FC_{it} X_{it} \quad (13.22)$$

The second objective is to minimize the total amount of SO_2 emissions (in tonnes) produced, from electricity generation, during the entire planning horizon.

$$\text{Minimize } Z_2 = \sum_{t=1}^{T} \sum_{i=1}^{I} ES_i \sum_{p=1}^{P} TD_p(OX_{ipt} + OE_{ipt}) \quad (13.23)$$

where ES_i is the SO_2 emission factor for the ith type units (tonne SO_2/MW-h)

The third objective is to minimize the total amount of CO_2 emissions (in tonnes) produced, from electricity generation, during the entire planning horizon.

$$\text{Minimize } Z_3 = \sum_{t=1}^{T} \sum_{i=1}^{I} EC_i \sum_{p=1}^{P} TD_p(OX_{ipt} + OE_{ipt}) \quad (13.24)$$

where EC_i is the CO_2 emission factor for the ith type units (tonne CO_2/MW-h)

The final form of the multi-objective model is as follows:

$$\text{Minimize } Z_1 = \sum_{p=1}^{P} \text{TD}_p \sum_{i=1}^{I} \sum_{t=1}^{T} \text{VC}_{it} (\text{OX}_{ipt} + \text{OE}_{ipt}) + \sum_{i=1}^{I} \sum_{t=1}^{T} \text{FC}_{it} X_{it}$$

$$\text{Minimize } Z_2 = \sum_{t=1}^{T} \sum_{i=1}^{I} \text{ES}_i \sum_{p=1}^{P} \text{TD}_p (\text{OX}_{ipt} + \text{OE}_{ipt})$$

$$\text{Minimize } Z_3 = \sum_{t=1}^{T} \sum_{i=1}^{I} \text{EC}_i \sum_{p=1}^{P} \text{TD}_p (\text{OX}_{ipt} + \text{OE}_{ipt})$$

Subject to

$$\begin{aligned}
&\text{OE}_{ipt} \le \text{AF}_{it}\text{CE}_{it} && \text{for all } i, p, \text{ and } t \\
&\text{OX}_{ipt} - \text{AF}_{it} X_{it} \le 0 && \text{for all } i, p, \text{ and } t \\
&\sum_{i=1}^{I} (\text{OX}_{ipt} + \text{OE}_{ipt}) \ge \text{PD}_{pt} && \text{for all } p \text{ and } t \\
&X_{it+1} - X_{it} \ge 0 && \text{for all } i \text{ and } t && \text{Model (13.3)} \\
&\text{CX}_i - X_{it} \ge 0 && \text{for all } i \text{ and } t \\
&\text{CX}_i - W_i \sum_{i=1}^{I} \text{CX}_i \ge 0 && \text{for selected } i\text{'s} \\
&\text{OX}_{ipt}, \text{OE}_{ipt}, X_{it}, \text{CX}_i \ge 0 && \forall i, p, t
\end{aligned}$$

13.4 A Water Supply Problem

Consider a water supply system in any big city. Suppose there are a number of sources for water supply such as underground wells, rivers, and rainwater collected in reservoirs. Besides meeting the total demand for water in the city, it is extremely important to maintain the water quality for human consumption as per the World Health Organization (WHO) standards. In the past, underground water used to be considered as the safest source of drinking water. Now, the major concern with this water is the high level of chlorides, nitrates, and arsenic content. As per WHO report in 1996, nitrate levels above 150 mg/L pose an extreme risk to infants' health in the form of blue baby syndrome. In addition, high nitrates may create carcinogenic effects in adults (Cantor, 1997). Arsenic, widely known as poison, is a common health hazard to living beings. Inorganic arsenic is a documented human carcinogen. Although the chlorides are not as harmful as nitrates, the salinity it causes, makes water unacceptable for drinking. As per WHO guidelines for drinking water quality (1993), the levels of chlorides, nitrates, and arsenic should be 250 mg/L, 50 mg/L, and 0.01 mg/L, respectively.

To reduce the above chemicals in the water supply, to an acceptable level, one can use sophisticated chemical treatments. However, it would make the drinking water unaffordable to most citizens due to its high treatment costs. So it is good neither for the water supply business nor for general consumers. An alternative is to appropriately mix water from different sources as the levels of the chemical content vary from one underground source to other and the surface water is much better in quality. So it is a problem of optimally mixing the water from different sources at minimum possible cost while satisfying the demand, supply, and quality constraints.

We assume that a water supply network has a number of water sources and several water-treatment (here mixing) facilities. Water is distributed, from the treatment facilities, to the individual consumers through regional nodes and storage facilities. We restrict our problem boundary to water sources and treatment facilities as our main problem is to appropriately mixing water.

13.4.1 Linear Programming Model

First we formulate the problem as an LP model.

Parameters:

I = number of sources (index i)
J = number of water treatment facilities (index j)
K = number of quality parameters (index k)
S_i = water supply from source i (liter/period)
D_j = water demand from treatment facility j (liter/period)
C_{ij} = supply cost per unit from water source i to treatment facility j
T_j = treatment cost per unit at facility j
Q_{ik} = amount of quality parameter k in water from source i (amount per liter)
$Q\max_k$ = maximum allowable amount of quality parameter k for water consumption (amount per liter)

Decision variables:
X_{ij} = the amount of water supplied from source i to treatment facility j

Objective function:
The objective is to minimize the total cost of water supply and water treatment in the facilities.

$$\text{Minimize } Z = \sum_i \sum_j C_{ij} X_{ij} + \sum_j T_j \sum_i X_{ij} \tag{13.25}$$

Constraints:
The amount of water supply from each source, at any period, must be less than or equal to a given amount or capacity.

$$\sum_j X_{ij} \leq S_i \quad \forall i \tag{13.26}$$

The amount of water treated in a facility, at any period, must be greater than or equal to the demand at that facility.

$$\sum_i X_{ij} \geq D_j \quad \forall j \tag{13.27}$$

The amount of any quality parameter k in supply water supplied to a facility j must be less than or equal to the maximum amount allowed for that indicator.

$$\sum_i Q_{ik} X_{ij} \leq Q\max{}_k \sum_i X_{ij} \quad \forall j, k \tag{13.28}$$

So the final model is

$$\text{Minimize} \quad Z = \sum_i \sum_j C_{ij} X_{ij} + \sum_j T_j \sum_i X_{ij}$$

Subject to

$$\begin{aligned} &\sum_j X_{ij} \leq S_i \quad \forall i \\ &\sum_i X_{ij} \geq D_j \quad \forall j \\ &\sum_i Q_{ik} X_{ij} - Q\max{}_k \sum_i X_{ij} \leq 0 \quad \forall j, k \\ &X_{ij} \geq 0 \quad \forall i, j \end{aligned} \qquad \text{Model (13.4)}$$

Agha (2006) developed a GP model for handling water quality management problem of the Gaza Strip where the author considered two quality parameters—the amount of nitrates and chlorides. In addition to water allocation plan, he determined the location of reservoirs as a separate case under the overall water management problem.

13.4.2 Goal Programming Model

The above model can be formulated as a GP model. The objective of the GP model is to minimize the weighted sum of all undesirable deviations. In this model, the deviational variables and weight parameters are defined as follows:

d_{jk}^{+} = overachievement deviational variables for facility j quality parameter k

d_{jk}^{-} = underachievement deviational variables for facility j quality parameter k

W_{jk} = weights of deviational variables for facility j quality parameter k

The final GP model is

$$\text{Minimize } Z = \sum_{j}\sum_{k} W_{jk} d_{jk}^{+}$$

Subject to

$$\sum_{j} X_{ij} \le S_i \quad \forall i$$

$$\sum_{i} X_{ij} \ge D_j \quad \forall j \qquad \text{Model (13.5)}$$

$$\sum_{i} Q_{ik} X_{ij} - Q\max{}_k \sum_{i} X_{ij} + d_{jk}^{-} - d_{jk}^{+} = 0 \quad \forall j,k$$

$$X_{ij} \ge 0 \quad \forall i,j$$

$$d_{jk}^{-}, d_{jk}^{+} \ge 0 \quad \forall j,k$$

13.5 A Supply Chain Problem

In this section, we will model a supply chain problem. The first step of any supply chain model is to develop a supply chain network representing the material sources, intermediate processes and storages, destinations, and flow paths. We consider a supply chain network as shown in Figure 13.2. In the network, there are a number of plants that supply finished goods to a number of silos and packaging lines. The silos store the products and deliver to the commercial customers throughout the year. On the other

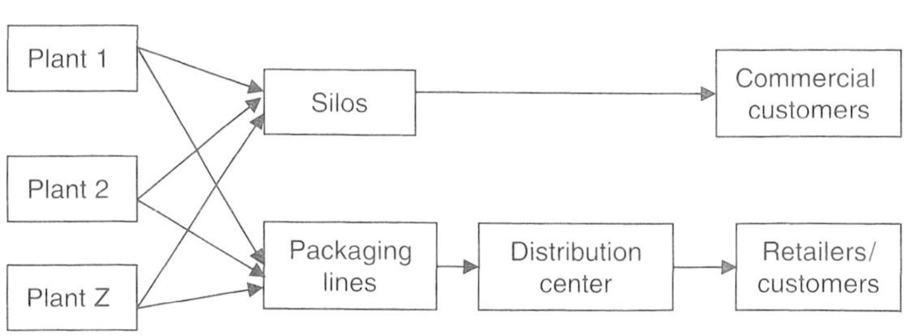

FIGURE 13.2
A sample supply chain network.

hand, the packaging lines pack the products for household consumption and subsequently the products are stored in warehouses or distribution centers. Finally, the packaged products are supplied to the retailers.

For formulating the above supply chain problem, we define the following nodes from the production facilities to the customers end:

N_{PF} = set of production facilities
N_{PL} = set of packaging lines
N_{DC} = set of distribution centers
N_{SB} = set of silos for bulk storages
N_{CC} = set of commercial customers require bulk delivery
N_{RD} = set of retailers require packaged goods
N = set of all nodes in the network

where, $N = N_{\text{PF}} \cup N_{\text{PL}} \cup N_{\text{DC}} \cup N_{\text{SB}} \cup N_{\text{CC}} \cup N_{\text{RD}}$

The parameters can be defined as follows:

D_{ij} = the distance between nodes i and j
C_{ij} = the cost per unit distance traveled between nodes i and j
SC_i = the storage capacity at node i for bulk storage, if $i \in N_{\text{SB}}$
SD_i = the storage capacity at distribution center i, if $i \in N_{\text{DC}}$
PC_i = the production capacity of plant i, if $i \in N_{\text{PF}}$
PL_i = the packaging line capacity of line i, if $i \in N_{\text{PL}}$
CD_i = the demand of commercial customer i, if $i \in N_{\text{CC}}$
RD_i = the demand of retailer i, if $i \in N_{\text{RD}}$

The decision variable of the supply chain problem can be defined as

X_{ij} = the quantity of goods transported between nodes i and j

The objective of the problem is to minimize the overall transportation cost between all nodes of the distribution network. We assume that the total cost between any two nodes is a linear function of the distance between these two nodes and the cost per unit distance.

$$\text{Minimize } Z = \sum_{i \in N} \sum_{j \in N} C_{ij} D_{ij} X_{ij} \tag{13.29}$$

Note that C_{ij} is zero if an arc (i, j) does not exist.

Constraints:
Production capacity: The goods supplied from a plant must be less than or equal to its production capacity.

$$\sum_{j \in N_{\mathrm{PL}} \cup N_{\mathrm{SB}}} X_{ij} \leq \mathrm{PC}_i \quad \forall i \in N_{\mathrm{PF}} \tag{13.30}$$

Packaging capacity: The quantity of goods supplied from all production facilities to any packaging line must be less than or equal to its packaging capacity.

$$\sum_{i \in N_{\mathrm{PF}}} X_{ij} \leq \mathrm{PL}_j \quad \forall j \in N_{\mathrm{PL}} \tag{13.31}$$

Silo capacity: The quantity of goods supplied from all production facilities to any silo must be less than or equal to its capacity.

$$\sum_{i \in N_{\mathrm{PF}}} X_{ij} \leq \mathrm{SC}_j \quad \forall j \in N_{\mathrm{SB}} \tag{13.32}$$

Distribution center capacity: The quantity of goods supplied from all packaging lines to any distribution center must be less than or equal to its capacity.

$$\sum_{i \in N_{\mathrm{PL}}} X_{ij} \leq \mathrm{SD}_j \quad \forall j \in N_{\mathrm{DC}} \tag{13.33}$$

Demand constraint: The total goods supplied to either commercial customers or retailers must be greater than or equal to their demand.

$$\sum_{i \in N_{\mathrm{SB}}} X_{ij} \geq \mathrm{CD}_j \quad \forall j \in N_{\mathrm{CC}} \text{ Commercial customers} \tag{13.34}$$

$$\sum_{i \in N_{\mathrm{DC}}} X_{ij} \geq \mathrm{RD}_j \quad \forall j \in N_{\mathrm{RD}} \text{ Retailers} \tag{13.35}$$

The flow of goods must be balanced in each intermediate node. The following equations represent the flow balance constraints of our supply chain network.

Plant-silo-customer Path: The amount of goods received by all commercial customers from a silo must equal to the total amount received by the silo from all production facilities.

$$\sum_{i \in N_{\mathrm{PF}}} X_{ij} = \sum_{k \in N_{\mathrm{CC}}} X_{jk} \quad j \in N_{\mathrm{SB}} \tag{13.36}$$

Plant-packaging-distribution center path: The amount of goods received by all distribution centers from a packaging line must equal to the total amount received by the packaging line from all production facilities.

$$\sum_{i \in N_{\mathrm{PF}}} X_{ij} = \sum_{k \in N_{\mathrm{DC}}} X_{jk} \quad j \in N_{\mathrm{PL}} \tag{13.37}$$

Packaging-distribution-retailer path: The amount of goods received by all retailers from a distribution center must equal to the total amount received by the distribution center from all packaging lines.

$$\sum_{i \in N_{\mathrm{PL}}} X_{ij} = \sum_{k \in N_{\mathrm{RD}}} X_{jk} \quad j \in N_{\mathrm{DC}} \tag{13.38}$$

Nonnegativity constraint: $X_{ij} \geq 0$

A slightly different supply chain model for a sugar industry can be found in Ioannou (2005). The company procures beets from farmers and processes them in their production facilities to produce white sugar in bulk form and a number of by-products. As the beets are seasonal agricultural products, the final product is either stored at appropriate silos for distribution to commercial customers throughout the year or transferred to packaging lines for production of packaged sugar. Packaged sugar is subsequently stored in warehouses or distribution centers. Finally the packaged sugar is sold to retailers. Ioannou (2005) considered five production facilities, three packaging lines of two different types and three bulk sugar storage facilities. The author showed that he is able to save \$1 million implementing his model to the sugar industry in Greece.

13.6 Coal Production and Marketing Plan

In this section, we introduce a real-world coal production and marketing problem. Coal is extracted from mines and upgraded on the surface for the end users. The raw coal is known as the run of mine (ROM) in the coal mining industry. Each coal type has its own typical quality specifications. Coal quality is measured in terms of percent of ash, sulfur, and moisture, and BTU (British thermal unit) content per pound. BTU content per pound expresses the heating value (calorific value) of coal. It is desirable to have high BTU content and low ash, sulfur, and moisture content. The customers specify the quality parameters (such as maximum percentage of ash and sulphur, and minimum BTU/pound) for their coals. We assume a coal company operates a number of mines. These mines differ greatly in their cost of production and coal quality. Preparation and blending are the two coal upgrading and processing facilities. Coal preparation (also known as washing) is a process of removing physical impurities from coal content. A coal preparation unit produces high-quality washed coal as well as low-quality middling product. Blending is a simple mixing process that allows upgrading the low-quality coals by mixing it with high-quality coals. The coals are sold either based on the total BTU supplied or total tonnes

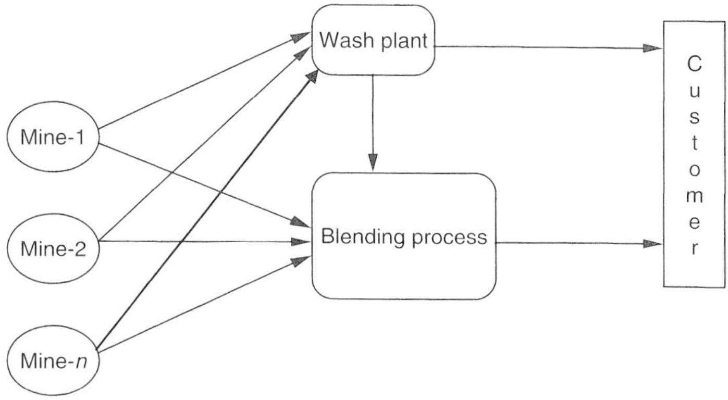

FIGURE 13.3
Single-period coal production process.

supplied within a given customer specification. The objective of coal production and marketing plan is to determine the coal-mix for maximizing the revenue from the coal sales while satisfying quality, demand, and other technical constraints.

A single-period LP model for coal production and marketing is presented below. It is designed to find an optimal strategy of coal blending, washing, and customer allocation so as to transform the available raw coal into final coal products within customer market specifications at maximum overall profit. This model can be used for any number of mines, washed coals, and blended products (see flow diagram in Figure 13.3). The constraints considered in this model are ash, sulfur, and calorific value limitations, production limit, and market requirements.

Variables:

bp_{jl} = tonnes of blended product for customer j made at location l

c_{mjl} = tonnes of raw coal from mine m used for blended product j at location l

wc_k = tonnes of washed product k produced

wb_{kjl} = tonnes of washed product k used for blended product j at location l

mb_{kjl} = tonnes of middling product k used for blended product j at location l

wp_{kc} = tonnes of washed product k sent to customer c

Data:

J = number of blended product customers

$L(j)$ = set of sites used for blended product for customer j

a_m^c, s_m^c, B_m^c = ash, sulphur, and BTU/lb analysis for coals from mine m

a_k^w, s_k^w, B_k^w = as received ash, sulphur, and BTU/lb analysis for washed product k

a_k^m, s_k^m, B_k^m = as received ash, sulphur, and BTU/lb analysis for middling product k

a_j^+, s_j^+ = maximum allowable ash and sulphur analysis for customer j

B_j^- = minimum allowable BTU/lb analysis for customer j

BTU_j^+, BTU_j^- = maximum and minimum BTU requirements for customer j

S_{NS}^+ = maximum allowable sulphur supplied

NS = set of blended product customers who correspond to local customers

ζ_{j} = amount of SO_2 per tonne of sulphur supplied to customer $j \in \text{NS}$

I = number of mines

r_{mk}^c = amount of raw coals from mine m used per tonne of washed product k

r_{mid}^k = ratio of middling in product k produced

MP_m^+, MP_m^- = maximum and minimum production from mine m

BCOST_{jl} = blending cost (\$/tonne) for customer j at location l

MPRO_m = mining cost (\$/tonne) for mine m

B_m^M = BTU content (million BTU/tonne) for raw coal from mine m

B_k^{Wb} = BTU content (million BTU/tonne) for washed product k

B_k^{Mb} = BTU content (million BTU/tonne) for middlings of washed product k

PB_j = price (\$/million BTU) offered by the blended customer j

PP_{kc} = price (\$/tonne) offered by customer c for washed product k

TCBC_{jl} = transportation cost (\$/tonne) to blended product customer j from blending location l

TCML_{ml} = transportation cost (\$/tonne) from mine m to blending location l

TCMW_m = transportation cost (\$/tonne) from mine m to the wash plant

TCWL_l = transportation cost (\$/tonne) from the wash plant to blending location l

$TCWC_c$ = transportation cost ($/tonne) from the wash plant to washed customer c

W_k = washing cost per tonne in the washing plant

The objective function of the LP model is to maximize profit. The mathematical expression is as follows:

$$\text{Minimize } Z = \sum_j \sum_l \left[-(\text{BCOST}_{jl} + \text{TCBC}_{jl})\text{bp}_{jl} \right]$$
$$+ \sum_m \sum_j \sum_l \left[B_m^M \text{PB}_j - \text{TCML}_{jl} - \text{MPRO}_m \right] c_{mjl}$$
$$+ \sum_k \left[-W_k - \sum_m a_{mk}(\text{MCMW}_m + \text{MPRO}_m) \right] \text{wc}_k$$
$$+ \sum_k \sum_j \sum_l \left[B_k^{Wb} \text{PB}_j - \text{TCWL}_l \right] \text{wb}_{kjl}$$
$$+ \sum_k \sum_j \sum_l \left[B_k^{Mb} \text{PB}_j - \text{TCWL}_l \right] \text{mb}_{kjl}$$
$$+ \sum_k \sum_c \left[\text{PP}_{kc} - \text{TCWC}_c \right] \text{wp}_{kc} \qquad (13.39)$$

The different components of the objective function are explained as follows:

Term 1: blending cost plus transportation costs to blended product customers

Term 2: revenue for BTU content of raw coals used in blending minus the mine production costs and transportation costs

Term 3: the washing costs, transportation costs, and mine production costs for raw coals used in washing plant

Term 4: revenue from washed coals used in blended product minus the transportation costs

Term 5: revenue from thermal middlings used in blended product minus the transportation costs

Term 6: revenue from washed coals sold as tonnes minus the transportation costs

The constraints of the model are discussed below.

Mass balance for blended products: The total blended product j produced in any location l must be equal to the total raw coal received from all mines plus all wash and middling coals from the wash plant.

$$-\text{bp}_{jl} + \sum_m c_{mjl} + \sum_k \text{wb}_{kjl} + \sum_k \text{mb}_{kjl} = 0 \quad j = 1, J, \quad l \in L(j) \qquad (13.40)$$

Ash limits in blended products: The percentage of ash in any blended product must be less than or equal to the allowable limit of ash.

$$-a_j^+ \mathrm{bp}_{jl} + \sum_m a_m^c c_{mjl} + \sum_k a_k^w \mathrm{wb}_{kjl} + \sum_k a_k^m \mathrm{mb}_{kjl} \leq 0 \quad j = 1, J, \quad l \in L(j) \tag{13.41}$$

Sulfur limits in blended products: The percentage of sulfur in any blended product must be less than or equal to the maximum allowable limit of sulphur.

$$-s_j^+ \mathrm{bp}_{jl} + \sum_m s_m^c c_{mjl} + \sum_k s_k^w \mathrm{wb}_{kjl} + \sum_k s_k^m \mathrm{mb}_{kjl} \leq 0 \quad j = 1, J, \quad l \in L(j) \tag{13.42}$$

Minimum BTU content in blended products: The BTU content per pound in any blended product must be greater than or equal to the minimum allowable limit of BTU content.

$$-B_j^- \mathrm{bp}_{jl} + \sum_m B_m^c c_{mjl} + \sum_k B_k^w \mathrm{wb}_{kjl} + \sum_k B_k^m \mathrm{mb}_{kjl} \leq 0 \quad j = 1, J, \quad l \in L(j) \tag{13.43}$$

Overall BTU supply to customers: The overall BTU supply to any customer must be between the customer's specified limit.

$$\mathrm{BTU}_j^- \leq \sum_l \left[\sum_m B_m^c c_{mjl} + \sum_k B_k^w \mathrm{wb}_{kjl} + \sum_k B_k^m \mathrm{mb}_{kjl} \right] \leq \mathrm{BTU}_j^+ \quad j = 1, J, \quad l \in L(j) \tag{13.44}$$

Overall sulfur supplied: To control the environmental pollution, the overall sulfur supplied to all local customers in a given year must be less than or equal to a set limit.

$$\sum_{j \in \mathrm{NS}} \xi_j \sum_l \left[\sum_m s_m^c c_{mjl} + \sum_k s_k^w \mathrm{wb}_{kjl} + \sum_k s_k^m \mathrm{mb}_{kjl} \right] \leq S_{\mathrm{NS}}^+ \tag{13.45}$$

Maximum and minimum mine production: The total raw coal used, from any mine, in the blending process and wash plant must within the given production limit.

$$\mathrm{MP}_m^- \leq \sum_j \sum_l c_{mjl} + \sum_k r_{mk}^c \mathrm{wc}_k \leq \mathrm{MP}_m^+ \quad i = 1, I \tag{13.46}$$

Mass balance in wash plant: The total wash product produced is equal to the wash product used in the blended process plus the wash product directly sent to customer. It is assumed that any excess amount of wash products can be sold in the market with the set price.

$$\mathrm{wc}_k - \sum_j \sum_l \mathrm{wb}_{kjl} - \sum_c \mathrm{wp}_{kc} = 0 \quad k = 1, K \tag{13.47}$$

Middlings ratio for washed products: The middling product produced maintains a certain ratio with the wash product. It is assumed that all the middling products will be used in blending process.

$$r_k^{\mathrm{mid}} \mathrm{wc}_k - \sum_j \sum_l \mathrm{mb}_{kjl} = 0 \quad k = 1, K \tag{13.48}$$

and nonnegativity constraints.

13.6.1 Multi-Objective Problem

The above problem can be formulated as a two-objective model where the objectives are

- Maximizing profit (current objective)
- Minimizing carbon dioxide emission

The mathematical expression for the second objective can be written as follows:

$$\text{Minimize } Z_2 = \sum_j \sum_l \left[\sum_m \mathrm{CE}_m^c c_{mjl} + \sum_k \mathrm{CE}_k^w \mathrm{wb}_{kjl} + \sum_k \mathrm{CE}_k^m \mathrm{mb}_{kjl} \right] \tag{13.49}$$

where CE_m^c, CE_k^w, and CE_k^m represent the amount of carbon dioxide emission per tonne of raw coals directly used in blending, washed products, and middlings, respectively, sold in a given year.

13.6.2 Multi-Period Problems

A single-period model can be used for annual or monthly planning, or for that matter, for any defined period. A single-period annual planning model is suitable for stable demand and production patterns. This model does not consider the demand, production, or cost fluctuations from period to period within the planning horizon. This may create severe problems in operational implementation for fluctuating cases. In case of monthly planning, a collection of 12 monthly planning models can be thought of as an annual planning model. However, a collection of monthly (single-period) models, which represents an annual planning process, may become infeasible in the case of fluctuating demand or production. If solving this model with zero initial and final inventories in each month leads to a feasible solution, this model provides a lower bound of the problem. A multi-period model is usually preferable over a collection of single-period models. A multi-period model is usually run with a rolling planning horizon, which allows updating the planning process at the end of every period with the availability of new information. Furthermore, it shows more savings than a collection of monthly models under seasonality and gives a better scenario of the planning process in an unstable planning environment.

A multi-period model allows the carrying of inventory from one period to another if it is profitable. A multi-period model is a collection of several single-period problems with some linking constraints from one period to another. Normally, inventory balance equations serve this purpose.

In the multi-period case, we can assume that the blended product inventory of period t would be an input to the blending process of period $t+1$. As indicated in Figure 13.4, the raw coals are supplied to the preparation plant and the blending process. The washed products produced in any period are sent to the customers and blending process or carried for future use. The inputs to a blending process in any period are raw coals and washed coals of that period, and the blended product inventory from the previous period. The blended product produced is sent to the customers or carried for use as an input in the next period. It is useful to carry inventory from one period to another where the demand or production pattern or cost figures are highly fluctuating. Alternatively, it can be assumed that the inventory of blended product of period t will be supplied to the customers in period $t+1$. This model will be less flexible in tackling the fluctuating situations.

In traditional multi-period formulation, the goods produced in any period t are carried as inventory either to supply customers in period $t+1$ or for further uses as inventory for period $t+2$ and so on (see Figure 13.5). These goods cannot (or are not allowed to) be used as raw material for the same kind of products. In case of coal blending, the quality of blended product is always greater than or equal to its minimum requirements. If the quality of blended product is very good then its use as an input to the blending process of the next period would be useful and make the model flexible. The mathematical models for these two alternative planning processes can be found in Sarker (2003) and Sarker and Gunn (1997).

13.7 General Blending Problem

Blending is a common process in many practical situations. All blending processes are required to meet a number of key quality parameters. Once the minimum quality parameters are met, the price is usually

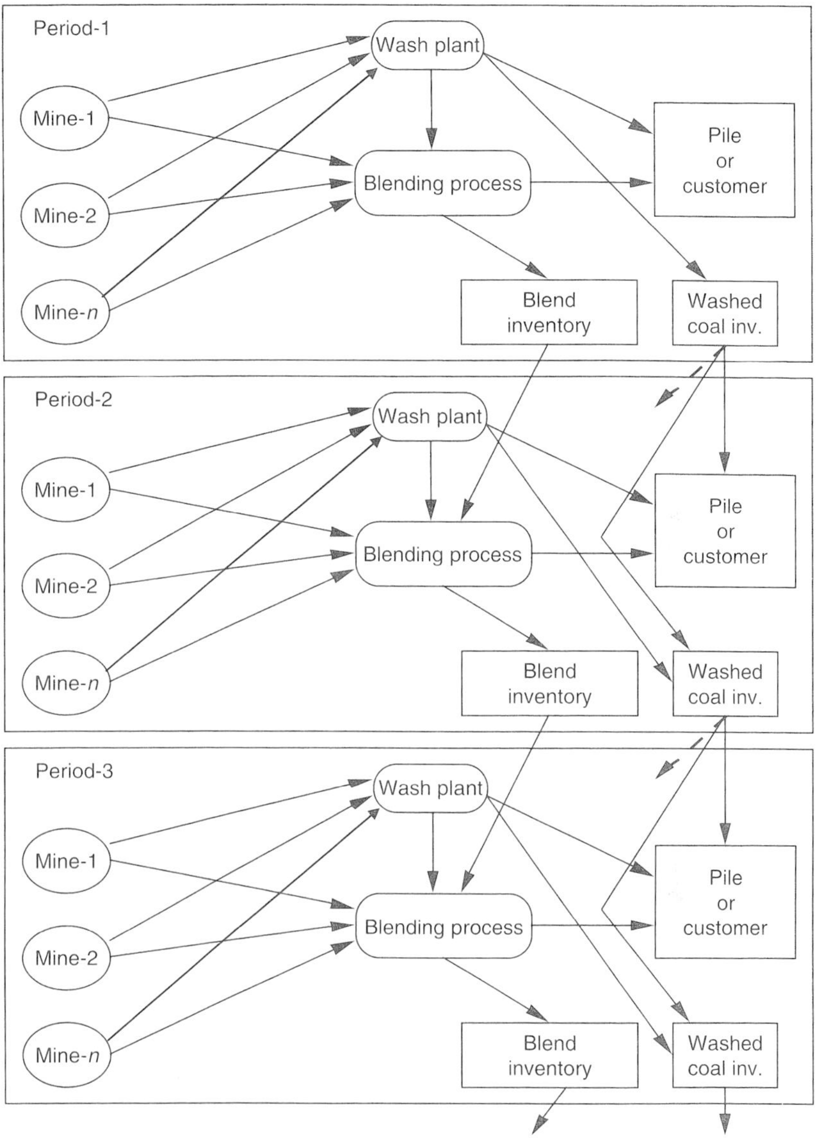

FIGURE 13.4
A multi-period coal production process. (Produced as a slightly modified version of Sarker, R. and Gunn, E., *Eur. J. Oper. Res.*, 101, 140, 1997. With permission from Elsevier.)

charged based on the amount of the blended product sold. However, there are situations where the customers accept both better and worse quality products, in relation to the contract specifications, with a revised price. In this case, the price is a function of the content of the quality parameters. That means, higher quality product will be rewarded and

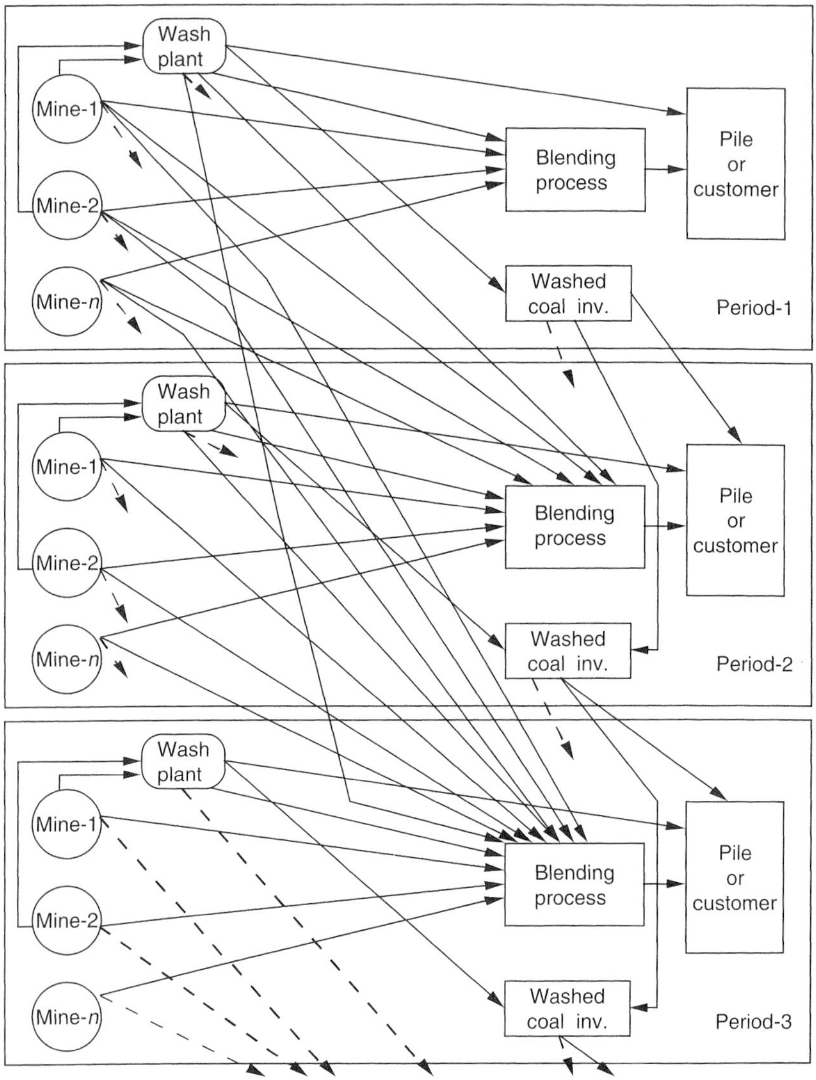

FIGURE 13.5
An alternative multi-period coal production process. (Produced as a slightly modified version of Sarker, R. and Gunn, E., *Eur. J. Oper. Res.*, 101, 140, 1997. With permission from Elsevier.)

lower quality product will be penalized. In this section, we will formulate two LP models for two situations of general blending problems.

Assumptions:

- A number of inputs, with known quality parameters, will be used in the process
- Only one blended product will be produced

- The contract specifications of the blended product are given
- The price of blended product is known which is set based on the contract specifications

Parameters:

I_1 = set of quality parameters in which higher the parameter better the quality

I_2 = set of quality parameters in which lower the parameter better the quality

Q_{ij} = quality parameter i in input j

CQ_i = acceptable level of quality parameter i as per contract specification

C_j = unit cost of input j

D = demand of blended product

P = unit selling price of blended product

Decision variables:
X_j = amount of input j used in producing the blended product

Objective Function:
The objective is to maximize the total profit from the blended product sales.

$$\text{Maximize} \sum_j (P - C_j)X_j \tag{13.50}$$

Constraints:
Demand Constraint: the amount of blended product produced must be *greater than or equal to* the demand of the product.

$$\sum_j X_j \geq D \tag{13.51}$$

Quality Restrictions: the quality level of the blended product must be *greater than or equal to* the acceptable level for I_1 parameters and *less than or equal to* the acceptable level for I_2 parameters.

$$\sum_j Q_{ij}X_j \geq CQ_i \sum_j X_j \quad \forall i \in I_1 \text{ and} \tag{13.52}$$

$$\sum_j Q_{ij}X_j \leq CQ_i \sum_j X_j \quad \forall i \in I_2 \tag{13.53}$$

So the first linear programming model:

$$\text{Maximize } \sum_j (P - C_j)X_j$$

Subject to

$$\sum_j X_j \geq D$$

$$\sum_j Q_{ij}X_j \geq CQ_i \sum_j X_j \quad \forall i \in I_1 \qquad \text{Model (13.6)}$$

$$\sum_j Q_{ij}X_j \leq CQ_i \sum_j X_j \quad \forall i \in I_2$$

$$X_j \geq 0 \quad \forall j$$

In the above model, there is no reward or credit for better quality than the contract specification CQ_i. For the second model, we assume that a blended product that has a better quality will attract a higher price. In addition, a lower quality product can also be sold but at lower price. To formulate this model, we need the following additional parameters and decision variables.

Parameters:

RP_i = Reward/penalty for improvement/deterioration of one unit of quality parameter per unit of blended product produced. Note that the improvement in quality parameter means above the specified level for I_1 and below the specified level for I_2, and deterioration means the opposite.

Variables:

Y_i = Total amount of deviation for quality parameter i. It equals quality parameter improvement (or deterioration) per unit of blended product multiplied by the amount of blended product produced. This variable is unrestricted in sign (i.e., can be either positive, zero, or negative). So the total reward plus penalty would be equal to $\sum_i RP_i Y_i$.

So the second LP model will be

$$\text{Maximize } \sum_j (P - C_j)X_j + \sum_i RP_i Y_i$$

Subject to

$$\sum_j X_j \geq D$$

$$\sum_j Q_{ij}X_j + Y_i = CQ_i \sum_j X_j \quad \forall i \in I_1 \qquad \text{Model (13.7)}$$

$$\sum_j Q_{ij}X_j + Y_i = CQ_i \sum_j X_j \quad \forall i \in I_2$$

$$X_j \geq 0 \quad \forall j$$

$$Y_i \text{ unrestricted} \quad \forall i$$

If the reward rate is different from the penalty rate, we need two variables representing the over-achievement and under-achievement deviations for each quality constraint. This is equivalent to replacing the unrestricted variable with the difference of two non-negative variables discussed in an earlier chapter. As the objective function of Model 13.7 is of maximization type, it would require additional constraints to make sure that at best one deviational variable in each quality constraint is positive.

13.8 Summary

In this chapter, we have introduced a number of LP-based practical problems and discussed their models referring to the published materials. These are crop planning, power generation planning, water supply management, supply chain, coal upgrading and marketing, and general blending problems. We have also discussed their alternative mathematical models and related issues. In the next chapter, we present a number of integer and nonlinear practical problems and their mathematical models.

References

Agha, S.R., Use of goal programming and integer programming for water quality management—a case study of Gaza Strip, *European Journal of Operational Research*, 174, 1991–1998, 2006.

Cantor, K.P., Drinking water and cancer, *Cancer Causes Control*, 8, 292–308, 1997.

Ioannou, G., Streamlining the supply chain of the Hellenic Sugar Industry, *Journal of Food Engineering*, 70, 323–332, 2005.

Mavrotas, G., Diakoulaki, D., and Papayannakis, L., An energy planning approach based on mixed 0–1 multiple objective linear programming, *International Transactions in Operations Research*, 6, 231–244, 1999.

Sarker, R., Operations Research Applications in a Mining Company, Dissertation.de-Publishing, Germany, 2003.

Sarker, R.A. and Gunn, E.A., A simple SLP algorithm for solving a class of nonlinear programs, *European Journal of Operational Research*, 101, 140–154, 1997.

Sarker, R. and Quaddus, M., Modelling a nationwide crop planning problem using a multiple criteria decision making Tool, *International Journal of Computers and Industrial Engineering*, 42, 541–553, 2002.

Sarker, R., Talukder, S., and Haque, A., Determination of optimum crop-mix for crop cultivation in Bangladesh, *Applied Mathematical Modelling*, 21, 621–632, 1997.

World Health Organization, *Guidelines for Drinking Water Quality*, 2nd ed., Volume 2, Health criteria and other supporting information, WHO, Geneva, 1996.

WHO guidelines for drinking water quality (1993) http://www.who.int/mediacentre/factsheets/fs210/en/index.html

14

Models for Practical Problems II

14.1 Introduction

In Chapter 13, we addressed a number of linear programming (LP)-based practical problems. In this chapter, in addition to one or two LP models, we briefly present various integer and nonlinear practical problems. The problems to be discussed are a combat logistics problem, a lot-sizing problem, a joint lot-sizing and transportation decision problem, a coal bank scheduling problem, an engineering design problem, a gas-lift optimization problem, and a multiple shifts planning problem. We also discuss their associated mathematical models and related issues.

14.2 A Combat Logistics Problem

For strategic reasons, modern war-fighting concepts emphasize small but highly mobile forces supported over the littoral, rather than from large, land-based supply points. In this situation, the objective of logistics planners is to support these forces with as little inventory on land as possible. In this section, we present a combat logistic model for a given scenario.

Let us consider sea-based combat and support units. Each combat unit is required to reach a particular set of targets on land. Combat units consume food, water, ammunition, and fuel. Supply units are free to deploy, move, and to build up and deplete inventories as necessary to meet the demand of the combat units. A fleet of vehicles are available (1) to transport combat units to targets or intermediate points, (2) to move entire supply units, or (3) to transport supplies between units. Definitely, the choices of vehicle types are dependent on the physical environments such as sea/water, and land with or without road connections. The problem is to determine the locations of supply units and the shipment of each commodity between units fulfilling the demands in each period while minimizing the land-based inventory.

The problem seems to resemble a multi-period, facility location and multi-commodity flow problem. Gue (2003) models such a situation as a network flow problem where the network consists of two types of nodes, combat and supply nodes. The combat nodes are given in the battle plan and a possible set of supply locations may be known or considered depending on the battle field and physical environment. The objective is to minimize the total inventory of land-based support units while maintaining the sea-based units as the origin of supplies. The solutions required for each period are (1) the locations of support units, (2) inventories held by the units, and (3) the amounts shipped between units. We assume that the sea base is the only node that is common to both supply and combat units. All other nodes are either combat or supply nodes. In this section, we present a mathematical model similar to Gue (2003). In modelling the problem, we use the following notations:

Decision variables:

X_{ijkt} = quantity of commodity k shipped from node i to j in period t

INV_{ikt} = inventory of commodity k held at node i in period t

$$Y_{ijt} = \begin{cases} 1 & \text{if a unit moves from node } i \text{ to node } j \text{ in period } t \\ 0 & \text{otherwise} \end{cases}$$

Sets:

I = set of all nodes

I_s = set of supply nodes

I_c = set of combat nodes

I_{sl} = set of land-based supply nodes

I_{cl} = set of land-based combat nodes

I_l = set of land-based nodes

I_b = set of beach nodes accessible by watercraft

I_i = set of inland (not beach) nodes

Parameters:

W_s = weight of support unit

W_c = weight of combat unit

S_{ij} = distance from node i to node j

B_{it} = maximum total inventory that can be held at node i in period t

D_{jkt} = demand of commodity k at node j in period t

N = maximum number of support units

L_a = available airlift in a period (in kg-km)

L_s = available ship-shore lift in a period (in kg-km)

T_{ijt} = indicates the movement of a combat unit from node i to j in period t

M = a large number

The mathematical model is presented below:

Objective function:
The objective is to minimize the total land-based inventory plus the sum of weights for all support units.

$$\text{Minimize } Z = \sum_{i \in I_L} \sum_{k} \sum_{t} \text{INV}_{ikt} + W_s \sum_{i \in I_{sl}} \sum_{j \in I_{sl}} \sum_{t} Y_{ijt} \tag{14.1}$$

Constraints:
Material balance constraints: For a land-based combat node, in a given period t, the commodity k coming from all other nodes plus the inventory from the previous period must be equal to its supplies to other nodes plus the inventory carried for the next period plus the demand of that period.

$$\text{INV}_{ikt} + \sum_{j \in I} X_{jikt} - \sum_{j \in I} X_{ijkt} - D_{ikt} = \text{INV}_{ik,t+1} \quad \forall i \in I_{cl}, k, t \tag{14.2}$$

Similarly, for a land-based supply node, in a given period t, the commodity k coming from all other nodes plus the inventory from the previous period must be equal to its supplies to other nodes plus the inventory carried for the next period plus the demand of that period.

$$\text{INV}_{ikt} + \sum_{j \in I_s} X_{jikt} - \sum_{j \in I} X_{ijkt} = \text{INV}_{ik,t+1} \quad \forall i \in I_{sl}, k, t \tag{14.3}$$

If any supply unit moves to a land-based supply unit in a period t, which is equivalent to the movement of a land-based supply unit to any supply unit in period $t + 1$. This constraint enforces the continuity of flow for supply units among nodes.

$$\sum_{j \in I_s} Y_{jit} - \sum_{j \in I_s} Y_{ij,t+1} = 0 \quad \forall i \in I_{sl}, t \tag{14.4}$$

It is required to prohibit shipments between the supply units unless the sending unit is stationary or the sending units are the shipping material that supports its own movement.

$$\sum_{k} X_{ijkt} - M(Y_{iit} + Y_{ijt}) \leq 0 \quad \forall i \in I_s, j \in I_s, t \tag{14.5}$$

It is necessary that a land-based supply unit to be stationary when shipping to a combat unit.

$$\sum_{j\in I_{\text{c}}}\sum_{k} X_{ijkt} - MY_{iit} \leq 0 \quad \forall i \in I_{\text{sl}}, t \tag{14.6}$$

Linking the unit location and inventory variables by only allowing inventory if a unit occupies a node.

$$\sum_{k} \text{INV}_{ikt} - M\sum_{j\in I_{\text{sl}}} Y_{ijt} \leq 0 \quad \forall i \in I_{\text{sl}}, t \tag{14.7}$$

The total shipping quantity from any land-based supply node must be less than or equal to its stock.

$$\sum_{j\in I_{\text{l}}} X_{ijk,t+1} - \text{INV}_{ikt} \leq 0 \quad \forall i \in I_{\text{sl}}, k, t \tag{14.8}$$

Specifying the maximum lift in a period from ship to shore via air

$$\sum_{j\in I_{\text{i}}}\sum_{k} S_{0j}X_{0jkt} + W_{\text{c}}\sum_{j\in I_{\text{l}}}\sum_{i\in I_{\text{l}}} S_{ij}T_{ijt} \leq L_{\text{a}} \quad \forall t \tag{14.9}$$

Specifying the maximum lift in a period from ship to beach location via air or ship

$$\sum_{j\in I_{\text{b}}}\sum_{k} S_{0j}X_{0jk1} + W_{\text{c}}S_{0j}T_{0j1} + \sum_{j\in I_{\text{l}}}\sum_{k} S_{0j}X_{0jk1} + W_{\text{c}}\sum_{i\in I_{\text{l}}} S_{0j}T_{ij1} \leq L_{\text{a}} + L_{\text{s}} \tag{14.10}$$

Imposing a limit on the amount of inventory that combat units can hold.

$$\sum_{k} \text{INV}_{ikt} \leq B_{it} \quad \forall i \in I_{\text{cl}} \tag{14.11}$$

Nonnegativity constraint and binary restrictions

$$\begin{aligned} &X_{ijkt}, \text{INV}_{ikt} \geq 0 \quad \forall i, j, k, t \\ &Y_{ijt} \in \{0, 1\} \quad \forall i, j, t \end{aligned} \tag{14.12}$$

The above model is a mixed-integer LP model. For further details on combat logistics modelling aspects, see Gue (2003) who implemented his model using GAMS with CPLEX-MIP Solver.

14.3 A Lot-Sizing Problem

Consider a manufacturing system where a manufacturer uses raw material to produce a product. In such a case, the raw material's ordering quantities are dependent on the batch quantity of the product. As a result, one should determine the optimum production batch size of a product and the ordering quantities of associated raw material simultaneously. This may be done by treating production and purchasing as components of an integrated system, minimizing the total cost of the system. The problem considered has the following attributes:

- A finite production-rate environment uses raw materials from outside suppliers.
- Only the product lot-sizing and its associated raw material supply quantities are under consideration.
- The supply policy for the product is to deliver an equal quantity at fixed intervals.
- The raw materials are received in lots.
- The product cannot be delivered until the whole lot is finished and quality certification is complete.

Here, the objective is to determine the optimum batch size of the product and the ordering quantities of raw materials minimizing the overall system cost. To simplify the model, we consider that a single raw material purchase provides stock for several production runs. The details of the process can be found in Sarker and Khan (1999) and it is briefly discussed below.

Figure 14.1 shows the finished product inventory system. The product is produced at a constant rate during the production uptime period t. The delivery of the finished product starts as soon as the level of the accumulated

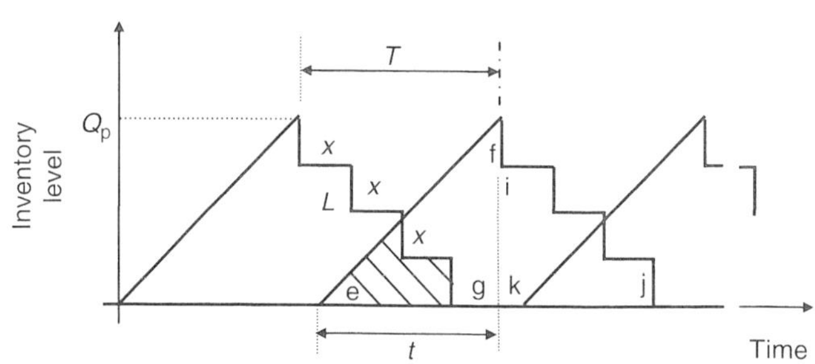

FIGURE 14.1
Finished product inventory level. (Reprinted from Sarker, R. and Khan, L., *Comput. Ind. Eng.*, 37, 711, 1999. With permission from Elsevier.)

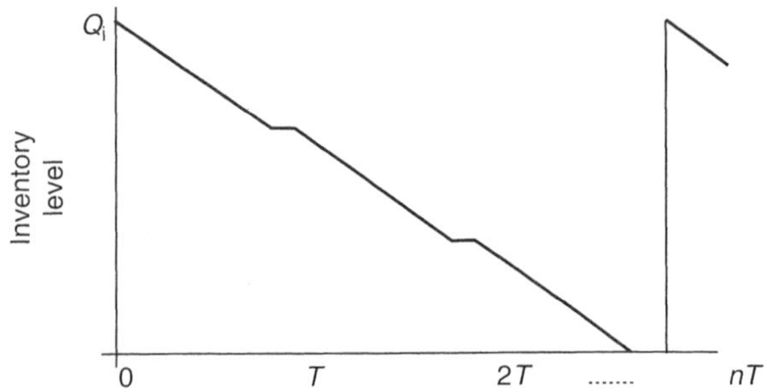

FIGURE 14.2
Raw material inventory. (Reprinted from Sarker, R. and Khan, L., *Comput. Ind. Eng.*, 37, 711, 1999. With permission from Elsevier.)

finished product inventory reaches the manufacturing batch quantity, Q_p. A fixed quantity of finished product (x) is delivered to the customers at the end of every L units of time. This forms a staircase pattern during the production downtime of the cycle. To fulfill the demand on time, there may be an overlapping of two consecutive finished product inventory cycles. The duration or amount of overlapping depends on the production rate, demand rate, and other relevant cost data. The shaded area in Figure 14.1 shows the overlapping of inventory depletion of a finished product inventory cycle with the inventory accumulation of the next cycle. Figure 14.2 presents the raw material inventory level of the system. At the beginning, the raw material of quantity nrQ_p is replenished for the purpose of providing stock for n production runs (here, r is the amount/quantity of raw material required in producing one unit of a product). The raw material is depleted at a constant rate during the production uptime of the finished product inventory cycle. The level of raw material inventory remains constant during the production downtime.

To find an ordering quantity for raw material and a production quantity for the production run, it is customary to consider the following costs:

- Raw material inventory carrying cost
- Finished goods inventory carrying cost
- Raw material ordering cost
- Manufacturing setup cost

To develop the mathematical model, the following assumptions have been made:

- The production rate is finite and constant.
- The production capacity is greater than the demand.
- No shortages are permitted.

- The time horizon is infinite.
- A fixed quantity of the product is delivered after a fixed interval of time.

The parameters and decision variables used in developing the cost functions are given below:

Parameters:

A_i = ordering cost of the raw material
A_p = setup cost for a product ($/setup)
r = amount/quantity of raw material required in producing one unit of a product
D_i = demand of raw material for the product in a year, $D_i = r\, D_p$
D_p = demand rate of a product, units per year
H_i = annual inventory holding cost for raw material, $/unit/year
H_p = annual inventory holding cost for product, $/unit/year
x = shipment quantity to customer at a regular interval (units/shipment)
L = time between successive shipments $= x/D_p$
t = production uptime in years in a cycle of length T
T = cycle time measured in years $= Q_p/D_p$, where $T > t$
P_p = production rate, units per year (here, $P_p > D_p$)
PR_i = price of raw material
Q_i^* = optimum ordering quantity of raw material

Decision variables:

Q_i = ordering quantity of raw material $= n\, r\, Q_p$
Q_p = production lot size
m = number of full shipments during the cycle time $= T/L = Q_p/x$
n = number of production run to use one lot of raw material

14.3.1 Finished Product Inventory

Since the product cannot be delivered to the customers until the whole lot is completed and quality certification is ready, there is a continuous build up of finished product inventory, at the rate equal to production, during the production uptime of a given lot (Figure 14.1: line ef). The delivery of the finished product is permitted during the production downtime of that lot only. In order to satisfy the demand in time, the production of a new lot may be started before finishing the delivery of the previous lot.

From Figure 14.1, $t = Q_P/P_P$ and $Q_P/D_P = \text{mL}$
Finished product inventory in a cycle = Area (efg + ikj)

$$\text{Area (ikj)} = x(m-1)L + x(m-2)L + \ldots + x\ 2L + x\ L = x\frac{(m-1)m}{2}L \tag{14.13}$$

$$\text{So, the finished product inventory in a cycle} = \frac{1}{2}Q_P t + x\frac{(m-1)}{2}mL \tag{14.14}$$

Hence, the average finished product inventory in a cycle is

$$\begin{aligned} &= \frac{\frac{1}{2}Q_P \cdot t + \frac{x(m-1)}{2}mL}{T} \\ &= \frac{1}{2}Q_P\frac{t}{T} + x\frac{m-1}{2}\frac{mL}{T} \\ &= \frac{1}{2}Q_P \cdot \frac{D_P}{P_P} + \frac{mx}{2} - \frac{x}{2} \\ &= \frac{1}{2}Q_P\left(\frac{D_P}{P_P} + 1\right) - \frac{x}{2} \end{aligned} \tag{14.15}$$

14.3.2 Raw Material Inventory

The raw material inventory is shown in Figure 14.2. For ease of understanding, it is redrawn for one replenishment cycle with more details in Figure 14.3.

The area of one replenishment cycle of raw material can be divided into n triangles and $(n-1)$ rectangles as shown in Figure 14.3.

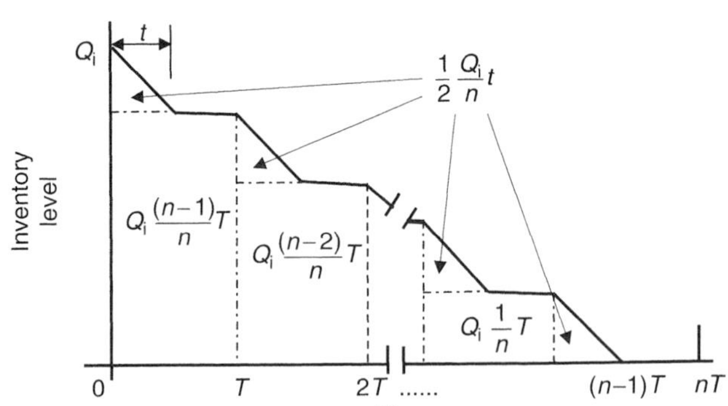

FIGURE 14.3
Raw material inventory system.

All the triangles have the same area equal to $\frac{1}{2}\frac{Q_i}{n}t$

$$\text{The total area occupied by } n \text{ triangles} = n\left(\frac{1}{2}\frac{Q_i}{n}t\right) = \frac{1}{2}Q_i t \quad (14.16)$$

The area of rectangles decreases as the cycle number n increases.

$$\text{The area of the rectangles} = \frac{(n-1)}{n}Q_i T,\ \frac{(n-2)}{n}Q_i T,\ \frac{(n-3)}{n}Q_i T,\ \ldots,\ \frac{2}{n}Q_i T,\ \frac{1}{n}Q_i T \quad (14.17)$$

The total area occupied by $(n-1)$ rectangles

$$= \frac{Q_i T}{n}\{1 + 2 + 3 + \cdots + (n-3) + (n-2) + (n-1)\} = \frac{(n-1)}{2}Q_i T \quad (14.18)$$

So the inventory for raw material in a cycle of nT periods is

$$\frac{1}{2}Q_i t + \frac{(n-1)}{2}Q_i T \quad (14.19)$$

For raw material, the average inventory per cycle

$$= \frac{\left[\frac{1}{2}Q_i t + \left(\frac{n-1}{2}\right)Q_i T\right]}{nT} = \frac{1}{2}rQ_p\left(\frac{D_p}{P_p} + n - 1\right) \quad (14.20)$$

14.3.3 Total Cost Function per Year

The total cost of the system per year is simply the sum of four cost components as discussed earlier.

Total cost of the system (TC) is

$$\mathrm{TC} = \frac{D_p}{Q_p}A_p + \left[\frac{1}{2}Q_p\left(\frac{D_p}{P_p} + 1\right) - \frac{x}{2}\right]H_p + \frac{D_i}{Q_i}A_i + \left[\frac{1}{2}rQ_p\left(\frac{D_p}{P_p} + n - 1\right)\right]H_i \quad (14.21)$$

The first term represents the total manufacturing setup cost in a year. The second term indicates the total product holding cost in a year. The third and fourth terms represent the raw material ordering cost and holding cost, respectively. After simplification, we can rewrite the total cost equation as follows:

$$\mathrm{TC} = \frac{D_\mathrm{P}}{Q_\mathrm{P}}\left(A_\mathrm{P} + \frac{A_\mathrm{i}}{n}\right) + \frac{Q_\mathrm{P}}{2}(\mathrm{kk}) - \frac{x}{2}H_\mathrm{P} \tag{14.22}$$

where

$$\mathrm{kk} = \left(\frac{D_\mathrm{P}}{P_\mathrm{p}} + 1\right)H_\mathrm{P} + rH_\mathrm{i}\left(\frac{D_\mathrm{P}}{P_\mathrm{p}} + n - 1\right) = \mathrm{CC} + nrH_\mathrm{i}$$

where

$$\mathrm{CC} = \left(\frac{D_\mathrm{P}}{P_\mathrm{p}} + 1\right)H_\mathrm{P} + rH_\mathrm{i}\left(\frac{D_\mathrm{P}}{P_\mathrm{p}} - 1\right)$$

Since, $Q_\mathrm{P} = mx$, then TC can be rearranged as follows:

$$\mathrm{TC} = \frac{D_\mathrm{P}}{mx}\left(A_\mathrm{P} + \frac{A_\mathrm{i}}{n}\right) + \frac{mx}{2}(\mathrm{kk}) - \frac{x}{2}H_\mathrm{P} \tag{14.23}$$

TC is a nonlinear function with integer variables m and n. The purpose is to determine the optimal m and n while minimizing the total cost. In optimization terms, this is an unconstrained nonlinear optimization model.

14.4 A Joint Lot-Sizing and Transportation Decision Problem

We consider a single retailer who purchases goods from an outside supplier and transports them via an external shipper (transport company). The amount of goods to be purchased by the retailer is usually determined using an economic replenishment policy (Khan and Sarker, 2002; Sarker and Newton, 2002), and then the transportation decision is made based on the quantity purchased. In this sequential decision process, the optimum purchasing quantity may not guarantee the minimum shipping cost incurred because of weight-dependent freight rates offered by the transport company and the limited vehicle capacity.

Motor carrier freight rates are a function of the total weight in a given shipment. The freight rates, usually form a step function, decrease as the shipping weight increases as shown in Figure 14.4. This figure shows a piecewise linear function, for price per unit, for quantities from Q_0 to Q_{n-1} where $p_1 > p_2 > p_3 > \cdots > p_{n-1}$. The prices for quantities 0 to Q_0 and Q_{n-1} to Q_n are fixed. In other words, the figure shows the price breaks for

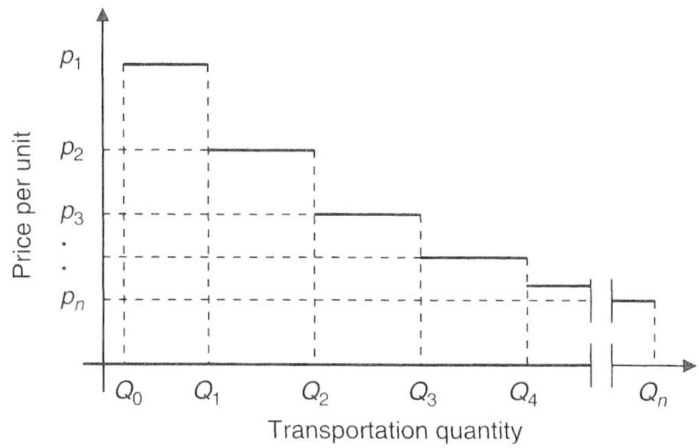

FIGURE 14.4
Transportation cost function.

different quantities. For a fixed unit weight, the upper and lower limits of these quantities can be converted to weights. Throughout this section, we use these figures as weight breaks. These price breaks (or weight breaks) reflect the economies of scale that accrue for larger shipping weights and the additional consolidation costs involved when determining load priorities for small shipments. Further information on freight rates can be found in Swenseth and Godfrey (2002) and Baker (1991).

Although some motor carriers provide software for rate lookups, it does not provide decisions based on simultaneous consideration of the inventory replenishment and the transportation decision. Suppose the optimal purchasing quantity is Q. Then there are three possible categories of shipping decisions: (1) shipments that result in true truckload (TL) shipping quantities (i.e., $Q = \text{TL}$), (2) shipments that are likely to be over-declared as TL (i.e., $Q < \text{TL}$ but declared as TL), and (3) shipments that are not likely to be over-declared as TL and are therefore shipped at less-than-truckload (LTL) rates (i.e., $Q < \text{TL}$ and declared as Q). Here, the problem is to determine the optimum purchasing quantities based on annual ordering, holding, and transportation costs. In the following section, we develop a mathematical programming model for simultaneous determination of purchasing quantities and the transportation decision.

Parameters:

D = annual requirements (units)

C_h = inventory holding cost (\$/unit/year)

C_o = ordering cost (\$/order)

w = weight per unit

I = number of weight breaks in freight rate

Tcst_i = transportation cost per hundred weight for weight break i (Tcst_I is fixed and independent of total weight)

U_i = upper limit of weight break i

L_i = lower limit of weight break i

M = a very big quantity (known as big M)

Mtc = minimum transportation charge ($)

TRC = total relevant cost ($)

Decision variables:

Q = ordering quantity (units)

X = quantity declared for transportation in weight (where $Q \leq X$)

$$y_i = \begin{cases} 1 & \text{if the quantity } X \text{ is in the weight break } i \\ 0 & \text{otherwise} \end{cases}$$

Total ordering cost per year is the number of orders per year multiplied by the ordering cost per order which is $C_o D/Q$. The total inventory holding cost per year is equal to the average inventory per cycle multiplied by the inventory holding cost per unit per year, which can be written as $(Q/2)C_h$. The total transportation cost is equal to the shipment size declared multiplied by the transportation cost and number of shipments per year.

$$\text{So the total transportation cost per year} = \left(\frac{D}{Q}\right)\left(\frac{X}{100}\right)\cdot\left(\sum_{i=1}^{I-1} y_i\,\text{Tcst}_i\right) + \left(\frac{D}{Q}\right) y_I\,\text{Tcst}_I \qquad (14.24)$$

The total relevant cost of the system is the sum of ordering cost, inventory holding cost, and transportation cost which is

$$\text{TRC} = C_o\frac{D}{Q} + \frac{Q}{2}C_h + \left(\frac{D}{Q}\right)\left(\frac{X}{100}\right)\cdot\left(\sum_{i=1}^{I=1} y_i\,\text{Tcst}_i\right) + \left(\frac{D}{Q}\right) y_I\,\text{Tcst}_I \qquad (14.25)$$

However, the following constraints must be satisfied in solving the total cost function (Equation 14.25).

The quantity for transportation (X) can be declared as a higher amount than the quantity purchased (Q) to take the advantage of cost bracket.

$$Q \leq X/w \qquad (14.26)$$

The quantity declared (X) must be within the range of each freight rate range.

$$L_i y_i \leq X \leq U_i + M(1 - y_i) \quad \forall i \tag{14.27}$$

The declared quantity must be under only one weight break.

$$\sum_{i=1}^{I} y_i = 1 \tag{14.28}$$

The last constraint confirms the minimum transportation charge requirement to hire a truck.

$$\frac{X}{100} \sum_{i=1}^{I-1} y_i \text{Tcst}_i + y_I \text{Tcst}_I \geq \text{Mtc} \tag{14.29}$$

The final integer programming model of the problem can be summarized as follows:

$$\text{Minimize TRC} = C_o \frac{D}{Q} + \frac{Q}{2} C_h + \left(\frac{D}{Q} \cdot \frac{X}{100}\right) \cdot \left(\sum_{i=1}^{I=1} y_i \text{Tcst}_i\right) + \left(\frac{D}{Q}\right) y_I \text{Tcst}_I$$

Subject to

$$Q \leq X/w$$
$$L_i y_i \leq X \leq U_i + M(1 - y_i) \quad \forall i$$
$$\sum_{i=1}^{I} y_i = 1$$
$$\frac{X}{100} \sum_{i=1}^{I-1} y_i \text{Tcst}_i + y_I \text{Tcst}_I \geq \text{Mtc}$$
$$X, Q \geq 0 \text{ and integer}$$
$$y_i = 0 \text{ or } 1 \quad \forall i$$

Model (14.1)

14.5 Coal Bank Scheduling

Coal banks are used to store coals for aging and for carrying normal inventory. We assume the production of each coal product in each period and the customer delivery schedule are known. Now the questions are how to build the banks in a given floor space and how large should each bank be and where should they be located. Because of different quality parameters and aging requirements, we cannot keep all the coals in a single bank. Coals that are produced with the same quality parameters in a period can be stored in the same bank. Interestingly, the total storage capacity of any floor space decreases with the number of banks on the same area built. For example, if the available storage space is 2625×145 sq. ft., for a 10° angle on one side for access by trucks and 45° on other side of the banks, and a height of up to 60 ft., the storage capacity of this bank is about 373,700 tonnes of coal. If the same area is used for two banks (of 1312.5×145 sq. ft. each), then total capacity will be $175{,}290 \times 2 = 350{,}580$ tonnes. The capacity is 327,460 tonnes for three banks and 304,340 tonnes for four banks. Bigger banks maximize the storage capacity utilization on a given space.

Assume, we need banks of different sizes and these banks need to be stored for different lengths of time. It would seem that the banks should be built in a cyclic order on a given floor space. In fact, this does not ensure the best possible utilization of floor space. For further details on bank management, see the paper by Sarker and Gunn (1994).

Withdrawal from smaller banks is more difficult than from large banks because of transport and handling facilities. In managing the storage area, our purpose is to determine the bank size and its location on a given floor space. We make the bank as large as possible to maximize capacity per unit area and to reduce the handling difficulties. Once a bank is built up, any movement of coal increases the handling cost and operational complexity. As a result, we determine the location of a bank with other banks, on a given floor space, in such a way that the bank movement is minimum over a given period. In building up a bank, we assume the height of a bank is limited due to handling facility or amount of coal to be stored in a bank, and the given area is one linear storage with known length and width.

In this section, we formulate two mathematical models for the bank location sequencing/scheduling problem. The first model is a static model and the second is a dynamic model. In the first model, once a bank is built it is not allowed to move in any direction until withdrawal. This model sequences a given number of banks for some periods. The other model allows movement of the bank at additional cost, if it provides a better utilization of the entire floor space. This model determines the location of new banks at the beginning of each week with other existing banks, over a planning horizon. This is sequencing and scheduling problem. The models are presented below:

14.5.1 Static Model

The static model* is useful when the bank movement is restricted.
Index:

$i =$ bank number ($i = 1, N$)
$t =$ time period ($t = 1, T$)

* Reproduced as slightly modified version from Sarker, R. and Gunn, E., *Appl. Math. Model.*, 18, 672, 1994. With permission from Elsevier.

Parameters:

Q_{it} = amount of coal in bank i at period t
A_i = maximum amount of coal to be stored in bank i
L = maximum length of floor space
M = a significantly big number
B_i = set of banks that could be adjacent to the bank i
$B_i = \{j;\ Q_{it} \times Q_{jt} > 0 \text{ for some } t \text{ and } i \neq j\}$

Variables:

s_i = starting location of bank i
l_i = length of bank i

$$y_{ij} = \begin{cases} 1 & \text{if bank } j \text{ is built after bank } i \\ 0 & \text{otherwise} \end{cases}$$

Note that we can eliminate half of these variables using $y_{ij} = 1 - y_{ji}$. This will help to solve the problem quickly.

The amount of coal in a bank i is the maximum amount over some time

$$A_i \geq \text{Max}\,(Q_{it},\ t = 1,\ T) \quad \forall i \tag{14.30}$$

The amount of coal in a bank can be expressed as a function of the length of that bank

$$A_i = f(l_i) \quad \forall i \tag{14.31}$$

Either bank i will be built after bank j or bank j will be built after bank i. Exactly one of these two constraints exists.

$$s_j + l_j \leq s_i + M\ y_{ij} \tag{14.32}$$

$$s_i + l_i \leq s_j + M(1 - y_{ij}) \quad \forall i \text{ and } j \in B_i \tag{14.33}$$

Either bank i or bank j will be built closer to the starting point of the storage space. This is one linear storage space with given length and width. The width of each bank is equal to the width of the storage. The starting position is the lengthwise leftmost position of the available storage space for building a bank.

The end of a bank must be less than the given length of bank space

$$0 \leq s_i \leq L - l_i \quad \forall i \tag{14.34}$$

and nonnegativity constraints

Two objective functions can be considered for this problem as follows:

- The first objective is to minimize L, where L is a decision variable. This objective minimizes the total storage space required.
- The second is to maximize $\sum_i (L - s_i)$, where L is a given data. This objective forces to build every bank as close to the starting point of the storage space as possible. The volume of a bank is determined based on its shape and then the quantity in tonnes is calculated by multiplying by a density factor.

14.5.2 Dynamic Model

For a dynamic model,* we need to redefine some of the parameters and variables as follows.

s_{it} = starting position of bank i at period t

l_{it} = length of bank i at period t

B_{it} = set of banks that could be potentially adjacent to the bank i

$B_{it} = \{j;\ Q_{it} \times Q_{jt} > 0 \text{ for } i \neq j\}$

Other parameters and variables are the same as for the static model.

The amount of coal in a bank can be expressed as a function of the length of that bank.

$$Q_{it} = f(l_{it}) \quad \forall i, t \tag{14.35}$$

In a period t, either bank i will be built after bank j or bank j will be built after bank i. Exactly one of these two constraints exists.

$$s_{jt} + l_{jt} \leq s_{it} + M\, y_{ij} \tag{14.36}$$

$$s_{it} + l_{it} \leq s_{jt} + M(1 - y_{ij}) \quad j \in B_i; \quad \forall i, t \ \exists\ Q_{it} \geq 0 \tag{14.37}$$

The end of a bank must be less than the given dimensions of the storage space

$$0 \leq s_{it} \leq L - l_{it} \quad \forall i, t \tag{14.38}$$

and nonnegativity constraints.

---

* Reproduced as slightly modified version from Sarker, R. and Gunn, E., *Appl. Math. Model.*, 18, 672, 1994. With permission from Elsevier.

The objective is to minimize the total movement of all banks during the planning horizon.

$$\text{Max } Z = \sum_{i} \sum_{t} x_{it} \tag{14.39}$$

Where x_{it} is the maximum movement of a bank i from period $t-1$ to period t. We have added few more constraints to define the variables x_{it} as given below:

$$x_{it} \geq s_{it} - s_{it-1} \quad \forall i, t \tag{14.40}$$

$$x_{it} \geq s_{it-1} - s_{it} \quad \forall i, t \tag{14.41}$$

$$x_{it} \geq s_{it} + l_{it} - s_{it-1} - l_{it-1} \quad \forall i, t \tag{14.42}$$

$$x_{it} \geq s_{it-1} + l_{it-1} - s_{it} - l_{it} \quad \forall i, t \tag{14.43}$$

14.6 A Scaffolding System

Consider a scaffolding system, a simple engineering design problem, consisting of three beams and six ropes as shown in Figure 14.5. Each of the ropes R1 and R2 can carry a load of L_1, each of the middle ropes R3 and R4 can carry a load of L_2, and each of the bottom ropes R5 and R6 can carry a load of L_3. Suppose the loads acting on beams 1, 2, and 3 are x_1, x_2, and x_3, respectively, formulate the problem of finding the maximum load ($x_1 + x_2 + x_3$) that can be supported by the system.

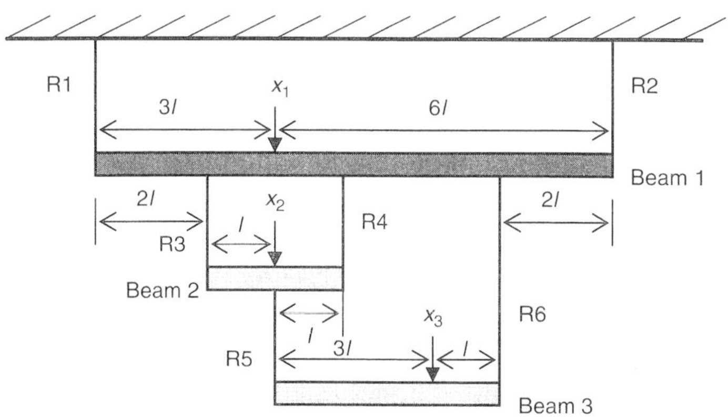

FIGURE 14.5
A scaffolding system.

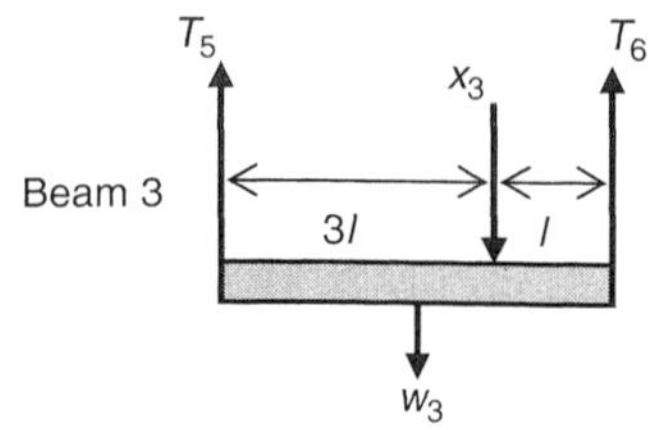

FIGURE 14.6
Beam 3 of scaffolding system.

Assume that the weights of the beams 1, 2, and 3 are w_1, w_2, and w_3, respectively, and the weight of the ropes is negligible.

We assume that the weights of the beams act through their respective middle points. Let, $T_1, T_2, \ldots, T_6$ denote the tension in ropes R1, R2, ..., R6, respectively. For ease of explanation, the beam 3 is separated from the system and all the forces acted on the beam are shown in Figure 14.6.

As shown in Figure 14.6, T_5 and T_6 are upward forces and x_3 and w_3 are downward forces. In such case, the equations of equilibrium for vertical forces in each beam and the moments for each beam can be expressed as follows.

Beam 3:

$$\text{Forces: } T_5 + T_6 = x_3 + w_3 \tag{14.44}$$

$$\text{Moments: } x_3(3l) + w_3(2l) - T_6(4l) = 0 \tag{14.45}$$

Beam 2:

$$\text{Forces: } T_3 + T_4 - T_5 = x_2 + w_2 \tag{14.46}$$

$$\text{Moments: } x_2(l) + w_2(l) + T_5(l) - T_4(2l) = 0 \tag{14.47}$$

Beam 1:

$$\text{Forces: } T_1 + T_2 - T_3 - T_4 - T_6 = x_1 + w_1 \tag{14.48}$$

$$\text{Moments: } x_1(3l) + w_1(4.5l) - T_2(9l) + T_3(2l) + T_4(4l) + T_6(7l) = 0 \tag{14.49}$$

The moment is measured as the force multiplied by its distance from a reference point. For example, the moment of x_3 from the left end of beam 3 is $x_3(3l)$. From the above equations, the tension in the ropes T_1 to T_6 can be represented as follows:

$$T_6 = (3/4)x_3 + (1/2)w_3 \tag{14.50}$$

$$T_5 = (1/4)x_3 + (1/2)w_3 \tag{14.51}$$

$$T_4 = (1/2)x_2 + (1/8)x_3 + (1/2)w_2 + (1/4)w_3 \tag{14.52}$$

$$T_3 = (1/2)x_2 + (1/8)x_3 + (1/2)w_2 + (1/4)w_3 \tag{14.53}$$

$$T_2 = (1/3)x_1 + (1/3)x_2 + (2/3)x_3 + (1/2)w_1 + (1/3)w_2 + (5/9)w_3 \tag{14.54}$$

$$T_1 = (2/3)x_1 + (2/3)x_2 + (1/3)x_3 + (1/2)w_1 + (2/3)w_2 + (4/9)w_3 \tag{14.55}$$

The optimization model is

$$\begin{aligned}
&\text{Maximize } Z = x_1 + x_2 + x_3 \\
&\text{Subject to} \\
&(2/3)x_1 + (2/3)x_2 + (1/3)x_3 + (1/2)w_1 + (2/3)w_2 + (4/9)w_3 \le L_1 && \text{For } T_1 \\
&(1/3)x_1 + (1/3)x_2 + (2/3)x_3 + (1/2)w_1 + (1/3)w_2 + (5/9)w_3 \le L_1 && \text{For } T_2 \\
&(1/2)x_2 + (1/8)x_3 + (1/2)w_2 + (1/4)w_3 \le L_2 && \text{For } T_3 \\
&(1/2)x_2 + (1/8)x_3 + (1/2)w_2 + (1/4)w_3 \le L_2 && \text{For } T_4 \\
&(1/4)x_3 + (1/2)w_3 \le L3 && \text{For } T_5 \\
&(3/4)x_3 + (1/2)w_3 \le L3 && \text{For } T_6 \\
&x_1, x_2, x_3 \ge 0
\end{aligned}$$

Model (14.2)

This is a simple linear programming model with three variables (x_1, x_2, and x_3) and six constraints. We must mention here that the weights of the beams (w_1, w_2, and w_3) are parameters (known) of the system.

14.7 A Gas-Lift Optimization Problem

In this section, we consider a crude oil-production system. In the system, there is a well-known underground oil reservoir and the reservoir has a number of wells for oil extraction. There are two basic methods of extracting oil from such reservoirs (Kosmidis et al., 2005): naturally flowing and gas lift. In the first method, the oil is able to flow naturally to the surface, whereas the second requires the injection of high-pressure gas to facilitate oil extraction. The gas lift is considered as the most economic method for artificial lifting of oil (Aaytollahi et al., 2004; Camponogara and Nakashima, 2006).

In this chapter, we consider the gas lift extraction method. For a given well, the oil production per day can be expressed as a nonlinear function of gas injected into the well on that day. The oil production per day increases with the increase of gas used to certain level and then decreases. That means an excessive use of gas may increase the gas cost, as well as production cost, without providing any benefit in terms of oil-production volume. For a given amount of gas used, the amount of oil extraction significantly varies from well to well. That means the nonlinear function of gas usage versus oil extracted varies from well to well. As a result, an inappropriate gas allocation to different wells, under limited gas availability,

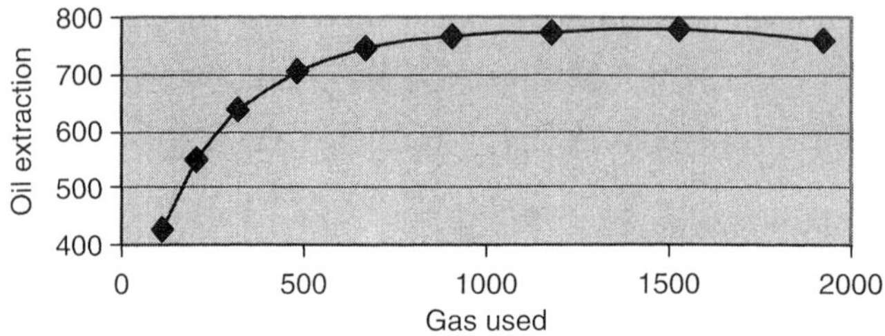

FIGURE 14.7
A sample oil-production function.

will reduce the overall production and hence profitability from the entire reservoir. So the gas-lift optimization problem is to allocate a limited amount of gas to a number wells in a reservoir while maximizing the total oil production in a day.

As discussed earlier, the oil production per day from a given well is a function of the gas injected into that well on that particular day. However, there is no standard function that can be used for determining the production level for all wells. The practice is to collect production data for a number of discrete points of gas use for each well and then generate an approximate function. A sample function for a given well is shown in Figure 14.7. Using these discrete data points, the researchers in the field generate the corresponding function either as a piecewise linear function (Kosmidis et al., 2005) or as a quadratic function (Camponogara and Nakashima, 2006). Although both functions have drawbacks in estimating the production level accurately, we can use the piecewise linear function as this method is widely used for such situations and easy to model the optimization problem. For further details of the data and alternative approach, see Ray and Sarker (2006).

The mathematical model of the problem is formulated as follows:

N = the number of wells

I_n = the number of line segments in the function in well n

G_{ni} = slope of the function at the line segment i in well n

GL = limit of gas usage in all wells per day

U_{ni} = upper limit of gas use at the individual line segment i in well n

Variables:

Y_n = gas used in well n

X_{ni} = gas usage segment variable in segment i in well n

$$Z_{ni} = \begin{cases} 0 & \text{if } X_{ni} \leq U_{ni} \\ 1 & \text{if } X_{ni} = U_{ni} \end{cases}$$

The relationships between Y_n and X_{ni} are as follows:

$$Y_n = \sum_i X_{ni} \quad \forall n \tag{14.56}$$

where

$$0 \leq X_{ni} \leq U_{ni} \quad \forall n, i \tag{14.57}$$

In addition, for a given value of Y_n (which is in segment i in well n), all $X_{n(i+1)}$ will be equal to zero and all $X_{n(i-1)}$ will be at the upper bound. However, the value of X_{ni} will be greater than zero and less than or equal to upper bound.

The objective of the problem is to maximize the total daily production.

$$\text{Maximize } Z = \sum_n \sum_i G_{ni} X_{ni} \tag{14.58}$$

Constraints:
Calculating the gas usage level in each well: the total gas used in a well n is the sum of gas used under all the segments of the piecewise linear function.

$$Y_n = \sum_i X_{ni} \quad \forall n \tag{14.59}$$

Suppose the point Y_n belongs to the segment i_n. The values of all segments (x_{ni}) on the left side of segment i_n must be at the upper bound and the value of the segment containing Y_n point must be greater than or equal to zero and less than or equal to its upper bound. All other segments must have zero value. To ensure these conditions, we need the following constraints.

$$U_{ni} Z_{ni} \leq X_{ni} \leq U_{ni} Z_{n(i-1)} \quad \forall n, i \tag{14.60}$$

Here $Z_{n0} = 1$ and $Z_{nI_n} = 0$. For each well, this constraint sets the point Y_n and ensures that for any $i < i_n$, $x_{ni} = U_{ni}$, for $i > i_n$, $x_{ni} = 0$ and for $i = i_n$, $0 \leq x_{ni} \leq U_{ni}$.

Gas limitation: The total use of gas in all wells on a given day must be less than or equal to the available gas.

$$\sum_n Y_n \leq GL \tag{14.61}$$

Nonnegativity constraints:

$$X_{ni} \geq 0,\ Y_n \geq 0,\ Z_{ni} \in (0, 1) \quad \forall n, i \tag{14.62}$$

So the overall mathematical model is

$$\text{Maximize } Z = \sum_n \sum_i G_{ni} X_{ni}$$

Subject to

$$\begin{aligned}
&Y_n = \sum_i X_{ni} \quad \forall n \\
&U_{ni} Z_{ni} \leq X_{ni} \leq U_{ni} Z_{n(i-1)} \quad \forall n, i \\
&\sum_n Y_n \leq GL \\
&Z_{n0} = 1 \text{ and } Z_{I_n n} = 0 \\
&X_{ni} \geq 0,\ Y_n \geq 0,\ Z_{ni} \in (0, 1) \quad \forall n, i
\end{aligned} \qquad \text{Model (14.3)}$$

14.8 Multiple Shifts Planning

A local organization is considering a second shift in its manufacturing unit for some working days to meet the increased demand. However, there are significant additional costs involved with the opening and closing of a second shift in any day. Suppose that all the costs, capacities, and demand data are given. Formulate the shift planning problem.

We define the variables and parameters as follows:

$X1_t$ = production volume in shift 1 of period t

$X2_t$ = production volume in shift 2 of period t

I_t = inventory carried from period t to $t + 1$

$$Y_t = \begin{cases} 1 & \text{second shift operating in period } t \\ 0 & \text{otherwise} \end{cases}$$

H = inventory carrying cost

$C1_t$ = first shift production cost per unit in period t

$C2_t$ = second shift production cost per unit in period t

COS = cost of opening the second shift

CCS = cost of closing the second shift

$CAP1_t$ = capacity of first shift in period t

$CAP2_t$ = capacity of second shift in period t

D_t = demand in period t

The following are the assumptions for the opening and closing of shift:

- A second shift can be opened in a period t, if it was not operated in period $t - 1$.
- A second shift can be closed in period t, if it was operated in period $t - 1$.

Under these assumptions, the cost function for second-shift opening and closing (SSOC) can be expressed as

$$f(\text{SSOC}) = (\text{COS})Y_t + (\text{CCS})Y_{t-1} - (\text{COS} + \text{CCS})Y_t Y_{t-1} \tag{14.63}$$

Let us analyze the second-shift cost function (Equation 14.63) as it would generate four different cases as follows.

Case	Y_t	Y_{t-1}	f(SSOC)	Remarks
1	0	0	0	Neither opening nor closing, and the second shift not operating in either period t or t–1
2	1	1	0	Neither opening nor closing, but second shift operating in both period t and t–1
3	1	0	COS	Second shift opening (operated) in period t, but not operated (closing) in period t–1
4	0	1	CCS	Second shift closing (or not operated) in period t and opening (operated) in period t–1

The objective is to minimize the sum of production, inventory holding, and second-shift opening and closing costs.

$$\text{Min } Z = \sum_t [C1_t X1_t + C2_t X2_t + HI_t + (\text{COS})Y_t + (\text{CCS})Y_{t-1} \\ - (\text{COS} + \text{CCS})Y_t Y_{t-1}] \tag{14.64}$$

In a given period t, the total production plus the inventory carried from previous period minus inventory to be carried for the next period must be equal to the demand of that period.

$$X1_t + X2_t + I_{t-1} - I_t = D_t \quad \forall t \tag{14.65}$$

The production in any period for any shift must be less than or equal to the production capacity in that period.

$$X1_t \leq \text{CAP1}_t \tag{14.66}$$

$$X2_t \leq \text{CAP2}_t Y_t \tag{14.67}$$

Finally, nonnegative constraints.

$$\begin{aligned} & X1_t,\ X2_t,\ I_t \geq 0 \text{ and integer} \\ & Y_t \text{ binary} \end{aligned} \tag{14.68}$$

We assume $L_t = Y_t Y_{t-1}$. To simplify the polynomial term in the objective function, the following additional constraints are required (for further details on this approach, see Section 5.14 and Glover and Woolsey (1974)).

$$\begin{aligned} & L_t - Y_t \leq 0 \\ & L_t - Y_{t-1} \leq 0 \\ & Y_t + Y_{t-1} - L_t \leq 1 \\ & 0 \leq L_t \leq 1 \end{aligned} \tag{14.69}$$

So the revised formulation:

$$\begin{aligned} \text{Minimize } Z = \sum_t & [C1_t X1_t + C2_t X2_t + HI_t + (\text{COS})Y_t + (\text{CCS})Y_{t-1} \\ & - (\text{COS} + \text{CCS})L_t] \end{aligned}$$

Subject to

$$\begin{aligned} & X1_t + X2_t + I_{t-1} - I_t = D_t \quad \forall t \\ & X1_t \leq \text{CAP1}_t \\ & X2_t \leq \text{CAP2}_t Y_t \\ & L_t - Y_t \leq 0 \\ & L_t - Y_{t-1} \leq 0 \\ & Y_t + Y_{t-1} - L_t \leq 1 \\ & 0 \leq L_t \leq 1 \\ & X1_t,\ X2_t,\ I_t \geq 0 \text{ and integer} \\ & Y_t \text{ binary} \end{aligned} \qquad \text{Model (14.4)}$$

14.9 Summary

In this chapter, we have briefly presented several integer and nonlinear practical problems that include a combat logistics problem, a lot-sizing problem, a joint lot-sizing and transportation decision problem, a coal bank scheduling problem, an engineering design problem, a gas-lift optimization problem, and a multiple shifts planning problem. The mathematical models and related issues of these problems have also been discussed.

References

Aaytollahi, S., Narimani, M., and Moshfeghian, M., Intermittent gas lift in Aghajari oil field, a mathematical study, *Journal of Petroleum Science and Engineering*, 42, 245, 2004.

Camponogara, E. and Nakashima, P., Solving a gas lift optimization problem using dynamic programming, *European Journal of Operational Research*, 174, 1220, 2006.

Glover, F. and Woolsey, E., Converting 0–1 polynomial programming problem to a 0–1 linear program, *Operations Research*, 22, 180, 1974.

Gue, K.R., A dynamic distribution model for combat logistics, *Computers and Operations Research*, 30, 367, 2003.

Khan, L. and Sarker, R., An optimal batch size for a JIT manufacturing system, *Computers and Industrial Engineering*, 42, 127, 2002.

Kosmidis, V., Perkins, J., and Pistikopoulos, E., A mixed integer optimization formulation for the well scheduling problem on petroleum fields, *Computers and Chemical Engineering*, 29, 1523, 2005.

Ray, T. and Sarker, R., Multiobjective evolutionary approach to the solution of gas lift optimization problems, IEEE Congress on Evolutionary Computation (CEC2006), Vancouver, BC, Canada, 10951, 2006.

Sarker, R. and Gunn, E., Coal bank scheduling using a mathematical programming model, *Applied Mathematical Modelling*, 18, 672, 1994.

Sarker, R. and Khan, L., An optimal batch size for a production system operating under a periodic delivery policy, *Computers and Industrial Engineering*, 37, 711, 1999.

Sarker, R. and Newton, C., Genetic algorithm for solving economic lot size scheduling problem, *Computers and Industrial Engineering*, 42, 189, 2002.

15

Solving Practical Problems

15.1 Introduction

In this chapter, we solve a number of practical problems and discuss their solutions. The mathematical models of most of the problems, to be discussed here, have already been formulated in earlier chapters. The size of these problems ranges from small to medium. The following are the problems discussed in this chapter:

- A sanitary-wares product-mix problem
- A crop planning problem
- A two-stage transportation problem
- A power generation and planning problem
- A gas-lift optimization problem

15.2 A Product Mix Problem

ABC sanitary-wares company produces a number of products using three existing production lines. The production lines A, B, and C produce 3, 4, and 6 products, respectively. All the products require the same basic raw materials but at different amounts per unit. The details of products and raw materials can be found in Sarker and Haque (1985). There is a limitation on the supply of raw materials in a given year and the working capital is also limited. The capacity of each production line is restricted by the skilled labor hours available which can be calculated easily. Initially, it was assumed that all products produced in a year can be sold as the prices are lower than their competitors when compared with their quality. So the problem is to find an optimal product-mix for the company.

The mathematical model of the problem is presented below.

Variables:
X_i = number of units of product i to be produced

Data:

W_i = raw material required for each unit of product i in kg
P_i = price of each unit of product i
C_i = variable cost per unit of product i
L_i = labor hours required for each unit of product i
RS = total raw material supply
TB = total working capital available
PC_j = capacity (labor hours available) of production line j

The objective function was set to maximize a payoff function ($P_i - C_i$) rather than profit maximization since the fixed cost component is not involved in the optimization process.

$$\text{Maximize } Z = \sum_i (P_i - C_i)X_i \tag{15.1}$$

Constraints:
Raw material limitation: the total amount of raw material required by all products must be less than or equal to the total amount available.

$$\sum_i W_i X_i \leq \text{RS} \tag{15.2}$$

Working capital restriction: the total amount of variable costs incurred by all products must be less than or equal to the total working capital available.

$$\sum_i C_i X_i \leq \text{TB} \tag{15.3}$$

Production line capacity: the total number of labor hours required by all products produced in a line j must be less than or equal to the total number of labor hours available in that line.

$$\sum_{i \in S_j} L_i X_i \leq \text{PC}_j \quad \forall j \tag{15.4}$$

Where S_j is a set of product produced in production line j

Nonnegativity constraint

$$X_i \geq 0 \quad \forall i \tag{15.5}$$

So the overall mathematical model is

$$\text{Maximize } Z = \sum_i (P_i - C_i)X_i$$

Subject to

$$\sum_i W_i X_i \leq \text{RS}$$

$$\sum_i C_i X_i \leq \text{TB} \qquad \text{Model (15.1)}$$

$$\sum_{i \in S_j} L_i X_i \leq \text{PC}_j \quad \forall j$$

$$X_i \geq 0 \quad \forall i$$

To solve the model, the data W_i, P_i, C_i, L_i, RS, TB, and PC_j were taken from Sarker and Haque (1985). The model contains 13 variables and 5 constraints. The LINGO codes for the model (Figure 15.1) are given below.

After running LINGO, it indicated that the smallest and largest elements in absolute value in the model are 0.300000 and 15000000, respectively. So LINGO provided a ** *WARNING* ** Problem is poorly scaled. The units of the rows and variables should be changed so that the coefficients cover a much smaller range. As discussed in an earlier chapter, it may create a

```
MODEL:
!Sanitarywares product-mix part-1;

!OBJECTIVE FUNCTION;
    [OBJ] MAX = 610*X1 + 885*X2 + 300*X3 + 725*X4 + 355*X5 + 80*X6 +
940*X7 + 210*X8 + 235*X9 + 600*X10 + 30*X11 + 20*X12 + 15*X13;

!CONSTRAINTS;

!Raw material;
    11.7*X1 + 17*X2 + 5.8*X3 + 13.95*X4 + 6.8*X5 + 1.5*X6 + 18*X7 +
4*X8 + 4.5*X9 + 11.5*X10 + 0.57*X11 + 0.39*X12 + 0.30*X13 <= 4000000;

!Working capital;
    94.32*X1 + 140.82*X2 + 47.67*X3 + 115.75*X4 + 56.57*X5 +
12.47*X6 + 149.07*X7 + 33.17*X8 + 39.22*X9 + 95.32*X10 + 4.79*X11 +
3.28*X12 + 2.5*X13 <= 15000000;

!Production line 1;
    32*X1 + 31*X4 + 15*X5 <= 2442600;

!Production line 2;
    45*X2 + 25*X3 + 16*X6 + 17*X7 <= 1486800;

!Production line 3;
    11*X8 + 20*X9 + 21*X10 + 15*X11 + 15*X12 + 15*X13 <= 743400;

END
```

FIGURE 15.1
LINGO codes for the product mix model.

```
MODEL:
!Sanitarywares product-mix part-2;

!OBJECTIVE FUNCTION;
     [OBJ] MAX = 610*X1 + 885*X2 + 300*X3 + 725*X4 + 355*X5 + 80*X6 +
940*X7 + 210*X8 + 235*X9 + 600*X10 + 30*X11 + 20*X12 + 15*X13;

!CONSTRAINTS;

!Raw material;
     11.7*X1 + 17*X2 + 5.8*X3 + 13.95*X4 + 6.8*X5 + 1.5*X6 + 18*X7 +
4*X8 + 4.5*X9 + 11.5*X10 + 0.57*X11 + 0.39*X12 + 0.30*X13 <= 40000;

!Working capital;
     94.32*X1 + 140.82*X2 + 47.67*X3 + 115.75*X4 + 56.57*X5 +
12.47*X6 + 149.07*X7 + 33.17*X8 + 39.22*X9 + 95.32*X10 + 4.79*X11 +
3.28*X12 + 2.5*X13 <= 150000;

!Production line 1;
     32*X1 + 31*X4 + 15*X5 <= 24426;

!Production line 2;
     45*X2 + 25*X3 + 16*X6 + 17*X7 <= 14868;

!Production line 3;
     11*X8 + 20*X9 + 21*X10 + 15*X11 + 15*X12 + 15*X13 <= 7434;

END
```

FIGURE 15.2
Rescaled LINGO model.

computational problem for the optimization algorithms. So it is recommended to scale the model properly. For this purpose, we define the variables in units of 100s. For such changes, we have divided the right-hand side (RHS) of all the constraints by 100 and made no changes to the objective function. The revised LINGO model (Figure 15.2) is presented above.

As can be seen, in the following output (Figure 15.3), LINGO stated that the smallest and largest elements in absolute value are 0.300000 and 150000, respectively, and it provided no warning message.

Because of the rescaling of the model, the solution must be read as shown in column 3 in Table 15.1.

As per the optimal product-mix, the company needs to produce only four products (*X*1, *X*6, *X*7, and *X*8) to maximize their payoff function. As per the solution, the company is using the full capacity of all three production lines and all of the working capital. The company management thinks it is not a good idea to produce a subset of the products if the company wants to increase its market share and grow in the future. Now the question is how to ensure that all products will appear in the final solution while maximizing the defined payoff function. By changing the parameters, one can get a maximum of five products at the positive level for this model as the number of basic variables is always less than or equal to the number of constraints when solving the model using a linear programming (LP) methodology.

```
Rows=      6 Vars=     13 No. integer vars=      0  ( all are linear)
 Nonzeros=     57 Constraint nonz=    39(     0 are +- 1)
Density=0.679
 Smallest and largest elements in abs value=   0.300000          150000.
 No. < :   5 No. =:   0 No. > :   0, Obj=MAX, GUBs <=   3
 Single cols=    0
 Global optimal solution found at step:              6
 Objective value:                             958863.7

                        Variable         Value        Reduced Cost
                              X1      763.3125           0.0000000
                              X2     0.0000000            5.691559
                              X3     0.0000000            2.430403
                              X4     0.0000000            19.39549
                              X5     0.0000000            8.728787
                              X6      585.0481           0.0000000
                              X7      323.9547           0.0000000
                              X8      675.8182           0.0000000
                              X9     0.0000000            14.06092
                             X10     0.0000000            2.322620
                             X11     0.0000000            1.779661
                             X12     0.0000000            2.274109
                             X13     0.0000000            2.363957
                             Row   Slack or Surplus     Dual Price
                             OBJ      958863.7            1.000000
                               2      21657.21           0.0000000
                               3     0.0000000            6.295067
                               4     0.0000000           0.5077886
                               5     0.0000000        0.9378178E-01
                               6     0.0000000           0.1084192
```

FIGURE 15.3
Output of rescaled model.

TABLE 15.1
LINGO Solutions vs. Actual Solution

Variable	LINGO Output (in 100 Units)	Actual Solution (in Units)
Z	958863.7	95886370.0
*X*1	763.3125	76331.25
*X*2	00	00
*X*3	00	00
*X*4	00	00
*X*5	00	00
*X*6	585.0481	58504.81
*X*7	323.9547	32395.47
*X*8	675.8182	67581.82
*X*9	00	00
*X*10	00	00
*X*11	00	00
*X*12	00	00
*X*13	00	00

So we need to add new constraints to have more than five products in the final solution. If there are no other constraints then it is recommended to set the variable bounds. In this problem, we are using the following constraints.

$$W_i X_i \geq \text{MD}_i \quad \forall i \tag{15.6}$$

Here, MD_i is the minimum level of production for product i in a year set by the management.

The corresponding LINGO model (Figure 15.4) is as follows.

```
MODEL:
!Sanitarywares product-mix part-3;

!OBJECTIVE FUNCTION;
     [OBJ] MAX = 610*X1 + 885*X2 + 300*X3 + 725*X4 + 355*X5 + 80*X6 +
940*X7 + 210*X8 + 235*X9 + 600*X10 + 30*X11 + 20*X12 + 15*X13;

!CONSTRAINTS;

!Raw material;
     11.7*X1 + 17*X2 + 5.8*X3 + 13.95*X4 + 6.8*X5 + 1.5*X6 + 18*X7 +
4*X8 + 4.5*X9 + 11.5*X10 + 0.57*X11 + 0.39*X12 + 0.30*X13 <= 40000;

!Working capital;
     94.32*X1 + 140.82*X2 + 47.67*X3 + 115.75*X4 + 56.57*X5 +
12.47*X6 + 149.07*X7 + 33.17*X8 + 39.22*X9 + 95.32*X10 + 4.79*X11 +
3.28*X12 + 2.5*X13 <= 150000;

!Production line 1;
     32*X1 + 31*X4 + 15*X5 <= 24426;

!Production line 2;
     45*X2 + 25*X3 + 16*X6 + 17*X7 <= 14868;

!Production line 3;
     11*X8 + 20*X9 + 21*X10 + 15*X11 + 15*X12 + 15*X13 <= 7434;

!Minimum production level;
     11.7*X1 >= 4294;
     17*X2 >= 1589;
     5.8*X3 >= 160;
     13.95*X4 >= 1166.9;
     6.8*X5 >= 3178;
     1.5*X6 >= 623.5;
     18*X7 >= 337.2;
     4*X8 >= 125;
     4.5*X9 >= 38.8;
     11.5*X10 >= 584.7;
     0.57*X11 >= 18.20;
     0.39*X12 >= 10.3;
     0.30*X13 >= 8.60;
END
```

FIGURE 15.4
LINGO codes for the revised model.

```
Rows=      19 Vars=      13 No. integer vars=       0  ( all are linear)
 Nonzeros=      83 Constraint nonz=     52(      0 are +- 1)
Density=0.312
 Smallest and largest elements in abs value=   0.300000         150000.
 No. < :    5 No. =:    0 No. > :  13, Obj=MAX, GUBs <=  13
 Single cols=     0
 Global optimal solution found at step:              13
 Objective value:                              952054.2

                     Variable           Value        Reduced Cost
                           X1        463.2061           0.0000000
                           X2        93.47059           0.0000000
                           X3        27.58621           0.0000000
                           X4        83.64875           0.0000000
                           X5        467.3529           0.0000000
                           X6        415.6667           0.0000000
                           X7        195.3825           0.0000000
                           X8        374.8743           0.0000000
                           X9        8.622222           0.0000000
                          X10        87.27798           0.0000000
                          X11        31.92983           0.0000000
                          X12        26.41026           0.0000000
                          X13        28.66667           0.0000000
                          Row    Slack or Surplus      Dual Price
                          OBJ        952054.2            1.000000
                            2        21767.11           0.0000000
                            3       0.0000000            6.222475
                            4       0.0000000           0.7217556
                            5       0.0000000           0.7303343
                            6       0.0000000           0.3273192
                            7        1125.511           0.0000000
                            8       0.0000000           -1.418467
                            9       0.0000000           -2.566161
                           10       0.0000000           -1.263504
                           11       0.0000000           -1.151725
                           12       0.0000000           -6.186407
                           13        3179.684           0.0000000
                           14        1374.497           0.0000000
                           15       0.0000000           -3.464854
                           16        418.9967           0.0000000
                           17       0.0000000           -8.272706
                           18       0.0000000           -13.63976
                           19       0.0000000           -18.21992
```

FIGURE 15.5
Output of the revised model.

The solution of the full model is shown in Figure 15.5.

Note that the revised model decreases the objective function value. However, it satisfies management's other qualitative goals.

15.3 A Two-Stage Transportation Problem

Consider the example of a two-stage transportation problem described in Chapter 10. As presented in Example 10.10 and Figure 10.14, the LINGO model for a problem with 15 plants, 10 warehouses, and 15 retailers is as follows. Although the problem is a representative practical problem, the data in the model do not represent any real scenario. They are chosen arbitrarily random, for demonstration, by ensuring feasibility of the model (Figure 15.6).

The model contains 300 integer variables and 50 constraints. The smallest and largest elements in absolute value are 1.00 and 4500.00, respectively. The objective function value is 91,300.00. As per LINGO, the solution summary is given in Tables 15.2 and 15.3.

```
MODEL:
!15 PLANTS, 10 WAREHOUSE AND 15 DIST CENTER ;

SETS:
   PLANTS/ P1 P2 P3 P4 P5 P6 P7 P8 P9 P10 P11 P12 P13 P14 P15/:
PLANT_CAPACITY;
   WAREHOUSES/ WH1 WH2 WH3 WH4 WH5 WH6 WH7 WH8 WH9 WH10/:
WHOUSE_CAPACITY;
   DISTCENTERS/ DC1 DC2 DC3 DC4 DC5 DC6 DC7 DC8 DC9 DC10 DC11 DC12
DC13 DC14 DC15/: DC_CAPACITY;
   LINKS1 (PLANTS, WAREHOUSES): COST1, VOLUME1;
   LINKS2 (DISTCENTERS, WAREHOUSES) : COST2, VOLUME2;
ENDSETS
!OBJECTIVE FUNCTION;
   [OBJ] MIN = PRODUCTION + DISTRIBUTION;
   PRODUCTION = @SUM ( LINKS1 (I, J): COST1 (I, J) * VOLUME1
(I,J));
   DISTRIBUTION = @SUM (LINKS2 (I, J): COST2 (I,J) * VOLUME2
(I,J));

!CONSTRAINTS;
   @FOR (WAREHOUSES(J):
        @SUM ( PLANTS(I): VOLUME1(I,J)) <= WHOUSE_CAPACITY(J));
   @FOR (PLANTS (I):
        @SUM ( WAREHOUSES (J): VOLUME1(I,J)) <= PLANT_CAPACITY
(I));

   @FOR (DISTCENTERS(I):
        @SUM (WAREHOUSES (J): VOLUME2(I,J)) >= DC_CAPACITY(I));

   @FOR (WAREHOUSES(J):
        @SUM ( PLANTS (I): VOLUME1(I,J)) - @SUM (DISTCENTERS
(I): VOLUME2(I,J)) >= 0);

   @FOR (PLANTS(I):
        @FOR (WAREHOUSES(J):
              @ GIN(VOLUME1(I,J)))) ;

   @FOR (DISTCENTERS(I):
        @FOR (WAREHOUSES(J):
              @ GIN(VOLUME2(I,J)))) ;
```

FIGURE 15.6
LINGO codes for a transportation model.

(continued)

```
!DATA ARE BELOW;
DATA:
   PLANT_CAPACITY = 1500 1400 1800.5 2200 1200 1700 1500 2000 2300
1950 2350 3700 1400 1600 2000;
   WHOUSE_CAPACITY = 3500 2500 4000 2000 3000 4500 4500 1000 2000
2000 ;
   DC_CAPACITY = 2000 2800 700 1400 1200 1600 1400 2100 2400 2050
2500 3450 1500 1700 1800 ;

   COST1 =      2  3  3  5  7  8  9  2  3  4
                6  4  2  3  9  7  4  5  6  7
                8  6  7  9  3  3  2  1  6  9
                3  5  7  9 10 12 14 12 16 18
                1 12  3  7  7  5  7  5  3  1
               15  2  8  6  2  2  6  2  1  1
                5  7  3  9  8 15  1  2  7  8
                3  2  8  6 12  4  9  1  1  1
               13 12  1 10  5  2  1  3  5  7
                7  3  2 14  1  1  2 14  6  3
               17  2  5  7  9  1  3  5  9  1
               10  5  6  4  3  1  2  3  7  9
                5  6  7  9  3  1  2  5  9  4
                1  5  9  1  1  4  7 14  5  6
                1  3  4  7  9 12  1 11 12 13;
      COST2 =  7  3  5  5  9  8  2  2  4  5
                5  4 12  7  3  7  5  7  6  7
                4  5  2  9  2  3  3  6 90  9
                3  6  4 11  5  2  4  1  1  3
                5  7  6  9  4  4  7 11 12  2
                2  9  8  7  8  6  8 15 16  1
                1 12 10  5 11  8  9  4  7  5
                2 15  3  3 12  7  1 14  8  7
                3  2  1  4  5  1 10  6  7  9
                7  7  5  6  1  4  5  7  7  3
               8  2  1  5  1  8 13 17  1  4
               9  3  3  7  3  9  1  2  3  4
               2  1 12  3  7  2  1 20  7  7
               3  1  5  3  9  5  3  3  9  3
               5  1  7  3  1  8  7  6  2  1;
 ENDDATA
 END
```

FIGURE 15.6 (continued)

15.4 A Crop Planning Problem

Consider the crop planning problem presented in Chapter 13. In this section, we solve an instance of this problem. To solve the problem, we need to collect and prepare data for the model. The basic data and relevant information are available in Sarker et al. (1997).

There are more than 100 crops in the country (a South Asian country) under consideration. The model developed in the previous chapter is an

TABLE 15.2

Transportation Plan from Plants to Warehouses

	To Warehouse									
From Plant	**1**	**2**	**3**	**4**	**5**	**6**	**7**	**8**	**9**	**10**
1	1200		300							
2			1400							
3							800	1000		
4	2200									
5										1200
6		950			750					
7							1500			
8									2000	
9			2300							
10					1950					
11		1550								800
12					300	3400				
13						1100	300			
14				1600						
15	100						1900			

aggregate planning model to make high-level planning decisions. Hence all the crops are divided into 10 major crop groups. According to the present cropping pattern, the number of crop combinations identified for single-, double-, and triple-cropped lands is 10, 17, and 6, respectively. Any of the 10 major groups/crops can be produced in a year in the single-cropped area. There are 17 pairs of crops that can be produced (one after another of the

TABLE 15.3

Transportation Plan from Warehouses to Retailers

From Warehouse										
1	**2**	**3**	**4**	**5**	**6**	**7**	**8**	**9**	**10**	**To Retailer**
						1000	1000			**1**
	1850			950						**2**
		700								**3**
								1400		**4**
					600				600	**5**
850									750	**6**
1400										**7**
1250		800				50				**8**
					2400					**9**
				2050						**10**
		2500								**11**
						3450				**12**
					1500					**13**
	100		1600							**14**
	550							600	650	**15**

pair) in a year in double-cropped areas while six combinations (three crops in each group, one after another in a year) in triple-cropped areas. The details of crop groups and their combinations are discussed below. Further details such as the individual crops under each crop group can be found in Sarker et al. (1997).

The production decision variables of the model with their coefficients are presented in Table 15.4. As indicated by Sarker et al. (1997), the data to prepare the coefficients were collected from different statistical year books.

TABLE 15.4

Model Parameters

Sl. No.	Variable X_{ijk}	A_{ijk} Million Tonnes/Acre	C_{ijk} Taka/Acre	Return Taka/Acre	B_{ijk} Taka/Acre
1	X_1_1_1	0.46	4313	5447	1134
2	X_2_2_1	0.41	3450	3566	116
3	X_3_3_1	0.70	4702	7475	2773
4	X_4_4_1	1.056	7557	9752	2195
5	X_5_5_1	0.75	4518	4809	291
6	X_6_6_1	0.658	5233	4531	–702
7	X_7_7_1	0.921	2090	10251	8125
8	X_8_8_1	0.935	1355	8445	7090
9	X_9_9_1	9.453	1680	49401	47722
10	X_10_10_1	0.506	12486	29247	16761
11	X_1_1_2	0.46	4313	5447	1134
12	X_3_1_2	0.70	4702	7475	2773
13	X_1_2_2	0.46	4313	5447	1134
14	X_4_2_2	1.056	7557	9752	2195
15	X_1_3_2	0.46	4313	5447	1134
16	X_5_3_2	0.75	4518	4809	291
17	X_1_4_2	0.46	4313	5447	1134
18	X_7_4_2	0.921	2090	10215	8125
19	X_2_5_2	0.41	3450	3566	116
20	X_4_5_2	1.056	7557	9752	2195
21	X_2_6_2	0.41	3450	3566	116
22	X_5_6_2	0.75	4518	4809	291
23	X_2_7_2	0.41	3450	3566	116
24	X_7_7_2	0.921	2090	10215	8125
25	X_3_8_2	0.70	4702	7475	2773
26	X_4_8_2	1.056	7557	9752	2195
27	X_3_9_2	0.70	4702	7475	2773
28	X_5_9_2	0.75	4518	4809	291
29	X_3_10_2	0.70	4702	7475	2773
30	X_6_10_2	0.658	5233	4531	–702
31	X_3_11_2	0.70	4702	7475	2773
32	X_7_11_2	0.921	2090	10215	8125
33	X_4_12_2	1.056	7557	9752	2195
34	X_6_12_2	0.658	5233	4531	–702
35	X_4_13_2	1.056	7557	9752	2195

(continued)

TABLE 15.4 (continued)

Model Parameters

Sl. No.	Variable X_{ijk}	A_{ijk} Million Tonnes/Acre	C_{ijk} Taka/Acre	Return Taka/Acre	B_{ijk} Taka/Acre
36	X_8_13_2	0.935	1355	8445	7090
37	X_5_14_2	0.75	4518	4809	291
38	X_6_14_2	0.658	5233	4531	–702
39	X_5_15_2	0.75	4518	4809	291
40	X_8_15_2	0.935	1355	8445	7090
41	X_6_16_2	0.658	5233	4531	–702
42	X_7_16_2	0.921	2090	10215	8125
43	X_7_17_2	0.921	2090	10215	8125
44	X_8_17_2	0.935	1355	8445	7090
45	X_1_1_3	0.46	4313	5447	1134
46	X_3_1_3	0.70	4702	7475	2773
47	X_4_1_3	1.056	7557	9752	2195
48	X_1_2_3	0.46	4313	5447	1134
49	X_3_2_3	0.70	4702	7475	2773
50	X_5_2_3	0.75	4518	4809	291
51	X_1_3_3	0.46	4313	5447	1134
52	X_3_3_3	0.70	4702	7475	2773
53	X_7_3_3	0.921	2090	10215	8125
54	X_3_4_3	0.70	4702	7475	2773
55	X_4_4_3	1.056	7557	9752	2195
56	X_6_4_3	0.658	5233	4531	–702
57	X_3_5_3	0.70	4702	7475	2773
58	X_5_5_3	0.75	4518	4809	291
59	X_6_5_3	0.658	5233	4531	–702
60	X_6_6_3	0.658	5233	4531	–702
61	X_3_6_3	0.70	4702	7475	2773
62	X_7_6_3	0.921	2090	10215	8125

Source: Reprinted from Sarker, R., Talukdar, S., and Haque, A., *Appl. Math. Model.*, 21, 621, 1997. With permission.

In the above table, taka represents the local currency. In addition to the above variables, there are six other variables representing the importation of crops. The LP model of the problem consists of 68 variables and 45 constraints. The RHS data are also available from Sarker et al. (1997). The corresponding detailed LP model is presented in Appendix-15A.

The country produces jute, kharif crops, fruits and juices, and drugs and narcotics in excess of their own demand. Importation is restricted for them. It is permitted only for cereals (rice and wheat) and rabi crops due to their shortage. This is why, the importation quantity variables, I_1, I_2, I_3, I_4, I_5, and I_7 are inserted whereas the others omitted.

15.4.1 Constraint and Variable Reduction

The constraints 15–43 state that one variable is equal to another variable. For example, the constraint 15 indicates that variable X_1_1_2 is equal to variable

X_3_1_2. So this constraint can be omitted by substituting one variable for the other in the remaining constraints and in the objective function. This way we can remove all the constraints from 15 to 43. It will also reduce the number of variables. After applying such a constraint and variable reduction method, the model consists of 39 variables and 16 constraints. For the convenience of variable tracking, we present all the variables in Table 15.5. For double-cropped areas, the variables in the first column will be substituted for the

TABLE 15.5

Variable Definition and Substitution

Variable Definition as per Model			Revised Definition	Solutions
X_1_1_1			X[1]	117.5200
X_2_2_1			X[2]	217.6824
X_3_3_1			X[3]	253.9847
X_4_4_1			X[4]	
X_5_5_1			X[5]	
X_6_6_1			X[6]	
X_7_7_1			X[7]	
X_8_8_1			X[8]	44.6538
X_9_9_1			X[9]	111.4678
X_10_10_1			X[10]	27.3913
X_1_1_2	X_3_1_2		X[11]	171.0008
X_1_2_2	X_4_2_2		X[12]	
X_1_3_2	X_5_3_2		X[13]	
X_1_4_2	X_7_4_2		X[14]	
X_2_5_2	X_4_5_2		X[15]	
X_2_6_2	X_5_6_2		X[16]	
X_2_7_2	X_7_7_2		X[17]	
X_3_8_2	X_4_8_2		X[18]	433.8992
X_3_9_2	X_5_9_2		X[19]	157.4304
X_3_10_2	X_6_10_2		X[20]	
X_3_11_2	X_7_11_2		X[21]	
X_4_12_2	X_6_12_2		X[22]	
X_4_13_2	X_8_13_2		X[23]	
X_5_14_2	X_6_14_2		X[24]	
X_5_15_2	X_8_15_2		X[25]	
X_6_16_2	X_7_16_2		X[26]	
X_7_17_2	X_8_17_2		X[27]	
X_1_1_3	X_3_1_3	X_4_1_3	X[28]	195.0805
X_1_2_3	X_3_2_3	X_5_2_3	X[29]	
X_1_3_3	X_3_3_3	X_7_3_3	X[30]	
X_3_4_3	X_4_4_3	X_6_4_3	X[31]	43.6195
X_3_5_3	X_5_5_3	X_6_5_3	X[32]	
X_6_6_3	X_3_6_3	X_7_6_3	X[33]	
I_1			X[34]	5.0000
I_2			X[35]	
I_3			X[36]	
I_4			X[37]	
I_5			X[38]	
I_7			X[39]	407.4360

second column, and for triple-cropped areas, the variables in the first column will be substituted for the second and third columns. Alternatively, we may redefine these variables as shown in column 4.

15.4.2 Scaling the Model

In the above model, the smallest and largest elements are 0.3333 and 1.7E + 11, respectively. As we discussed earlier, it will create a computational problem when solving the model using a software package. To ignore that difficulty we may redefine the decision variable in 10,000 units.

After solving with LINGO, we provide the summary solutions in Table 15.5. The fifth column of the table shows solutions of each variable of the model.

15.4.3 Working with Solutions

The area-used variables are defined in 10,000 units in the revised model. In Table 15.6, column 4 shows the area used (in 10,000 acres) under different crop combinations as per the optimum solutions. The total area used (in million acres) by each crop is summarized in column 5. The total crop production (in million tonnes) is calculated using the yield rate presented earlier and they are reported in column 6.

TABLE 15.6

Working with Solutions

[1] Crop	[2] Variable	[3] Land Type	[4] Land Used (in 10,000 Acres)	[5] Total Land Used (Million Acres)	[6] Total Production (Million Tonnes)
Aus rice	X_1_1_1	Single	117.5200	4.836	2.225
	X_1_1_2	Double	171.0008		
	X_1_1_3	Triple	195.0805		
Amon rice (B)	X_2_2_1	Single	217.6824	2.177	0.892
Amon rice (T)	X_3_3_1	Single	253.9847	12.55	8.785
	X_3_1_2	Double	171.0008		
	X_3_4_3	Triple	43.6195		
	X_3_8_2	Double	433.8992		
	X_3_9_2	Double	157.4304		
	X_3_1_3	Triple	195.0805		
Boro rice	X_4_8_2	Double	433.8992	6.73	7.103
	X_4_1_3	Triple	195.0805		
	X_4_4_3	Triple	43.6195		
Wheat	X_5_9_2	Double	157.4304	1.574	1.181
Jute	X_6_4_3	Triple	43.6195	0.436	0.287
Kharif crops	X_8_8_1	Single	44.6538	0.447	0.418
Fruits	X_9_9_1	Single	111.4678	1.115	10.537
Drugs and narcotics	X_10_10_1	Single	27.3913	0.274	0.139

TABLE 15.7

Import Summary

Importation of Crop	Variables	Quantity (Million Tonnes)
Rice	I_1	0.05000
Rabi crops	I_7	4.0744

From the above solutions, one can easily find the total area used under different types of land. The solutions of import variables are recorded in Table 15.7.

15.4.4 Multi-Objective Crop Planning Problem

Let us now consider two objectives (contribution maximization and cultivation cost minimization) for the above crop planning problem. To solve the two-objective problem, we use the ε-constraint method. In this method, one of the objective functions is selected to be optimized and all other objective functions are converted into constraints by setting an upper bound to each of them. The problem to be solved is now of the form:

$$\begin{aligned} &\text{Minimize } f_l(x) \\ &\text{Subject to} \\ &\qquad f_j(x) \le \varepsilon_j \quad \text{for all } j = 1, \ldots, k, \quad j \neq l \qquad \text{Model (15.2)} \\ &\qquad x \in S \\ &\qquad \text{where } l \in \{1, \ldots, k\} \end{aligned}$$

So the method is basically solving a single-objective model each time. The level of ε_i is altered to generate the entire Pareto-optimal set. We used LINGO/PC Release 10.0 for solving various problems based on ε-constraint method. For the Pareto frontier shown below, the total profit and the cultivation costs are expressed in million units of currency. It is clear that the total profit increases with increase of cultivation costs (Figure 15.7).

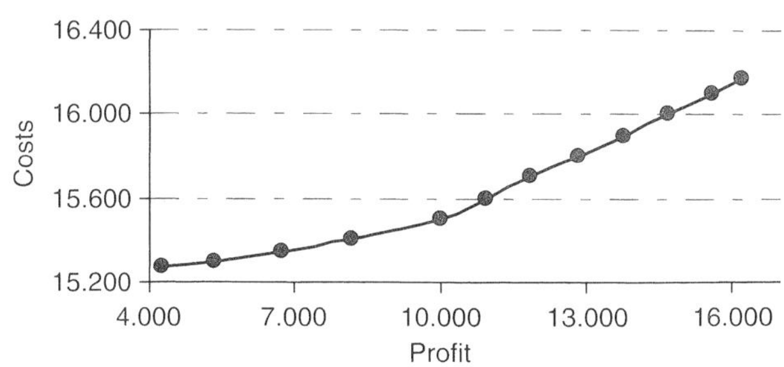

FIGURE 15.7
Pareto frontier for a bi-objective crop-planning problem.

TABLE 15.8

Plant Type and Technology

Plant Type i	Technology
1	Steam turbine
2	Gas turbine
3	Hydro
4	Combined cycle
5	Diesel cycle

15.5 Power Generation Planning Problem

Recall the power generation planning model presented in Chapter 13. The information/data for the problem are taken from Begum (2004). We consider five different plants (Table 15.8) and 6 years planning horizon (Table 15.9). Each period is divided into four demand blocks as shown in Table 15.10.

The existing plant capacities (CE_{it}) are presented in Table 15.11. The variable costs and fixed costs in different plant types for each period are given in Tables 15.12 and 15.13, respectively.

15.5.1 Model Validation

In this section, we validate the model by analyzing the solutions and comparing them with our expected outcomes and conditions. In the power generation problem, at the beginning of the modelling, one may think that the capacity constraint can be set as follows:

$$OX_{ipt} + OE_{ipt} \leq AF_{it}(X_{it} + CE_{it}) \quad \text{for all } i, p, \text{ and } t \tag{15.7}$$

It indicates that the output required from each type of power generation unit cannot exceed the total capacity of the existing units plus the planned units of this type, multiplied by the corresponding availability factor in each load block of a given year t. However, the above constraints with $p=2$, 3,

TABLE 15.9

Plant Operational Period

Period t	Year of Operation
1	2005
2	2006
3	2007
4	2008
5	2009
6	2010

TABLE 15.10

Time Block Details

Demand Block, p	Time Block in Hours of Any Year, t	Duration in Hours (TD_p)
1	0–910	910
2	910–2190	1280
3	2190–5840	3650
4	5840–8760	2920

TABLE 15.11

Existing Plant Capacity

		Period, t					
Plant Type i	Technology	1	2	3	4	5	6
1	Steam turbine	2438	340	210	210	0	0
2	Gas turbine	869	120	0	100	0	0
3	Hydro	230	0	0	0	100	0
4	Combined cycle	180	340	310	0	450	450
5	Diesel cycle	18	0	0	0	0	0

TABLE 15.12

Variable Cost for Different Periods (Value × 10^3 Taka/MW-h)

		Period					
Plant Type i	Technology	1	2	3	4	5	6
1	Steam turbine	1.423	1.437	1.452	1.466	1.481	1.496
2	Gas turbine	4.485	4.530	4.575	4.621	4.667	4.714
3	Hydro	0	0	0	0	0	0
4	Combined cycle	0.667	0.674	0.680	0.687	0.694	0.701
5	Diesel	4.177	4.219	4.261	4.304	4.3477	4.390

TABLE 15.13

Fixed Cost Terms (Value × 10^3 Taka/MW-h)

		Period					
Plant Type i	Technology	1	2	3	4	5	6
1	Steam turbine	54.568	60.025	62.576	69.66		
2	Gas turbine		33.472		34.013		
3	Hydro	533.978				593.978	
4	Combined cycle	2512.93	2889.87	3323.35		4395.13	5054.40
5	Diesel	2912.641					

and 4 would not affect the capacity determination as they would always have a much lower demand. So this type of constraint can be reduced to one-fourth, in number of constraints, by redefining it as follows.

$$\mathrm{OX}_{ipt} + \mathrm{OE}_{ipt} \leq \mathrm{AF}_{it}(\mathrm{X}_{it} + \mathrm{CE}_{it}) \quad \text{for } p = 1 \text{ only, and all } i \text{ and } t \tag{15.8}$$

The above constraints are technically correct and it may sound a great idea. After solving the model, you may check to see whether it serves its purpose properly or not. To report our experience to the readers, we have solved the model with the above capacity constraints. We have observed that although the above constraints are satisfied properly, there are a number of occasions where the outputs of existing plants exceed their available capacity. To remove this concern, we have included the following constraint which added 120 new constraints.

$$\mathrm{OE}_{ipt} \leq \mathrm{AF}_{it}\mathrm{CE}_{it} \quad \text{for all } i, p, \text{ and } t \tag{15.9}$$

It ensures that the output of any individual existing unit would not exceed their available capacity in any load block in a given period. These constraints were able to remove the above concern. However, it is now exceeding the available capacity of a new plant (type 3), in load block 1 and period 6. So, it is necessary to incorporate the following constraint, which would add another 120 constraints.

$$\mathrm{OX}_{ipt} \leq \mathrm{AF}_{it}\mathrm{X}_{it} \quad \text{for all } i, p, \text{ and } t \tag{15.10}$$

The constraint ensures that the output of any individual new unit would not exceed their available capacity in any load block in a given period. The model is now working well with no violation of conditions. However, if the latter two constraints (Equations 15.9 and 15.10) are satisfied, the earlier constraint (Equation 15.8) would automatically be satisfied. In other words, the constraint (Equation 15.8) would now be redundant, which can be verified by resolving the model. So we have removed the constraint (Equation 15.8) to provide an efficient solution to the model.

An alternative way of validating the model or solutions is to compare the solutions with existing practices. In reality, the practitioners usually try both.

The corresponding LP model contains 275 variables and 336 constraints. LINGO requires 218 iteration to reach the optimum solution with an objective function value of 48.098 million taka. The solutions of the model can be summarized as shown in Table 15.14.

In the Table 15.14, the values of variable OX_{ipt} (for new plant type) and OE_{ipt} (for existing plant type) are taken from the LINGO output. Column 1 indicates plant type, column 2 either new plant or existing plant, column 3 shows load block, and columns 4–9 indicate the periods from 1 to 6. To explain the results, let us consider row 1 in Table 15.14. It indicates that a new plant of type 1 (that is a steam turbine type power plant) requires to generate a power of 1121.97, 1192.09, 974.04, 827.77, 516.85, and 55.00 MW for

TABLE 15.14

Existing and New Capacity Utilization

Plant Type i	Existing/ New	Load Block	Period 1	2	3	4	5	6
1	New	1	1121.97	1192.09	974.04	827.77	516.85	55.00
3	New	1	27.50	16.15		1.34	27.50	27.50
4	New	1	354.17	354.17	354.17	354.17	354.17	354.17
5	New	1	2.43					
1	Existing	1	2106.43	2417.13	2610.50	2819.34	3044.89	3228.48
2	Existing	1	608.30	741.75	801.09	865.18	934.39	1009.15
3	Existing	1	207.00					23.00
4	Existing	1	153.00	229.50	612.00	918.00	1377.00	2065.50
5	Existing	1	16.20	16.20	16.20	16.20	16.20	16.20
3	New	2	27.50		19.74	17.76		
1	Existing	2	2106.43	2417.13	2610.50	2819.34	2699.15	2340.85
2	Existing	2	608.30	565.72	157.38			
3	Existing	2	76.62	73.37				
4	Existing	2	153.00	229.50	612.00	918.00	1377.00	2065.50
5	Existing	2	16.20	16.20	16.20	16.20		
3	New	3	27.50	27.50				
1	Existing	3	2106.43	2417.13	2340.40	2273.10	2072.05	1662.95
2	Existing	3	139.74					
3	Existing	3	85.48	41.52				
4	Existing	3	153.00	229.50	612.00	918.00	1377.00	2065.50
5	Existing	3	16.20	16.20				
3	New	4					17.50	27.50
1	Existing	4	1915.65	2005.65	1803.60	1692.90	1427.45	853.55
3	Existing	4						104.00
4	Existing	4	153.00	229.50	612.00	918.00	1377.00	2065.50

years 1–6, respectively, during the load block 1. The summary of the new capacity required (X_{it}) by plant type for a 6 year period is given in Table 15.15.

15.6 Gas-Lift Optimization

Consider the gas-lift optimization model presented in Section 14.7. Representative gas used vs. oil production data for a 6 wells practical problem and

TABLE 15.15

Summary of Capacity Required

Plant Type	New Capacity (MW)
Steam Turbine	1402.467
Gas Turbine	0.00
Hydro	55.00
Combined Cycle	416.667
Diesel Cycle	2.700

other data can be found in Buitrago et al. (1996). From those data, the slopes of line segments in each well can be calculated easily. The LINGO codes for this model are presented in Figure 15.8.

```
MODEL:
!There are 6 wells, each has 8 segments in their gas-production function;

SETS:
        WELL: GASWELL, OILPWELL;
        SEGMENT;
        WELLSEG (WELL, SEGMENT): LIMIT, SLOPE, X, Y, Z;
ENDSETS

!Objective function;
MAX = @SUM (WELLSEG(I,J): X(I,J) * SLOPE(I,J));

!Total gas used - additional equation to calculate total gas used;
GasUsed = @SUM (WELL(I):
                (@SUM (SEGMENT(J): X(I,J))));

!Oil produced in each well - additional equation to find oil produced;
@FOR (WELL(I):
        OILPWELL(I) = @SUM (SEGMENT(J): X(I,J)* SLOPE(I,J)));

!Gas used in each well - constraint (14.60);
@FOR (WELL(I):
        GASWELL(I) = @SUM (SEGMENT(J): X(I,J)));

!Other conditions (14.61) to be satisfied;
@FOR (WELL(I):
        @FOR (SEGMENT(J)| J #GT# 1:
                (X(I,J)- LIMIT(I,J)*Z(I,J-1))<=0));

@FOR (WELL(I):
        @FOR (SEGMENT(J)| J #EQ# 1:
                (X(I,J)- LIMIT(I,J))<=0));

@FOR (WELL(I):
        @FOR (SEGMENT(J)| J #GT# 1:
                (LIMIT(I,J)*Z(I,J) - X(I,J))<=0));

!Gas use limitation (14.62);
@SUM (WELL(I):
        (@SUM (SEGMENT(J):
                X(I,J)))) <= 12000;

!Binary declaration of variables;
@FOR (WELL(I):
        @FOR (SEGMENT(J):
                @BIN(Z(I,J))));

DATA:
        WELL = W1    W2   W3   W4   W5   W6;
        SEGMENT = S1  S2  S3  S4   S5   S6  S7   S8;
```

FIGURE 15.8
LINGO codes for gas-lift optimization model.

(continued)

```
    LIMIT =    32.1   61.1   93.5   129.6  173.9  230.9  292.3  358.4
               113.2  90.9   121.1  154    191.8  231.9  280.2  347.6
               157    131.5  183.4  236.1  295.6  346.4  439.5  530.1
               141.5  106.4  146    196    242.6  306.8  388.3  455.4
               32.1   84     127.3  203.4  295.8  400.6  524.1  697.5
               742.6  400.6  524.1  697.5  888.8  1144.3 1446.5 0;

    SLOPE =    6.735  0.936  0.457  0.229  0.124  0.071   0.028 -0.0003
               3.778  1.303  0.760  0.4214 0.223  0.094   0.026  0.005
               3.749  1.395  0.841  0.476  0.259  0.093   0.055  0.009
               2.500  0.874  0.521  0.3193 0.159  0.086   0.045  0.002
               5.0    2.233  1.004  0.594  0.339  0.178   0.085  0.045
               0.0    0.243  0.136  0.064  0.035  0.049  -0.004 -0.001;
ENDDATA
END
```

FIGURE 15.8 (continued)

TABLE 15.16

Summary of Solutions

Well	Gas Used	Oil Produced
1	1,013.400	391.9329
2	1,183.100	774.8991
3	2,255.800	1,175.803
4	1,527.600	667.8241
5	2,364.800	844.2198
6	3,655.300	300.4421
Total =	**12,000.00**	**4,155.121**

For a limit of 12,000 units of gas, the solution of the model suggests 4155.121 units of oil production. The well-wise gas consumption and oil production can be summarized as shown in Table 15.16.

15.7 Summary

In this chapter, we solved a number of representative practical problems and discussed their solutions. The problems considered were a sanitary-wares product-mix problem, a crop planning problem, a two-stage transportation problem, a power generation and planning problem, and a gas-lift optimization problem. The data preparation, LINGO codes, model solving difficulties, model validation, and practical model solving issues were also discussed.

References

Begum, S., Determination of optimum generation mix for the power sector of Bangladesh, Unpublished M.Sc. Thesis, Industrial and Production Engineering, Bangladesh University of Engineering and technology, Dhaka, 2004.

Buitrago, S., Rodriguez, E., and Espin, D., Global optimization techniques in gas allocation for continuous flow gas lift systems, SPE Gas Technology Conference, Calgary, Canada, SPE35616, 375, 1996.

Sarker, R. and Haque, A., Industrial engineering (IE) techniques in production decision, *Journal of the Institution of Engineers, Bangladesh*, 13, 67, 1985.

Sarker, R., Talukder, S., and Haque, A., Determination of optimum crop-mix for crop cultivation in Bangladesh, *Applied Mathematical Modelling*, 21, 621, 1997.

Appendix-15A Crop Planning Linear Programming Model

Objective function:

Maximize Z = 1134 X_1_1_1 + 116 X_2_2_1 + 2773 X_3_3_1 + 2195 X_4_4_1 + 291 X_5_5_1 − 702 X_6_6_1 + 8125 X_7_7_1 + 7090 X_8_8_1 + 47722 X_9_9_1 + 16761 X_10_10_1 + 1134 X_1_1_2 + 2773 X_3_1_2 + 1134 X_1_2_2 + 2195 X_4_2_2 + 1134 X_1_3_2 + 291 X_5_3_2 + 1134 X_1_4_2 + 8225 X_7_4_2 + 116 X_2_5_2 + 2195 X_4_5_2 + 116 X_2_6_2 + 291 X_5_6_2 + 116 X_2_7_2 + 8125 X_7_7_2 + 2773 X_3_8_2 + 2195 X_4_8_2 + 2773 X_3_9_2 + 291 X_5_9_2 + 2773 X_3_10_2 − 702 X_6_10_2 + 2773 X_3_11_2 + 8125 X_7_11_2 + 2195 X_4_12_2 − 702 X_6_12_2 + 2195 X_4_13_2 + 7090 X_8_13_2 + 291 X_5_14_2 − 702 X_6_14_2 + 291 X_5_15_2 + 7090 X_8_15_2 − 702 X_6_16_2 + 8125 X_7_16_2 + 8125 X_7_17_2 + 7090 X_8_17_2 + 1134 X_1_1_3 + 2773 X_3_1_3 + 2195 X_4_1_3 + 1134 X_1_2_3 + 2773 X_3_2_3 + 291 X_5_2_3 + 1134 X_1_3_3 + 2773 X_3_3_3 + 8125 X_7_3_3 + 2773 X_3_4_3 + 2195 X_4_4_3 − 702 X_6_4_3 + 2773 X_3_5_3 + 291 X_5_5_3 − 702 X_6_5_3 − 702 X_6_6_3 + 2773 X_3_6_3 + 8125 X_7_6_3 − 5940 I_1 − 5940 I_2 − 5940 I_3 − 5940 I_4 − 4609 I_5 − 12341 I_7

Constraints:

(1) + 0.46 X_1_1_1 + 0.46 X_1_1_2 + 0.46 X_1_2_2 + 0.46 X_1_3_2 + 0.46 X_1_4_2 + 0.46 X_1_1_3 + 0.46 X_1_2_3 + 0.46 X_1_3_3 + I_1 ≥ 2274566

(2) + 0.41 X_2_2_1 + 0.41 X_2_5_2 + 0.41 X_2_6_2 + 0.41 X_2_7_2 + I_2 ≥ 892498

(3) + 0.7 X_3_3_1 + 0.7 X_3_1_2 + 0.7 X_3_8_2 + 0.7 X_3_9_2 + 0.7 X_3_10_2 + 0.7 X_3_11_2 + 0.7 X_3_1_3 + 0.7 X_3_2_3 + 0.7 X_3_3_3 + 0.7 X_3_4_3 + 0.7 X_3_5_3 + 0.7 X_3_6_3 + I_3 ≥ 8785105

(4) + 1.056 X_4_4_1 + 1.056 X_4_2_2 + 1.056 X_4_5_2 + 1.056 X_4_8_2 + 1.056 X_4_12_2 + 1.056 X_4_13_2 + 1.056 X_4_1_3 + 1.056 X_4_4_3 + I_4 ≥ 7102647

(5) + 0.75 X_5_5_1 + 0.75 X_5_3_2 + 0.75 X_5_6_2 + 0.75 X_5_9_2 + 0.75 X_5_14_2 + 0.75 X_5_15_2 + 0.75 X_5_2_3 + 0.75 X_5_5_3 + I_5 ≥ 1180728

(6) + 0.658 X_6_6_1 + 0.658 X_6_10_2 + 0.658 X_6_12_2 + 0.658 X_6_14_2 + 0.658 X_6_16_2 + 0.658 X_6_4_3 + 0.658 X_6_5_3 + 0.658 X_6_6_3 ≥ 287016

(7) + 0.921 X_7_7_1 + 0.921 X_7_4_2 + 0.921 X_7_7_2 + 0.921 X_7_11_2 + 0.921 X_7_16_2 + 0.921 X_7_17_2 + 0.921 X_7_3_3 + 0.921 X_7_6_3 + I_7 ≥ 4074360

(8) + 0.935 X_8_8_1 + 0.935 X_8_13_2 + 0.935 X_8_15_2 + 0.935 X_8_17_2 ≥ 417513

(9) + 9.453 X_9_9_1 ≥ 5942200

(10) + 0.506 X_10_10_1 ≥ 138600

(11) + X_1_1_1 + X_2_2_1 + X_3_3_1 + X_4_4_1 + X_5_5_1 + X_6_6_1 + X_7_7_1 + X_8_8_1 + X_9_9_1 + X_10_10_1 ≤ 7727000

(12) 0.5 X_1_1_2 + 0.5 X_3_1_2 + 0.5 X_1_2_2 + 0.5 X_4_2_2 + 0.5 X_1_3_2 + 0.5 X_5_3_2 + 0.5 X_1_4_2 + 0.5 X_7_4_2 + 0.5 X_2_5_2 + 0.5 X_4_5_2 + 0.5 X_2_6_2 + 0.5 X_5_6_2 + 0.5 X_2_7_2 + 0.5 X_7_7_2 + 0.5 X_3_8_2 + 0.5 X_4_8_2 + 0.5 X_3_9_2 + 0.5 X_5_9_2 + 0.5 X_3_10_2 + 0.5 X_6_10_2 + 0.5 X_3_11_2 + 0.5 X_7_11_2 + 0.5 X_4_12_2 + 0.5 X_6_12_2 + 0.5 X_4_13_2 + 0.5 X_8_13_2 + 0.5 X_5_14_2 + 0.5 X_6_14_2 + 0.5 X_5_15_2 + 0.5 X_8_15_2 + 0.5 X_6_16_2 + 0.5 X_7_16_2 + 0.5 X_7_17_2 ≤ 9615000

(13) + 0.333333 X_1_1_3 + 0.333333 X_3_1_3 + 0.333333 X_4_1_3 + 0.333333 X_1_2_3 + 0.333333 X_3_2_3 + 0.333333 X_5_2_3 + 0.333333 X_1_3_3 + 0.333333 X_3_3_3 + 0.333333 X_7_3_3 + 0.333333 X_3_4_3 + 0.333333 X_4_4_3 + 0.333333 X_6_4_3 + 0.333333 X_3_5_3 + 0.333333 X_5_5_3 + 0.333333 X_6_5_3 + 0.333333 X_6_6_3 + 0.333333 X_3_6_3 + 0.333333 X_7_6_3 ≤ 2387000

(14) + 4313 X_1_1_1 + 3450 X_2_2_1 + 4702 X_3_3_1 + 7557 X_4_4_1 + 4518 X_5_5_1 + 5233 X_6_6_1 + 2090 X_7_7_1 + 1355 X_8_8_1 + 1680 X_9_9_1 + 12486 X_10_10_1 + 4313 X_1_1_2 + 4702 X_3_1_2 + 4313 X_1_2_2 + 7557 X_4_2_2 + 4313 X_1_3_2 + 4518 X_5_3_2 + 4313 X_1_4_2 + 2090 X_7_4_2 + 3450 X_2_5_2 + 7557 X_4_5_2 + 3450 X_2_6_2 + 4518 X_5_6_2 + 3450 X_2_7_2 + 2090 X_7_7_2 + 4702 X_3_8_2 + 7557 X_4_8_2 + 4702 X_3_9_2 + 4518 X_5_9_2 + 4702 X_3_10_2 + 5233 X_6_10_2 + 4702 X_3_11_2 + 2090 X_7_11_2 + 7557 X_4_12_2 + 5233 X_6_12_2 + 7557 X_4_13_2 + 1355 X_8_13_2 + 4518 X_5_14_2 + 5233 X_6_14_2 + 4518 X_5_15_2 + 1355 X_8_15_2 + 5233 X_6_16_2 + 2090 X_7_16_2 + 2090 X_7_17_2 + 1355 X_8_17_2 + 4313 X_1_1_3 + 4702 X_3_1_3 + 7557 X_4_1_3 + 4313 X_1_2_3 + 4702 X_3_2_3

$+ 4518\ X\_5\_2\_3 + 4313\ X\_1\_3\_3 + 4702\ X\_3\_3\_3 + 2090\ X\_7\_3\_3 + 4702\ X\_3\_4\_3 + 7557\ X\_4\_4\_3\ 5233\ X\_6\_4\_3 + 4702\ X\_3\_5\_3 + 4518\ X\_5\_5\_3 + 5233\ X\_6\_5\_3 + 5233\ X\_6\_6\_3 + 4702\ X\_3\_6\_3 + 2090\ X\_7\_6\_3 \leq 1.7E+11$

(15) $X\_1\_1\_2 - X\_3\_1\_2 = 0$

(16) $X\_1\_2\_2 - X\_4\_2\_2 = 0$

(17) $X\_1\_3\_2 - X\_5\_3\_2 = 0$

(18) $X\_1\_4\_2 - X\_7\_4\_2 = 0$

(19) $X\_2\_5\_2 - X\_4\_5\_2 = 0$

(20) $X\_2\_6\_2 - X\_5\_6\_2 = 0$

(21) $X\_2\_7\_2 - X\_7\_7\_2 = 0$

(22) $X\_3\_8\_2 - X\_4\_8\_2 = 0$

(23) $X\_3\_9\_2 - X\_5\_9\_2 = 0$

(24) $X\_3\_10\_2 - X\_6\_10\_2 = 0$

(25) $X\_3\_11\_2 - X\_7\_11\_2 = 0$

(26) $X\_4\_12\_2 - X\_6\_12\_2 = 0$

(27) $X\_4\_13\_2 - X\_8\_13\_2 = 0$

(28) $X\_5\_14\_2 - X\_6\_14\_2 = 0$

(29) $X\_5\_15\_2 - X\_8\_15\_2 = 0$

(30) $X\_6\_16\_2 - X\_7\_16\_2 = 0$

(31) $X\_7\_17\_2 - X\_8\_17\_2 = 0$

(32) $X\_1\_1\_3 - X\_3\_1\_3 = 0$

(33) $X\_3\_1\_3 - X\_4\_1\_3 = 0$

(34) $X\_1\_2\_3 - X\_3\_2\_3 = 0$

(35) $X\_3\_2\_3 - X\_5\_2\_3 = 0$

(36) $X\_1\_3\_3 - X\_3\_3\_3 = 0$

(37) $X\_3\_3\_3 - X\_7\_3\_3 = 0$

(38) $X\_3\_4\_3 - X\_4\_4\_3 = 0$

(39) $X\_4\_4\_3 - X\_6\_4\_3 = 0$

(40) $X\_3\_5\_3 - X\_5\_5\_3 = 0$

(41) $X\_5\_5\_3 - X\_6\_5\_3 = 0$

(42) $X\_6\_6\_3 - X\_3\_6\_3 = 0$

(43) $X\_3\_6\_3 - X\_7\_6\_3 = 0$

(44) $I\_1 + I\_2 + I\_3 + I\_4 + I\_5 \leq 0$

(45) $X\_9\_9\_1 + X\_10\_10\_1 \leq 1969350$

Index

L

M